MW00445426

Second Edition

International Law for Energy and the Environment

Second Edition

International Law for Energy and the Environment

Patricia Park

CRC Press is an imprint of the
Taylor & Francis Group, an **informa** business

CRC Press
Taylor & Francis Group
6000 Broken Sound Parkway NW, Suite 300
Boca Raton, FL 33487-2742

Printed in the United States of America on acid-free paper
Version Date: 20120813

International Standard Book Number: 978-1-4398-7096-9 (Hardback)

Library of Congress Cataloging-in-Publication Data

Park, Patricia D.
International law for energy and the environment / Patricia Park. -- Second Edition.
pages cm
Includes bibliographical references and index.
ISBN 978-1-4398-7096-9
1. Power resources--Law and legislation. 2. Energy industries--Law and legislation. 3. Environmental law. I. Title.

K3981.P37 2013
343.09'2--dc23 2012031893

Visit the Taylor & Francis Web site at
http://www.taylorandfrancis.com

and the CRC Press Web site at
http://www.crcpress.com

Contents

SECTION I

SECTION II

SECTION III

Preface

The first edition of this book, *Energy Law and the Environment*, was published in 2002 and was the first one in the field that considered the legal environmental imperatives for the energy sector. The first edition was written to be accessible to both lawyers and non-lawyers alike and the same theme continues in this, the second edition. However, the interest and concerns of policymakers, multinational companies, and international jurists have expanded to consider not only what should be done to protect global resources for sustainable development but also why. To this end, the second edition considers how the international law engages with multinationals in respect of energy sources, ownership of those resources, and state sovereignty. The book also considers issues of regulation of the energy sector within an economic context and the need for protection of the intellectual property rights of those companies who develop the technology that will help both the developed and the developing nations mitigate any environmentally damaging emissions. The scope of the book has enlarged and so the title has been changed to more realistically reflect the development in the area to *International Law for Energy and the Environment.*

This edition has been written for all the actors within the energy sector. Policymakers, CEOs, and senior managers of any company, working either upstream or downstream, within the energy sector will benefit from greater knowledge and understanding of the legal issues. Engineers working in research and development will have an understanding of the intellectual property rights and the broader legal context within which they and their company operate. The book will also benefit those who are studying at undergraduate level, but more so those at postgraduate level on an MBA or a legal course.

The book itself is divided into three specific sections that build one upon the other. Section I looks at the more general issues concerning the energy sector. Chapter 1 sets into context the interrelationship between international law, environmental law, and the energy sector. It considers international policy and the regulatory bodies, international standard setting, and the influence of science; it also looks towards the more ethical issues of corporate social responsibility within environmental management.

Chapter 2 is more theoretical in nature and considers regulatory theory within an economic context, as no regulated sector can operate without consideration of the economic imperatives.

Chapter 3 discusses the interrelationship between international regulation and state rules that companies must comply with. It also discusses the duties and liabilities of multinational companies operating in the energy sector who have subsidiaries operating in different jurisdictions. This chapter brings together all of the global imperatives with regard to the energy sector.

Chapter 4 completes the general section of the book by considering trade, competition, and environmental law with reference to the energy sector.

Section II of the book addresses the individual regulation of the various energy sectors. This section of the book looks at international law as it affects the different energy sectors. Unless an energy resource is found wholly within the territory of a single state, international law will affect what any one state may do with regard to the regulation of a resource that may cross the borders of more than one state. If the development of that resource has a trans-boundary effect on a third-party state, then international law will come into play.

Chapter 5 looks at the international law with regard to the regulation of the oil and gas sector, considering the different types of state regulation and ownership, liability in both the criminal and civil areas, and issues of decommissioning.

In Chapter 6, the international regulation of the nuclear industry, control of nuclear risk, and liability for nuclear waste are considered.

Section III of the book considers some of the main energy producer/user jurisdictions within which any energy company may operate. It looks at the more developed systems around the world and identifies some areas of good practice. It also looks at the new and emerging economies where the regulation of the energy sector is evolving but is not yet complete. A number of case studies are used to identify problems and solutions that any enterprise operating in the energy sector needs to be aware of.

Chapter 7 considers the regulation of the energy sector in the United States, one of the major producers as well as users of energy. It considers the interrelationship between federal and state policy and regulations. Federal regulatory bodies and the different rules for the different energy sectors are considered. The chapter also uses case studies to illustrate some areas of the law.

Chapter 8 looks at the European Union as it represents another large block of energy producers/users. There are European laws concerning the energy sector and climate change, which in some ways reflect the federal jurisdiction of the United States, and these are issues by way of Directives and Regulations. The European competition legislation and the use of 'state aid' for environmental purposes are a major influence.

Chapter 9 considers the United Kingdom as a major producer and representative of a Member State of the European Union as these laws must be transposed into national laws of the European Member States. Again, issues of energy policy, the regulatory bodies, and the different energy sectors are considered.

Chapter 10 considers Norway as a large hydropower producer and a significant exporter of oil and gas. Norway is also a member of the European Economic Area (EEA), which although not a full member of the European Union, follows most of the Directives and Regulations with regard to trade and competition, through the European Economic Agreement.

Chapter 11 turns to Australia, the major producer and user of energy in the Southern Hemisphere. Australia also has very strong environmental credentials but has some difficulty in signing up to the Framework Convention on Climate Change lest it damage their international trade with the far eastern countries.

Chapter 12 considers the first of two major emerging economies, namely, India. India has a traditional and cultural empathy with nature and the environment. It has signed up to the major international treaties and has policies to support renewable energy sources. However, India does have a problem with the enforcement of the law.

Chapter 13 considers the emerging policies of the second of these economies, namely, China. Policies are being developed to consider the environment; as the main natural energy source is coal, the mitigation of any environmentally detrimental effects of coal-fired power stations is currently taken seriously by the Chinese government.

Chapter 14 provides a summary and discussion on the regulation of energy and the environment as discussion in greater depth in the earlier chapters.

Patricia Park
Southampton, United Kingdom

INTERNATIONAL TREATIES AND INTERNATIONAL DOCUMENTS

IEA (2011), Regulatory Review, 2nd edition.
Task 2–Policy Options paper v FINAL 27 April 2007.
(1998) O.J. L69/1; Trade amendment, Council Decision 98/537/EC (1998) O.J.
16th Conference of the Parties and 6th Conference of the Parties serving as the meeting of the Parties to the Kyoto Protocol to the United Nations Framework Convention on Climate Change.
1958 Convention Article 5, paras 2–6.
1958 Convention on the Continental Shelf, Article 2.
1958 Geneva Convention on the Territorial Sea and Contiguous Zone Article 5(1) and the 1982 Convention on the Law of the Sea Article 8(1).
1959 Antarctic Treaty, Article 5.
1963 Convention Supplementary to the Paris Convention, with additional Protocols.
1963 Treaty Banning Nuclear Weapons Tests in the Atmosphere, in Outer Space and Under Water.
1968 and 1978 Conventions on Civil Jurisdiction and the Enforcement of Judgements.
1986 Noumea Convention for the Protection of the Natural Resources and Environment of the South Pacific, Article II.
1989 Protocol on the Protection of the South East Pacific against Radioactive Pollution.
1991 ICJ, *ILR* (1994) 94, 446.
1992 Convention for the Protection of the Marine Environment of the North East Atlantic.
(1994) O.J. L336/100.
(1994) O.J. L336/213.
Annex V to the Montreal Protocol on Substances That Deplete the Ozone Layer.
Article 1 of the 1972 Convention on the Prevention of Marine Pollution by the Dumping of Wastes and Other Matter.
Article 1 of the EC Framework Directive on Waste (Council Directive 75/442/EEC as amended by Council Directive 91/156/EEC).

Article 3 of the 1958 Convention on the Continental Shelf.
Article 33 of the CCS Directive, amending Directive 2001/80/EC.
Communication on Implementing Community Environmental Law, COM (96) 500, p. 9.
Commission press release IP/05/1517, 'State Aid; Commission closes formal investigation on CO_2 taxation system in Slovenia following changes to legislation', 1 December 2005.
Committee on the Peaceful Uses of Outer Space, *Report of the Legal Subcommittee*, UN Doc. A/AC. 105/430, Annex III; Safety of Life at Sea Establishments sites at Aldermaston and Burghfield, at the press release on 10 May 2000.
Communication on a Community Strategy to Limit Carbon Dioxide Emissions and to Improve Energy Efficiency. COM (91) 249 final,[61] (1994) O.J. L33/13.
Community Guidelines on State Aid for Environmental Protection, (2001) O.J. C37/3.
Convention for the Protection of the Ozone Layer, 26 *ILM* 1529, 1985.
Convention on Assistance in the Case of a Nuclear Accident or Radiological Emergency. Legal Series no. 14. IAEA: Vienna, 1987, 61.
Convention on Early Notification of a Nuclear Accident. Legal Series no. 14. IAEA: Vienna, 1987.
Convention on Environmental Impact Assessment in a Transboundary Context (Espoo), 1991.
Convention on Long-Range Transboundary Air Pollution (Geneva), UKTS 57 (1983), Cmd. 9034; *ILM* (1979), 1442.
Cop 7. fccc/CP/2001/13 Marrakesh.
Council Directive 86/61/EC concerning integrated pollution prevention and control, O.J. L257.
Council Directive 92/43, (1992) O.J. L206/7.
Council Directive 96/29/EURATOM, Article 6.
Decision 15/CP.7, Preamble, para. 5.
Decision 24/CP.7, Annex, Section XV, para. 4.
Decision 24/CP.7.
Decision VII/9: Basic Domestic Needs, 7th Meeting of the Parties, 5–7 December 1995.
Directive 2004/35/CE of European Parliament and Council on environmental liability with regard to the prevention and remedying of environmental damage; (2004) O.J. L143/56.
Directive 93/12/EEC relating to the sulphur content of certain liquid fuels.
EC Habitats Directive 92/43.
EC Wild Birds Directive 79/409.
EU CCS Directive 2009/31.
Framework Directive on Combating Air Pollution from Industrial Plants, 84/360/EEC/L188.
General Assembly Declaration of the Principles on the Seabed and Ocean Floor, GA Res. 2749(XXV), 1970.
IAEA Conference on Civil Liability, 67.
IAEA General Conference, Special session, 1986, IAEL/GC (SPL,1)4.

IMO Assembly Resolution A.414 (XI), 15 November 1979.
IPCC Guidelines for National Greenhouse Gas Inventories 1966 as revised in 2006.
Law of the Sea Convention 1982.
LDC Resolution 14(7), 1983; LDC Resolution 21(9), 1985.
OECD (1996), *Pollutant Release and Transfer Registers: A Tool for Environmental Policy* (n 20), pp. 94–95.
OECD Report, 1989.
OECD Report, 1989, p. 13.
OECD (1982), *Combating Oil Spills*. OECD: Paris.
OPEC Resolution XV/90, 1968.
OPRC Convention, Articles 3 and 4.
OPRC Convention, 30 November 1990, IMO-55OE, 1991st edition. 30 *ILM* (1991), p. 733.
OSPAR Decision 98/3, Annex 2, 12.
OSPAR Decision 98/3.
OSPAR Guidelines for Risk Assessment and Management of CO_2 Streams in Geological Formations (Annex 7).
Statute of the International Court of Justice, Article 34.
The Annual Report of the International Oil pollution Compensation Funds, 1997, 86–94.
The Convention on the Continental Shelf was part of the outcome of the First Law of the Sea Conference, 1958.
The High Seas Convention, 1958, Article 25(1); the Roraonga Treaty, 1985, Article 7.
The New EU ETS Directive.
The Noumea Convention, 1986, Article 10(1).
UN Doc. A/5217, 1962.
UN Doc. A/7750, Part 1, p3, 10 November 1969.
UN Doc. A/Res/1803 (XVII), 14 December 1962.
UN Doc. A/RES/3201 (S-VI), 1974; UN Doc. A/RES/3281) (XXIX), 1974.
UN Resolution 2574 (XXIV), 1970.
UN Resolution 2749 (XXV), 17 December 1970.
UN/ECE, *Air Pollution Studies*, nos 1–12, 1984–1996.
UNCTAD, *Formulation of a Strategy for the Technological Dependence of Developing Countries*, (1980) UNCTAD Doc. TD/B/779.
UNEP, Ozone Secretariat, http://www.unep.org/ozone. Accessed December 2010.
United Nations Framework Convention on Climate Change, 1992.

Acknowledgments

During the writing of this book I have visited many places and have spoken to many of those who operate within the energy sector and have a particular concern about the environment. I have engaged with academics, engineers, policymakers, and practitioners from both the legal and management sectors. I have visited the United States, Belgium, Norway, India, and Australia. In addition, of course, I have found many in the United Kingdom from the legal field as well as from multinational companies who have given freely of their time and experience to inform the content of this book. There are too many to mention by name, but I must thank in particular the Climate Change team and the Head of the Legal Section from Statoil in Norway, two advocates from the Superior Court in India, the environmental manager from General Motors, the Environmental Law group, University of Sydney, and my friends and colleagues from the University of Leuven, Belgium. As Head of the Law Research Centre, I gained much knowledge from my international PhD candidates and also from my friends in the local government in the United Kingdom. Last but not least, a big 'thank you' must go to all the librarians who are usually the unsung heroes, but without whom much of the desk-based research would not be possible; in particular, my thanks go to Hannah Young, the Law Librarian in the Mountbatten Library at Southampton Solent University, Southampton, United Kingdom. A very big thank you to you all. I have only mentioned one by name but the rest of you know who you are. Any mistakes herein, of course, remain mine alone.

Patricia Park
Southampton, United Kingdom

Authors

Patricia Park is a research professor at Southampton Solent University, Southampton, United Kingdom. She was previously head of the Law Research Centre for 12 years. Patricia has carried out research for international oil companies, the United Kingdom, and German governments, and specialises in the regulation of industry, in particular, the energy sector. She has worked and carried out research in the United States, Sweden, Norway, Germany, Belgium, and India. She is a fellow of the Energy Institute and a member of the International Bar Association, and was appointed as an original member of the Regional Environmental Pollution Advisory Committee to the newly formed Environment Agency in 1996. She was appointed a member of the Commission on Environmental Law of the IUCN in 1995. Patricia has published and lectured extensively in both the environmental and energy areas. Patricia has been a magistrate since 1982.

Duncan Park is the intellectual property manager at Leogriff AS, Oslo, Norway. He was recently leading counsel at Statoil ASA, Oslo for four years and prior to that was IP Counsel at Schlumberger WesternGeco. Duncan has worked both in private practice as a patent attorney and as in-house counsel, both in the United Kingdom and in Norway. He specialises in the strategic development of intellectual property rights, mainly within the energy sector, providing advice to technical and commercial departments. Duncan's first degree was in mechanical engineering.

Section I

It is impossible to perform many socially beneficial functions . . . . . without some adverse environmental consequences (Pearce & Gellhorn, *Regulated Industries* [1994], p. 352)

1 Introduction to International Law

Section I of this book provides an introduction to and setting in context of the interrelationship between environmental law and the energy sector. It is concerned with the international regulation of energy activities. In particular, it considers the general principles of jurisdiction and control that states may exercise over their territory and their nationals, both natural and legal entities, including companies carrying out energy activities.

1.1 INTERNATIONAL LAW

International law is the law that states make to regulate matters among themselves. Initially, it was the League of Nations established after World War I that condemned external aggression against territorial integrity and political independence. After the attainment of a minimum peace order, other matters including economic development, exchange rates, trade, the environment, and intellectual property rights were considered, and a number of organisations were developed to deal with such matters. The international system also developed a number of instruments that focus on the protection of rights. In addition to the rights under the Universal Declaration of Human Rights,[1] which presents what could be called the *traditional human rights*, such as the right to life, the right to property, and the right to freedom from discrimination, other rights have been considered. These additional rights include the right to development[2] and the right to a decent environment.[3] Human rights articulate the demands for a maximum order of law, but this order goes beyond the achievement of elementary peace and incorporates the aspiration for a better quality of life. Human rights shape the notion of human dignity, which gives the direction for the future development of international law, with the protection of human dignity being the ultimate goal of the international law process.[4]

International environmental law was initially conceived as an institution that would establish rules for the management of environmental problems that started to become all too obvious in the late 1970s.[5] States gradually became aware that these problems were not amenable to easy solutions, and the best way to deal with them was as follows:

- Define the problem while there was scientific uncertainty.
- Devise management solutions that are economic.
- Deal with the distributive issues involved.

The foundations of international environmental law are the pursuit of a minimum order and sound environmental management through the satisfaction of perceived equitable outcomes. Without the maintenance of peace, there is no meaning to an international legal order, and most international environmental issues involve the management of common resources on an allocative basis. The tragedy associated with the use of common property resources is the inability to exclude others from the use of these resources. Inasmuch as all users of a common property resource are wealth maximisers, they try to get as much of the resource as they can in their effort to outperform other users. The resource will eventually collapse. The rationale of a polluter is the same as that of an extractor, because each in turn is a wealth maximiser. Environmental management, therefore, becomes a collective action problem. If an individual polluter or extractor takes measures to diminish the impact on a resource, others would continue to behave as profit maximisers, leading eventually to the collapse of the resource. The remedy for such common property problems is, therefore, government control of private property, in the form of taxes and property rights.[6] Common pool problems are essentially collective action problems, and collective action problems in the management of common pool resources involve distributive decisions in terms of the following:

- Who is to be included in the management of a resource
- How to distribute the benefits among those included
- How to compensate those who are excluded

The international law as it affects the environment and the energy sector addressed in this book are essentially common pool problems. Mineral resources under the seabed are common pool resources because they constitute global resources that could be accessed by everyone. The use of these resources by some will subtract from the use of others. Air quality is a common pool resource because the pollution of the air by some affects the utility of the air for others. The high seas are also a common pool resource because the pollution of waters by some would disadvantage the use by others.

Waste is not a common pool resource because waste is normally a material of low value and is frequently disposed of in a haphazard way. Generators, transporters, and disposers of waste are eager to get rid of the waste in either a legal or sometimes an illegal manner. Almost all economic activity will generate waste, which is an externality for any business; and the business would be content to move the responsibility for that externality to society. Most states have chosen to deal with waste products under the 'polluter pays' principle[7] and so businesses have been forced to take responsibility for the waste they produce, transfer, and dispose of.

Air is generally conceived as a public good as the consumption of air by one person does not affect the availability of the air for others. However, with industrialisation, it was quickly understood that the use of air by new technologies affects air quality. Polluted air can cause any number of physical ailments and the increase in polluting emissions transforms the air from a public good to an open access resource leading to its degradation. In the absence of government regulation or property rights, a tragedy of the commons thus became inevitable.

The first efforts to control air pollution were regulatory in nature but it was soon realised that all polluters were not equally effective in reducing their pollution. It was then decided to introduce a tradeable quota system.[8]

To consider the international law with regard to natural resources is to discover that it is not a single topic. Almost everything depends, if not on the actual resource that one is considering, then on the category of natural resource, for example, the mineral resources of the deep seabed; water, including water as it is carried along by the great international rivers; and petroleum found on shore, or beneath a state's territorial waters, or on its continental shelf. The mineral resources of the deep seabed are located in an area beyond national jurisdiction, so who may exploit them? In fact, legally, how is it to be determined who may exploit them? How is it to be done, and within what constraints? Under whose jurisdiction is water that flows through many countries fall? Does international law place any constraints upon its use?

1.2 THE RESOURCES OF THE DEEP SEABED

From the earliest days of international law, the idea of the 'freedom of the high seas' has been accepted.[9] This principle was first codified in the 1958 UN Convention on the High Seas. However, although specific treaty articles were not binding on parties, the basic status of the high seas undoubtedly represented what was already customary international law. The assumption was that the freedom of the high seas also meant the freedom to remove the resources that were found in those waters. Fishing on the high seas does not require anyone's permission, because no one has title over the resources of the high seas. It was beginning to be realised in the 1950s that fish were not the only valuable resources to be found in the high seas, beneath the waters, and under the seabed itself. Mineral resources such as manganese, cobalt, and nickel were present in clear evidence, but it was also likely that petroleum too lay within certain parts of the deep seabed. In addition, it would not be long before technology would prove it possible to exploit these resources for human benefit and commercial gain. In 1967, the government of Malta introduced into the UN General Assembly an idea that was to have a profound effect on the way international law developed in relation to these perceived realities.[10] It was suggested that the resources of the deep seabed were 'common heritage of mankind', and that this idea of common heritage over a resource was different from the perception of fish in waters of the high seas, because fish were *res nullius* (belonging to no one), and therefore could be exploited by anyone who wished to. But a resource that was termed as *common heritage* apparently meant that a resource that could, in principle, be exploited by anyone, but only with permission of the world community and upon such conditions as the institutions representing that community would lay down.

The economic consideration was that the resources of the deep seabed were likely to be of profoundly more commercial value than the fish that swim above them. The political considerations at play were the dwindling *res nullius* approach to commonality, which meant that, although anyone could exploit these resources, in fact only a few did so. To allow any state to explore and exploit the mineral resources of the deep seabed was not as open as it seemed, because only a handful of states would have the material wealth and the technical knowhow to be able to engage in such activities.

What was in principle open to all, in reality would become a resource in the hands of a few.

It was these underlying policy purposes that the principle of 'common heritage of mankind' served, and it was interesting to see how this principle was, in a short space of time, accepted as having a normative quality. In fact the principle was being sought within the wider framework of the multilateral negotiation for the comprehensive law of the sea treaty. The proposals for the United Nations Convention on the Law of the Sea (UNCLOS) contained various separate parts, on the territorial sea, the continental shelf, the new concept of the exclusive economic zone, scientific matters, and the deep seabed. These were to be a 'package deal', negotiated together over a period of 15 years, and they were to stand or fall in their entirety. No state could expect to be fully satisfied with every article of this huge treaty but the hope was that every state would find that their interests were sufficiently represented in most areas that they would be content to accept the parts they did not like. The advanced industrial states that had the potential capacity to exploit the seabed expressed deep reservations about the legal basis of the principle of 'common heritage of mankind', but the Assembly recommended that all states concerned exercised their domestic legal powers to exploit the deep seabed when the moment was right.[11]

These proposals envisaged a UN licensing authority, which would, upon the payment of a fee, grant a licence to explore and exploit certain parts of the proposed licence area, and that certain parts of the proposed area were to be set aside for future exploitation by a different international body, which would sit on behalf of the poorer nations that did not have the financial or territorial base to explore and exploit the area themselves.[12]

Almost inevitably, the advanced industrial states found this unacceptable, and although they found, for the most part, the UNCLOS satisfactory, they could not accept the section on the deep seabed and so refused to accept the treaty as a whole. It took until 1994 before the treaty came into force, without both the United Kingdom and the United States. After more negotiations and an agreement on part XI, the United Kingdom finally signed 2 years later, but the United States is yet to do so. The issues that surround the natural resources of the deep seabed largely concern questions of jurisdiction and law development.

1.3 PETROLEUM DEPOSITS ON THE CONTINENTAL SHELF

Virtually all known commercial petroleum deposits are on shore, under territorial waters, or in the continental shelf. It is now accepted that the initially unilateral acts by which states asserted jurisdiction, for the purpose of exploitation of resources, over the continental shelf rather rapidly became a permissive rule of customary international law. Such claims were then recognised in the 1958 Geneva Convention on the Continental Shelf.[13] The formula under this convention was that the shelf be measured either by a designated depth of 200 m or by exploitability, and the assumption was that these were practically identical. However, rapidly evolving technology made it apparent that it was possible to exploit these resources at deeper levels. Nevertheless, no other state has any entitlement to the exploration for and exploitation of resources on the continental shelf. The conventions of both 1958 and 1982 provide that the coastal state has sovereign jurisdiction for purposes of the exploration and exploitation of these resources, but this is a functional sovereignty and is of great legal importance. Firstly, states must take care that any legislation they pass that purports to have application on the continental shelf—criminal, civil, or tax legislation—is limited to matters relating to the exploitation and exploration of shelf resources. The issue of criminal jurisdiction and civil liability has led to interesting case laws in various jurisdictions, as has the question of taxation of activities on the continental shelf. The issue of whether a state may tax a company that operates on its continental shelf without having any office within its jurisdiction, and which is not itself engaged in exploring and exploiting, has created particular difficulties in the North Sea. Service companies, with vessels providing food or entertainment on oil platforms, have resisted their liability to tax at the hands of the coastal states. Again, the local jurisdiction will have to interpret the tax legislation against the specific jurisdiction authorised under international law.[14]

There is a more important consideration. If onshore mineral resources have been vested in the state, it is clear that it owns them, but it has no right of ownership in the resources *in situ* in the continental shelf; it merely has jurisdiction for the purposes of exploration and exploitation. This means that no one else may explore or exploit without the coastal state's permission. The coastal state may give that permission by the grant of a licence for the purpose of either or both exploring and exploiting the mineral wealth within the continental shelf. The holder of such an offshore licence does not get title to a resource over which the licensing authority does not itself have title. The holder of the licence gets an entitlement to explore and exploit and to reduce the mineral resource into possession. It is the actual reduction into possession that gives the licensee title.

This is of great importance because of the implications for lending institutions. Loans to the licensee cannot be secured against collateral in the form of resources owned by the licensee. Any government could always stop the intended reduction into possession by unilateral action, for example, revoking the licence for any variety of reasons. Inevitably, investors and other non-licence-holding partners have evolved a variety of financial techniques, whereby payment for a loan, or the recouping of profit on an investment, are made by reference to the barrels of oil produced.[15]

The exact entitlement of a licensee under international law in the event of termination, alteration, or the indirect taking of the benefit of the licence by a government is not exactly clear. The question becomes harder to answer when what is altered or taken is not the title to petroleum, but the entitlement to reduce the petroleum into possession. The usual situation is that the resource is located within the jurisdiction of one state. It could be that a foreign company, either private or a public sector company of another state, secures permission to search for and produce the petroleum under a licence. Although it could be considered that the agreement is under contract between the state and the investor, in fact, international law remains ever present in any transnational relationship, thus bringing into play all the norms relating not only to the treatment of foreigners, denial of justice, and good faith, but also to the taking of property.[16]

In an old-style concession, the investor secured title to the petroleum *in situ*. In offshore petroleum, as we have seen, he does not, but only has the right to take into possession. However, it is now accepted that contract rights are themselves a form of property,[17] and international law has much to say on property. In the *Texaco v Libya* case, the international arbiter found that the very fact that there was to be international arbitration made it inevitable that international law would have a role to play. The reasoning was that a foreign investor would be nervous to put himself at risk of a domestic law, which the government could change.

Given that international law has a role to play, a few short points need to be addressed.

The concept of permanent sovereignty over natural resources, which forms part of various resolutions, and which is invoked in tribunals and writings, was challenged in a number of industrialised countries. However, this challenge has now settled down with the acceptance that states do own rights over their natural resources, and tribunals look sympathetically at ways to liberate a state from a disadvantageous contract. In particular, clauses that seek to 'freeze' the situation at the moment of contracting are being given less and less credence by the court. Either the court/tribunal will not regard such clauses as 'stabilizing clauses' as in the *Khemco case*[18] or it may be said that they must explicitly prohibit nationalisation for the clause to have that effect as in the *Aminoil case*.[19]

Attempts to secure a negotiated change will be tolerated but nationalisation will require compensation, and will only be lawful if it is not discriminatory and serves a public purpose. The concept of permanent sovereignty over natural resources does not leave a state free to ignore contracts it has voluntarily entered into.

1.4 ENERGY ACTIVITIES UNDER INTERNATIONAL LAW

The regulation of energy activities at the global and regional levels is addressed under public international law, which is a system of law governing relations between states.[20] International relations relating to the energy sector are found embedded in the rules of international law applicable to a wide range of activities carried out by states in both the energy sector and other commercial activities. Of particular interest is how international law sets standards in respect of energy activities and, in particular, in the environmental sphere. International law sets standards in respect of energy activities and their consequences, and the enforcement of those standards by or against states.[21] It is not practicable to regulate industrial activities to the extent of requiring 0% emissions. Instead, the standard regulatory approach has been to set permissible emission levels and/or stipulate the use of particular technologies, that is, the command-and-control method of regulation. Specific standards and/or techniques are mandated and enforced. Increasingly, however, the importance of markets in controlling environmental problems is being stressed both at the domestic and international levels. The use of market-based instruments such as taxes or marketable emission permits increases the internalisation of costs and reduces negative externalities. Externalities are 'the systematic by-products of the method of production and supply of energy . . . services, which are effectively unaccounted for in the conventional pricing mechanisms of the market'.[22] One negative externality is environmental pollution. Although encouraging the internalisation of negative externalities is typically described as an application of 'the polluter pays' principle, it is of course consumers who ultimately pay through the increased price. This is the 'cost pass through' phenomenon,[23] which in turn has the effect of influencing consumer behaviour and generating awareness of the 'true' environmental cost of their activities.[24]

State government intervention is a necessary corollary to eliminating or reducing externalities, often expressed as the need for public regulation to correct the so-called market failure. The method of regulation chosen will determine the extent to which the burden of a polluting activity is borne equitably. Command-and-control standards and technology forcing, for example, are designed to cause the absorption of externalities, the philosophy being that it is more equitable for the polluter to pay. However, the form that government regulation should take can be controversial.

1.4.1 The Impact of Privatisation

The precise impact of privatisation on environmental regulation of the energy sector is difficult to measure. Although an assumption may be made that privatisation, especially coupled with deregulation, will adversely affect levels of environmental protection, Lucas illustrates from a Canadian perspective that it is difficult in fact to generalise about the effects on environmental regulation of commercialisation and privatisation processes.[25] Perhaps the most discernible effect upon environmental regulation of policies of liberalisation and privatisation is the phenomenon of 're-regulation'. This is manifested in a shift away from the standard regulatory approach of setting permissible emission levels and/or stipulating the use of particular technologies—that is, the command-and-control method of regulation. This shift is observable not only at state level but also at the EU and international levels.

1.5 THE INFLUENCE OF SCIENCE ON ENVIRONMENTAL REGULATION IN THE ENERGY SECTOR[26]

The role of science has been prominent in the development of international law and environmental regulation. Unfortunately, scientists do not always agree on the definition of an environmental problem or the prescriptions for its solution. This can be the result of scientists using different data as in the establishment of quotas for fisheries, or the use of different models as in the study of climate change. Therefore, given the lack of certainty, most decisions in the environmental field have to be made based on political considerations. Despite that, it is the scientific communities who inform and frame the environmental discourse. The United Nations Environmental Programme (UNEP) and national scientific institutions such as the US National Aeronautics and Space Administration (NASA) have executed much research on ozone depletion. The Intergovernmental Panel on Climate Change (IPCC) was organised by the UNEP to provide insights into the climate change debates.

Scientists and their expressed positions are used by both proponents and opponents of environmental issues and action groups to advance their own views. In fact, the climate change debate has been influenced by scientists and their inability to reach consensus on certain areas of concern. This lack of scientific consensus has led some states to delay constructive action for the abatement of carbon dioxide emissions, which is known to be the main culprit of climate change. By contrast, the scientific evidence on the depletion of the ozone layer over Antarctica produced robust international action to phase out ozone-depleting substances.[27]

The international environmental debate could be characterised as a series of dialectic interactions between opposing groups[28]: those who wish to continue with 'business as usual' and those who wish to inject a principle of sustainability into development. It is the role of the scientific community to facilitate this dialectic interaction.

1.5.1 Definition of Pollution

Under the UK Environmental Protection Act 1990, pollution is defined as 'a release into any environmental medium from any process of substances which are capable of causing harm to man or any other living organisms supported by the environment'. Alternatively, scientific definitions of pollution often contain phrases that relate to the ecological injury caused by pollution such as 'the introduction by man into the environment of substances or energy liable to cause hazards to human health, harm to living resources and ecological damage, or interferences with legitimate uses of the environment'.[29]

The procedures used to evaluate the environmental risk of a given pollutant are primarily based on the physiochemical properties of the pollutant and single-species or multi-species toxicity tests.[30] The results from such assessments are then incorporated into environmental regulation by way of standards, as they give a relatively rapid and reproducible assessment of safe environmental concentrations of any given pollutant. One reason for this has been the advances in science and technology with toxicology and ecotoxicology now established as disciplines. More and more synthetic substances have been brought into use and become of interest to those responsible for protecting the environment. Numerical standards seemed to be the most obvious and convenient way of summarising and codifying scientific understanding of human impacts on the environment in order to make it readily usable by policymakers and regulators.

As we shall see, the growth of European legislation with its predominance of the use of standards reflecting the traditions of regulation in some Member States also contributes to the importance of numerical standards in order to secure consistent implementation of policies across several tiers of government.

1.5.2 Setting Environmental Standards

The term 'standard' has sometimes been used in the environmental field in the narrow sense of a legally enforceable numerical limit, but the Royal Commission on Environmental Pollution has redefined an environmental standard to also include those that are not mandatory, but are contained in guidelines, codes of practice, or sets of criteria for deciding individual cases. Further, the definition also includes standards not set by governments but which carry authority for reasons of the scientific eminence of those who set them.[31]

The procedure for setting standards first requires the recognition and definition of the problem. Policy aims then have to be formulated by rigorous and dispassionate investigation and analysis, deliberation and synthesis, all of which should be informed by people's values. The decision then must be taken whether to set a standard, and if so, what type of standard. Once the standard is set by specifying the content of the standard, it should be followed up with monitoring and evaluating its effectiveness.

The analytical stage of the policy process has several complementary and closely interrelated components.[32]

1.5.3 Scientific Assessment

The primary source of evidence for assessing new and existing chemical substances is the toxicity data required to be submitted by the manufacturer or importer when notifying a new substance to the regulator. For these new substances, the aim is to prevent those that are potentially hazardous, posing an actual risk. For existing substances, the aim is to identify any need for better management of the risks posed by a substance. This assessment of new and existing chemical substances is a typical example of the scientific assessments carried out to provide the basis for decisions on environmental standards.

The methods for testing for toxicity and ecotoxicity have become increasingly standardised across the world with the Organisation for Economic Cooperation and Development (OECD) producing guidelines for tests in order to avert an escalation of testing to meet different data requirements by individual governments for the same substance. The International Organisation for Standardisation (ISO) has also been active in this field.

It has long been a central theme of UK environmental policies that decisions should be based on what has frequently been called *sound science*. The 1990 Environment White Paper[33] stated that 'We must base our policies on fact not fancy, and use the best evidence and analysis available . . . the need, in environmental decisions as elsewhere, to look at all the facts and likely consequences of action on the basis of the best scientific evidence'.[34] Statutory guidance given by Ministers to the Environment Agency affirmed that they should 'operate to high professional standards, based on sound science, information and analysis of the environment and of processes which affect it'.[35] Therefore, when considering the process of scientific assessment and its output, two separate issues need to be addressed. Firstly, is the science well done with recognition of any uncertainties, and are limitations in the data properly recognised? Secondly, does the science provide a firm basis for policy decisions, or will the policymaker have to take decisions in the face of uncertainty?

1.5.4 The Technical Options

Analysis of technical options is an essential component in the analyses needed to underpin decisions about environmental policies and standards. Depending on the circumstances, the emphasis may be on the environmental acceptability of a technological development, or it could be on the extent to which new technology can reduce the environmental effects of an activity. However, increasingly, the emphasis is on an integrated consideration of both.

Emission standards are often set at a level that it is known an available technology can achieve. Thus, the limit values in EC stage II legislation on emissions from cars[36] presupposed that manufacturers would comply by fitting catalytic converters and engine management systems to all the new petrol-fuelled cars. Another relevant technological consideration is the precision of available methods for measuring the concentrations at which substances are present as in the determination of the limit value for individual pesticides set in the EC Drinking Water Directive in 1980.

The use of technology standards has not been widespread in the United Kingdom as they would prescribe the use of a particular technology to limit emissions. The type of standard in which technology is most directly taken into account in the United Kingdom is the process standards set by the Environment Agency. This is somewhat similar to the use of emission standards, in that plants have legal limits placed on their emissions in accordance with benchmark levels contained in the guidance notes.[37] The key difference is that those benchmark levels are based on a comprehensive view of the relevant industrial process, covering in particular the release of substances to air, water, and land.

The desirability, in principle, of taking a comprehensive view of the environmental implications of each option, however, is sometimes constrained by practicability, both in terms of the analyses that can be carried out and in terms of the decisions that can be taken on the findings.

1.5.5 Best Available Techniques Not Entailing Excessive Cost (BATNEEC)

The operator of any prescribed industrial process in England, Wales, and Scotland is required to prevent the release into any environmental medium of substances prescribed for that medium or, where that is not practicable, using best available techniques not entailing excessive cost (BATNEEC), to use such techniques to reduce releases of substances to a medium and render them harmless. Where a prescribed industrial process is likely to involve the release of substances into more than one environmental medium, the conditions the regulator attaches to the authorisation for the process must have the objective of ensuring that BATNEEC 'will be used for minimising the pollution . . . to the environment taken as a whole . . . having regard to the best practicable environmental option available as respects the substances which may be released'.[38]

The use of this technique first appeared in legislation in the context of emissions to air from industrial plants in the EC Framework Directive in 1984.[39] The draft of this directive required the use of 'state-of-the-art' technology, and was amended following pressure from the United Kingdom. In the original draft directive, the 'T' stood for 'technology', whereas in the 1990 Act, the United Kingdom used the word 'techniques' to ensure that BATNEEC could be given as wide an interpretation as the earlier term of best practicable means (BPM).

An important advantage claimed for BATNEEC as a regulatory principle is that it is dynamic rather than static and is capable of different interpretations over time as techniques improve. Irrespective of the limit values contained in any authorisation, the operator is under a general obligation to use BATNEEC.

The UK Environment Agency, as regulators, enjoy wide discretion as to what constitute BATNEEC, but one factor in determining whether the cost of a particular set of techniques is excessive is the damage to the environment that the process is causing, or would otherwise cause. The Agency also considers the resources a typical operator has available for capital expenditure and the extent to which market conditions allow costs to be passed on to customers, passed back to suppliers, or absorbed through lower profitability.

1.5.6 Best Practicable Environmental Option (BPEO)

The purpose of the duty to have regard to the best practicable environmental option (BPEO) is to impose an obligation to take into account local conditions if they point to the need for more stringent limits on emissions than would be required by the obligations to use BATNEEC taken on its own. The obligation is site specific as local conditions will vary from site to site. It is, therefore, for the operators of a process to identify what they regard as BATNEEC, having regard to the BPEO, and to justify their conclusions to the regulator.

The statutory requirement to have regard to the BPEO has two serious limitations. Firstly, it applies only with regard to authorisations of those processes subject to integrated pollution control and not across all the functions of the environment agencies, neither to any form of pollution that they do not regulate. Secondly, even in the context where it is applicable, it is given a very narrow meaning. In particular, differences in energy requirements between different options are not taken into consideration.[40] Further, although any solid waste arising from a process is taken into account, the regulator is not entitled to take into account how it is dealt with if it is disposed of at another site.

1.5.7 Best Available Techniques (BAT) and Life Cycle Assessment

The EC Directive on Integrated Pollution Prevention and Control (IPPC)[41] embraces, as an overall view of emissions from a process, its decommissioning and key off-site aspects of the life cycle in the form of the energy requirements for the process and disposal of its wastes. The Directive's use of best available techniques (BAT) as the criterion, without the not entailing excessive cost (NEEC) part, is not as important as may at first appear. The definition of 'available' includes consideration of costs and refers to 'implementation . . . under economically . . . viable conditions'.[42]

Life cycle assessment is a formal technique that brings into consideration all the environmental impacts associated with the delivery of a service or a product. It involves identifying and quantifying the emissions and resources used at all stages of the life cycle.[43] However, looking at the entire material and energy supply required to make a product or provide a service may lead to different conclusions about environmental impacts. The greatest pollution may occur, not from the manufacturing process, which is subject to other regulatory controls, but either upstream[44] or downstream.[45] The value of life cycle perspectives is not confined to the differences between technologies, but can be used to identify 'hot spots' in the supply chain where any environmental impacts may be particularly significant.

The addition of pollution abatement devices to an industrial process to comply with the regulatory requirements inevitably involves economic cost or reduced efficiency.

As emission standards are progressively tightened, the requirement to capture pollutants before emission, usually by reaction with some appropriate reagent, becomes more expensive.[46] Additionally, clean-up technology usually involves transforming pollutants to a different form. The problem may be alleviated but is rarely eliminated, an example being provided by the generation of power from coal. In a conventional combustion plant, the established form of clean-up technology to abate acid gas emissions is flue gas desulphurisation (FGD), which makes the acid components react with lime. The limestone required for this is obtained by mining, and produces calcium sulphate in quantities that may be too large for economic use and may, therefore, have to be disposed of. Under the EC IPPC Directive, which takes into account energy requirements, a regulator could query the use of FGD as necessarily BAT, as operators of FGD equipment significantly reduce thermal efficiency.

On whatever basis technological options are compared, there remains a fundamental conceptual and methodological problem. How can different kinds of environmental impact, which are not directly commensurable with each other, be included in an overall assessment? This may arise in determining which option should be regarded as 'best' in terms of BATNEEC or BPEO just as much as in deciding which of the themes considered in life cycle assessment should be accorded most significance. The determination of what represents BPEO for a given process, for example, involves assessing the relative environmental impact each of the substances given off by the process would have if released to each medium.

1.6 INTERNATIONAL POLICY AND REGULATORY BODIES

Energy-producing states will normally have energy in excess to their industrial and domestic needs, whereas other states have a demand for energy as they do not have the natural resources to produce sufficient energy for their internal needs. This produces an interdependence between the states and a necessity for them to cooperate with one another. Although traditionally the regulation of energy issues has been regarded as a domestic matter of a sovereign state, the move further towards international trade between states for energy products requires greater international regulation of energy. In order to secure a continuing and self-sustaining system of energy relations, an institutional framework has been established to provide a forum of cooperation, consultation, and negotiations. This consists of a number of organisations that may be classified according to their functions.

1.7 ENERGY ACTIVITIES WITHIN THE UNITED NATIONS

The UN Charter established six principle organs: the General Assembly (GA), the Security Council, the Trusteeship Council, the Secretariat, and the International Court of Justice in The Hague. Entities within the United Nations undertake a wide range of energy activities that are mainly dictated by their mandates. The main activities as far as this book is concerned are the development of environmental concerns within the energy sector, particularly with regard to the conservation of valuable fossil resources and the promotion of new and renewable sources of energy.

They concern the security of energy supplies and analysis of current and future supply and demand. Over the last 20 years, legally binding environmental standards have been developed by international organisations, notably Principle 21 of the UN Conference on the Human Environment.

The GA unanimously adopted the resolutions on permanent sovereignty over natural resources.[47] The UN Conference on Environment and Development at Rio de Janeiro endorsed the Nairobi Programme of Action for the Development and Utilisation of New and Renewable Sources of Energy.[48] The Assembly also reviews the reports of the International Atomic Energy Agency (IAEA) by adopting resolutions. The Security Council may well deal with energy-related questions if they concern world peace.[49]

The UNEP promotes environmental science by coordinating activities within the UN system. It encompasses six divisions that include the Division of Technology, Industry and Economics, within which the Energy and Ozone Action Unit promotes good management practices and use of energy. The Collaborating Centre on Energy and Environment aids UNEP as a research and technical support unit through energy and planning policy.[50]

A major organ of the UN system with regard to the energy sector is the International Maritime Organisation (IMO) whose mandate is to ensure safe and clean seas. With continuing exploration for and exploitation of oil and gas on the continental shelf of coastal states and the transportation of oil by tanker, the IMO is responsible for setting up and monitoring a number of international maritime conventions that cover the energy sector.

1.7.1 The International Atomic Energy Agency (IAEA)

The IAEA was set up as an organ of the United Nations in 1956 and the statute came into force in July 1957. It has the duty to report to the GA and the Security Council in special situations.[51] It also has a duty to cooperate by agreement with the work of EURATOM.[52] The main area of activity is to support the contribution of nuclear energy to ensure peace, health, and prosperity throughout the world. The IAEA cannot support military purposes.[53] Its main duty is to 'safeguard' the use of nuclear materials and facilities in Agency projects and carry out peaceful activities in the nuclear field under the Non-Proliferation Treaty (NPT).[54]

1.7.2 The Organisation for Economic Cooperation and Development (OECD)

Although the role of the OECD under the Convention is primarily 'to strive for the highest sustainable economic growth while maintaining financial stability',[55] the energy crises of the 1970s provoked action, and both energy and environmental policies were added to its ambit. There are over 200 specialised bodies that cover inter alia environment and energy, scientific and technological policies.

The International Energy Agency (IEA) was founded in 1974 and is the forum for coordinating the energy policies of the industrial countries who are members of the OECD. It is one of the few organisations that deal exclusively with energy.

The Nuclear Energy Agency (NEA) was established by the OECD Council in 1957 to further the peaceful uses of nuclear energy. The NEA provides a databank that enables participating countries to share computer programmes used in nuclear data applications. It makes regulations for major nuclear accidents and regulates radioactive waste disposal.

1.8 THE INTERNATIONAL SCOPE OF STANDARD SETTING

Under the international law of treaties and customary law, there lies a further dimension of regulatory measures that are followed by international organisations. This group of regulatory measures is collectively known as *transnational administrative law.*[56] One such technique is standard setting, for example, for permits or licences that are not commonly issued at international level but are left to the individual state to administer. Regulations require the sovereign state to issue such licences in accordance with internationally accepted standards. There is much work undertaken on a global scale to set environmental standards that, although not binding in international law, is nevertheless of practicable significance, in particular in the area of the protection of humans. Additionally, there is much scientific activity on environmental issues at the global level that can be differentiated from the international conventions as not having a direct regulatory purpose. Much of it is, however, directly linked to the work of the regulatory bodies; for example, the United Nations Scientific Committee on the Effects of Atomic Radiation carries out regular reviews of the scientific evidence that provides the basis for the work of the International Commission on Radiological Protection (ICRP). It does not itself, however, make recommendations about acceptable levels of exposure or put forward guidelines. The work of the IPCC, established by the World Meteorological Organization (WMO) and UNEP, led to the Climate Change Convention, and its scientific assessments continue to provide a major input into the implementation of the convention.

Many international conventions have been drawn up on a regional basis by the group of nations most affected by, or able to influence, the type of pollution in question. The Convention on the Protection of the North Sea and North-East Atlantic (the Oslo/Paris [OSPAR] Convention, 1992) replaced the 1972 Oslo Convention on dumping from ships and aircraft and the Paris Convention 1974 on land-based sources of marine pollution. It regulates those forms of marine pollution that are not covered by other global conventions. The High Contracting Parties to the Convention are required to implement control programmes to reduce the amounts of a wide range of substances reaching the sea, including heavy metals, oil and hydrocarbons, and organic halogen compounds. Parties to the Convention are required to have regard to the 'best environmental practice'. This concept includes the use of the BAT and other dimensions such as information to the public, product labelling, and collection and disposal systems. The OSPAR Commission responsible for implementing the Convention has the duty to draw up documents specifying BAT for disposal reduction and elimination of a range of substances from various industrial sectors.

1.9 ISO, EUROPEAN ECO-MANAGEMENT AND AUDIT SCHEME (EMAS), AND SECTOR STANDARDS

The ISO has set a series of voluntary standards for all industry sectors as well as for a variety of cross-sector, horizontal themes. The Organisation has a membership of 160 national standard bodies from countries large and small; industrialised, developing, and in transition; and has a presence in most regions of the world. ISO has developed a portfolio of over 18,600 standards and provides business, government, and society with practical tools for the three dimensions of sustainable development, namely, economic, environmental, and social. The aim of the ISO is to establish standards that provide solutions for almost all sectors of activity and represent a global consensus on the state of the art in the subject of that standard.

In July 2011, the ISO published the ISO 50001, which gives organisations the requirements for energy management systems (EnMS). This was in response to a request from the United Nations Industrial Development Organisation (UNDO) to develop an international energy management standard. The United Nations had recognised industry's need to mount an effective response to climate change and the proliferation of national energy management standards. ISO had, in turn, identified energy management as one of the top five fields for the development of international standards and, in 2008, created a project committee, ISO/PC 242, Energy Management, to carry out the work. ISO/PC 242 was led by ISO members for the United States (American National Standards Institute—ANSI) and Brazil (Associação Brasileira de Normas Technicas—ABNT). These were joined by 44 ISO participating member countries. The standard was developed under the auspices of the UNDO and the World Energy Council (WEC).

ISO 50001 is based on the ISO management system model familiar to most organisations worldwide who implement standards such as ISO 9001 on quality management, ISO 14001 on environmental management, and ISO/International Electrotechnical Commission (IEC) 27001 on information security. In particular, ISO 50001 follows the plan–do–check–act process for continual improvement of the energy management system. This enables organisations to integrate energy management with their other ISO management systems.

The ISO 14000 is a series of standards on environmental management that provides a framework for the development of an environmental management system (EMS) and a supporting audit programme. The main thrust for its development came as a result of the world summit at Rio de Janeiro held in 1992.

As a number of national standards emerged, such as the British Standard 7750,[57] the ISO created a group to investigate how such standards might benefit business and industry. As the ISO 14000 series was developed, 14001 became the cornerstone. It specifies a framework of control for an environmental management system against which an organisation can be certified by a third party.

Other standards in the series are actually guidelines that include the following:

- ISO 14004 provides guidance on the development and implementation of EMSs.

- ISO 14010 provides general principles of environmental auditing (now superseded by ISO 19011).
- ISO 14012 provides guidance on qualification criteria for environmental auditors and lead auditors (now superseded by ISO 19011).
- ISO 14013/5 provides audit programme review and assessment material.
- ISO 14020+ deals with labelling issues.
- ISO 14030+ provides guidance on performance targets and monitoring within an environmental management system.
- ISO 14040+ covers life cycle issues.

Of all of these, ISO 14001 is not only the most well known, but is also the only ISO standard against which it is currently possible to be certified by an external certification authority.

1.9.1 The European Eco-Management and Audit Scheme (EMAS)

The EMAS[58] is a regional environmental standard that is similar but more extensive than the ISO 14000. The EC Regulation 1836/93 on Eco-Management and Audit came into force in April 1995, having been agreed in 1993 and piloted in a number of EU Member States. As a regulation has direct effect on all Member States, it is now automatically part of a uniform EC-wide system.

The EMAS is generally a site-based registration system with due consideration provided to off-site activities that may have a bearing upon the products and services of the primary site. Within the United Kingdom, an extension to the scheme has been agreed for local government operations, who may also register their environmental management systems with the EMAS Regulations. EMAS requires an environmental policy to be in existence within the organisation, fully supported by senior management and outlining the policies of the company not only to the staff but also to the general public and other stake holders. The policy needs to clarify compliance with environmental regulations that may affect the organisation and stress a commitment to continuous improvement. Emphasis has been placed on policy as this provides the direction for the remainder of the management system.

Those corporations who have witnessed ISO 9000 assessments will know that the policy is frequently discussed during assessment, with many staff asked if they understand or are aware of the policy, and any problems associated with the policy are seldom serious. However, the environmental policy is different as this provides the initial foundation and direction for the management system and will be more stringently reviewed than a similar ISO 9000 policy. The statement must be publicised in non-technical language so that it can be understood by the majority of readers. It should relate to the sites within the organisation encompassed by the management system in addition to providing an overview of the company's activities on the site and a description of those activities. This should provide a clear picture of the company's operations.

In addition to a summary of the process, the statement requires quantifiable data on current emissions and environmental effects emanating from the site, waste generated, raw materials utilised, energy and water resources consumed, and any other environmental aspect that may relate to operations on the site. The preparatory review is part of an EMAS assessment, which is not the case for the British Standard 7750. The environmental review must be comprehensive in consideration of all input processes and output on the site. This control process is designed to identify all relevant environmental aspects that may arise from the site, which may relate to current operations, future operations, or indeed even unplanned future activities. They will certainly relate to any past activities on the site that may have caused, for instance, contamination of the land. This preparatory review must also include a wide-ranging consideration of all legislation that may affect the site, and of whether the legislation is currently being complied with. The company must declare its primary environmental objectives and those that can have the most environmental impact.

The programme will be a plan to achieve specific goals or targets along the route to a specific goal and describe the means to reach those objectives. The EMS establishes procedure, work instructions, and controls to ensure that implementation of the policy and achievement of the targets can become a reality.

As with ISO 9000, the environmental management system requires a planned comprehensive periodic audit to ensure that it is effective in operation, is meeting specified goals, and that the system continues to perform in accordance with relevant regulations and standards. Under EMAS, the minimum frequency for an audit is at least once every 3 years.

The peculiarity with EMAS is that the policy statement, the programme, the management system, and audit cycles are reviewed and validated by an external accredited EMAS verifier. The verifier not only provides a registration service but is also required to confirm the company's periodic environmental statements.

1.10 CORPORATE SOCIAL RESPONSIBILITY AND MULTINATIONALS

The globalisation of production and services by international corporations has heightened the sense of awareness concerning the potential for transnational environmental harm that the activities of such companies may create. Alternatively, a responsible international corporation may also help to raise the environmental standards in developing countries, whose environmental regulation may be somewhat lower than that which would be expected in a developed state, by acting as the main repository of modern, environmentally friendly technology and as the most advanced experts on environmentally sound management practices.[59] This section considers environmental issues in business, in particular for the multinational corporation and how the responsible company may use corporate social responsibility (CSR) to promote environmentally sustainable development.

Legal regulation of the environmental activities of international companies is based on the concept of 'sustainable development' with more specific elements of environmental protection. With regard to the specific responsibilities of a multinational corporation, attention must be given to other concepts such as 'the polluter pays'; the 'preventative', and the 'precautionary' principles, the first two being enforceable in law, whereas the third one will go to 'due diligence' if things go wrong. The most widely recognised definition of 'sustainable development' is that of the Bruntland Commission, which called for development that 'meets the needs of the present without compromising the ability of future generations to meet their own needs'.[60] The emphasis is on *needs* rather than *wants*, and on inter and intra-generational justice. This approach is based on an accommodation between economic growth, environmental concerns, and the wider social effects of economic activity. CSR would support methods and processes of economic growth that ensure the survival of a sustainable ecosystem that can last for generations. Equally, the social effects of environmental protection, or damage, if that is so, need to be taken into account as part of the complex range of interactions that characterise the concept of sustainable development.

Following on from the concept of sustainable development is the idea of environmental protection. This is a wide concept that cannot be exhaustively defined, but includes key issues such as the preservation of the quality of the air, water, and soil; the sustainable use of natural resources; the preservation of human, animal, and plant life and health, and of the ecosystem more generally (United Nations Conference on Trade and Development [UNCTAD] 1972).[61] These goals form the bedrock of environmental regulation policy, both at national and international levels. They not only apply to states but to non-state actors such as corporations as well. These goals have been progressively developed in both national legal instruments and international conventions relating to, inter alia, the atmosphere, pollution of the sea, the protection of freshwater resources, the preservation of biological diversity, the control of hazardous substances and activities, the regulation of waste creation and its disposal, and the preservation of the Arctic regions.[62]

1.11 ENVIRONMENTAL REGULATION OF INTERNATIONAL CORPORATIONS

The main point to make here is that environmental regulation does not distinguish between international corporations and domestic companies as a matter of principle; however, certain international regimes concerning environmental protection are emerging, which focus on the international corporation in particular. In the case of developing countries, the activities of international companies may, in some cases, be the main or significant area of industrial activity that can lead to environmental harm. Therefore, international companies can be seen as major subjects of responsibilities in this field.[63]

Although there are differing kinds of environmental regulations including informal or self-regulation, which includes CSR, these will be addressed later in the section on environmental management and auditing. Here we address the more enforceable aspects of formal regulation of the environmental practices of international corporations. These 'formal' regulations refer to those mandatory requirements imposed by governments and regional organisations with law-making powers, as well as standard-setting environmental agreements aimed at both environmental protection and the furtherance of sustainable development, commonly referred to as *multinational environmental agreements* (MEAs). Equally, the interaction of international investment agreements (IIAs) with national regulations, and with MEAs, should be considered.

1.11.1 National and Regional Regulations

National and regional regulations are the principal way in which the standards within the MEAs are established. Furthermore, it is at the national level where corporations actually operate, and where legal claims are brought against the firm. National standard setting and enforcement is an essential part for ensuring that corporations actually meet the levels of environmentally sound practice needed to protect the environment efficiently.[64] Explicit standards, backed up by effective enforcement measures, change corporate behaviour. The more stringent carbon dioxide emission standards in national legislation have led to the development of new technologies in the car industry that reduce such emissions in addition to developing more efficient cars. Fiona Harvey[65] notes that, in the first 3 months of 2005, companies in countries that have ratified the Kyoto Protocol, including EU Member States, Canada, and Japan saw their shares rise by an average of 21.9%, whereas shares in companies specialising in renewable energy in Australia, a non-signatory of the Protocol, rose by only 4.2%. Shares in the United States (also a non-signatory of the Protocol) renewable energy companies actually fell 13.8% on average. More stringent environmental standards, therefore, should give rise to cleaner forms of commercial activities. As to corporate environmental governance, the European Union has taken a lead with the eco-audit and management scheme. Equally, liability rules, especially in the United States, under their Comprehensive Environmental Response, Compensation and Liability Act 1980 (CERCLA), can be particularly harsh, with strict liability for environmental damage extending to corporate directors, officers, lenders, and shareholders. In 2004, the principles applicable for EU Member States were laid down in the Directive on Environmental Liability.[66]

Turning to the substantive content of international instruments covering the responsibilities of international corporations in the environmental field, two voluntary codes are of particular importance. Firstly, the continuing importance of Agenda 21 in this area was affirmed at the World Summit on Sustainable Development (WSSD) in Johannesburg in 2002. Secondly, the OECD brought out their *Guidelines for Multinational Enterprises*. Both of these instruments emphasise the furtherance of sustainable development through the transfer of environmentally sound technology and management practices.

1.12 CARBON CAPTURE AND STORAGE (CCS)

In 'The Carbon Capture and Storage Legal and Regulatory Review', the IEA considered that carbon capture and storage (CCS) was a crucial part of worldwide efforts to limit global warming by reducing green house gas (GHG) emissions. It was estimated that the broad deployment of low-carbon energy technologies could reduce projected 2050 emissions to half of the 2005 levels, and that CCS could contribute about one-fifth of those reductions in a least-cost emissions reduction portfolio. However, that would mean that about 1000 CCS projects must be implemented by 2020 and over 3000 by 2050.[67]

One of the major issues would be the long-term liability for stored carbon dioxide. Within the CCS industry, liability tends to be used as a generic term for any legal liabilities arising from a storage site,[68] including the responsibility for undertaking and bearing the cost of putting right any damage caused by activities associated with the storage site. This also covers leakage of CO_2 from the site to the atmosphere, where CCS operations are undertaken as part of a CO_2 emissions reduction scheme. 'Long-term liability' normally refers to any liabilities arising after the permanent cessation of CO_2 injection and active monitoring of the site.

Most discussions on long-term liability have been based on who should retain this residual liability—the operator or the regulating government in whose jurisdiction the site is, through the public purse. A number of emerging legislations from different jurisdictions have shown a trend towards the government taking on this long-term liability after a period of about 15 years post closure.[69] However, there is no clear consensus on this issue as some of the emerging legislations are silent on the matters that would imply that the liability remains with the operator in perpetuity.

Generally, prior to liability being transferred, there are three requirements to be fufilled:

- There is evidence that there is no significant risk of physical leakage or seepage of the stored CO_2.
- A minimum time period has elapsed from the cessation of the injection operations.
- There is financial contribution to long-term stewardship of the site that would minimise the financial exposure of the (government) entity designated to take on the long-term liability.

However, there are some marked differences between jurisdictions on how these requirements are interpreted in legislation and the processes under which the operator may demonstrate that they have been met. Even the way in which the liability is transferred differs from one jurisdiction to another.

For Member States of the European Union, the long-term liability arrangements are set by the EU CCS Directive. Article 18 of the Directive requires that all legal obligations under the Directive relating to monitoring and corrective measures, surrender of allowances if leakage occurs, and preventative and remedial action shall be transferred to the competent authority where:

- All available evidence suggests that stored CO_2 will be completely and permanently contained.
- A minimum period as specified by the competent authority has elapsed (this is to be no shorter than 20 years, except where the competent authority deems that the evidence of complete and permanent containment of the stored CO_2 is available before such period has elapsed).
- Certain financial obligations have been fulfilled.
- The site has been sealed and injection facilities removed.

If there is negligence or wilful deceit on behalf of the operator, the transfer will not take place. All Member States are obliged to transpose these provisions of the EU CCS Directive into their national laws. Although Norway is not a member of the European Union, she is a member of the European Economic Area and has undertaken to follow the EU CCS Directive.[70]

In Australia, the Commonwealth and state legislations generally follow the principles of the EU Directive. In most of these jurisdictions, the transfer of liability extends to all those associated with a storage site. The exception is the State of Victoria, where common law liability remains with the operator in perpetuity. Considering this difference regarding long-term liability between certain states and the Commonwealth, there may well be implications for any cross-boundary storage projects. This is an issue currently being considered in Australia.

In both the United States and Canada, CCS is regulated at federal and regional levels and long-term liability is still being considered in these jurisdictions. A number of federal bills addressing long-term liability for geological storage have been introduced but none have yet been passed by Congress. The Environmental Protection Agency is currently considering a conditional exemption for CO_2 from federal hazardous waste legislation requirements, which could affect the amount of long-term liability.[71] The US Interagency Task Force for CCS recommended to the US President in August 2010 four approaches on long-term liability and stewardship for further consideration:

- Reliance on the existing framework for long-term liability and stewardship.
- Adoption of substantive or procedural limitations on claims.
- Creation of an industry-financed trust fund to support long-term stewardship activities and compensate parties for various types and forms of losses or damages that occur after site closure.
- Transfer of liability to the federal government after site closure (with certain contingences).

The task force have recommended that open-ended federal indemnification should not be considered. Six of the states have already enacted legislation relevant to long-term liability for stored CO_2, with five of them providing for transfer of liability to the government. The exception is Wyoming, but Illinois, Louisiana, Montana, North Dakota, and Texas are all transferring long-term liability to the government.

In Canada, Alberta has provided for long-term liability being transferred to the government once a storage site has been properly closed and the operator has demonstrated that stored CO_2 is stable. However, in Saskatchewan, it is the individual well licence owner who is liable for the long term. British Columbia is still considering the issues of long-term liability within its broader CCS framework.

Not only these individual governments but also international organisations are giving serious thought to long-term liability. In the United States, CCSR, for example, is advocating the development of a federal programme to manage and limit long-term liabilities, as well as special liability arrangements for first-mover projects.[72]

1.12.1 Measuring Stability

A critical aspect of the viability and environmental efficacy of CCS is the long-term security of such sites. This must be demonstrated through monitoring and verification for some time after cessation of CO_2 injection.[73] If a jurisdiction provides for transfer of liability to government or allows the operator to discontinue monitoring and verification, then the government will need to be confident that the site is behaving in a predictable and safe manner. This will ensure that the government will not bear an unacceptable risk of storage site failure.

All jurisdictions require a minimum specified time period to have elapsed after cessation of injection before liability is transferred. During this period, the behaviour of the site and injected CO_2 are closely monitored. These periods currently range from 20 to 50 years. However, this period can be modified at the discretion of the relevant designated authority.

Covering the costs of liability once any government has accepted its transfer will require a contribution by the operator, if the public purse, by way of the tax payer, is to reduce the limit of government exposure. The way any contribution is collected, the amount of that contribution, and how the funds are managed vary throughout the different jurisdictions. Some mechanisms for accruing contributions may include royalties, fees, trust funds, and insurance. The trust fund is widely used in Alberta, Louisiana, North Dakota, and Germany, where the fund benefits from being able to build up over the course of the project and may be used to pool risks across several projects. Alternatively, a combination of royalties and insurance is used by Victoria in Australia to cover different parts of the CCS chain. Insurers find it difficult to contemplate open-ended insurance, so this is only considered to be appropriate for the period of operation.

1.13 DEVELOPMENTS ON THE INTERNATIONAL CCS SCENE

1.13.1 The 2009 London Protocol Amendment

Article 6 of the London Protocol sets out a general prohibition on the export of wastes or other matter to other countries for dumping or incineration at sea. The CCS legal and technical working group under the London Protocol determined, in 2008, that Article 6 prohibits the export of CO_2 from a contracting party to other countries for injection into sub-seabed geological formations irrespective of any commercial basis for the movement of CO_2, and that an amendment would be required to facilitate the development of CCS activities. In October 2009, the Contracting Parties adopted a resolution proposing an amendment to Article 6 to provide an exception for the export of CO_2 streams in certain specified cases.

1.13.2 The OSPAR Convention

Annexes II and III of the OSPAR Convention[74] were amended in 2007 to enable the injection of CO_2 into the sub-seabed. These amendments were ratified by Norway, the United Kingdom, the European Union, Germany, Luxemburg, and Spain. Others are well advanced in the process, and as only one further party to the convention is needed for the amendments to come into force, it is anticipated that this will happen later this year.

At the Conference of Parties (COP) 16, climate change negotiations in Cancun, Mexico in 2010,[75] it was determined that CCS should be included as an eligible clean development mechanism (CDM) project activity. However, issues identified in Decision 2/CMP, 5, paragraph 29, as well as Decision 7/CMP, 6 ‘Carbon dioxide capture and storage in geological formations as clean development mechanism project activities' must be addressed and resolved in a satisfactory manner prior to inclusion.

1.14 HOW DO YOU DEVELOP A CCS REGULATORY FRAMEWORK?

This is the question that the IEA attempted to address in its report in 2011.

The IEA came to the conclusion that it is difficult to establish a 'one-size-fits-all' regulatory framework but they did lay down a model framework including the potential steps that a jurisdiction might take in implementing CCS regulation, such as the following:

- Identifying the purpose behind CCS framework development (e.g. will the resulting framework regulate a small number of demonstration projects, or is it intended to regulate large-scale deployment?).
- Developing an understanding of how existing regulatory frameworks address issues associated with CCS (e.g. to what extent do existing oil and gas, mining, waste, industrial permitting, health and safety, property rights and transportation laws already cover aspects of the CCS chain? Are any international laws, policy, or commitments relevant?).
- Undertaking a 'gap and barrier' analysis to compare how existing frameworks match the aims of future CCS legislation (e.g. are any existing provisions likely to impact on CCS deployment? What additional provisions are required to regulate CCS?).
- Determining whether existing regulation should be amended or dedicated legislation developed (e.g. will existing frameworks effectively regulate CCS or is a dedicated framework likely to be more suitable?).
- Potentially undertaking a review of proposed regulatory approaches (e.g. is CCS regulation fit for purpose?).[76]

Part of the methodology of preparing to develop a CCS regulatory framework is to gain a clear understanding of the extent to which existing frameworks at a national, regional, or international level already cover aspects of the CCS chain and to analyse how these frameworks compare with comprehensive CCS legislation. Once this analysis has been undertaken, then any areas where parts of the CCS chain are not addressed represent a 'gap' and any regulation that conflicts with a part of the CCS chain is a 'barrier'. The IEA suggests that any existing framework must be reviewed in respect of the following:

- Scope and coverage; that is are CCS operations likely to fall within the scope of the framework?
- Suitability, with or without modification, to appropriately regulate CCS.
- Whether specific derogations are required to remove any barriers to CCS and their potential consequences of any modifications for existing activities and operations.
- Any potential conflicts.

Once the context is understood, it is suggested that any gaps in which aspects of the CCS chain are not addressed by existing laws can be identified. This process will help to determine whether existing frameworks should be amended or new frameworks developed to regulate CCS.

1.15 POLICY INCENTIVES ESSENTIAL TO DEPLOY CCS

The World Bank published a study in September 2011,[77] which examines various barriers to CCS deployment, looking at the Southern Africa region and the Balkans as case studies, as both depend heavily on fossil fuels for their electricity generation. The study presents the impacts of different policies on CCS deployment such as electricity prices. The report also explores the legal and regulatory issues in addition to the potential of climate finance for CCS projects. The report concludes as follows:

- Combining CCS with enhanced hydrocarbon recovery, such as enhanced oil or coal bed methane recovery could make CCS technology in the power sector economically competitive in parts of the Balkans and Southern Africa, without additional policies.
- High carbon prices could lead to the deployment of CCS in the power sector in both regions in some cases.
- There are grounds to recommend a platform for countries in the Southern African and the Balkan regions to discuss and agree on multilateral and regional treaties for important CCS related issues, such as compliance, enforcement, and dispute resolution mechanisms, in case these countries decide to move towards using CCS technology in the future.
- There is a great deal of uncertainty about the future structure and specific features of climate finance instruments and channels. It is likely, however, that in a highly ambitious GHG Emission Mitigation Scenario, market-based climate finance instruments, as part of a mix of funding sources, will have to play an important role in attracting finance at the international level.
- Broad CCS deployment depends on the viability of a mix of sources of finance, including public funds, carbon market finance, and concessional finance. Concessional financing will significantly lower the cost of electricity for power plants with CCS.[78]

This report emphasises the need for domestic and international mechanisms to pump prime CCS projects, in particular in the developing countries, to reduce investor risk and thereby open up any private capital investment in CCS.

ENDNOTES

1. The Universal Declaration of Human Rights, 10 December 1948.
2. The Rio Declaration, 1992.
3. Shelton (1991), 'Human rights, environmental rights and the right to a decent environment', *Stanford Journal of International Law* 28, 103–138.
4. Myres, McDougal & Lasswell (1959), 'The identification and appraisal of diverse systems of public order', *American Journal of International Law* 53, 1–30.

5. Louka (2006), *International Environmental Law: Fairness, Effectiveness, and World Order*. Cambridge University Press: Cambridge, UK.
6. Hardin (1968), 'The tragedy of the commons', *Science* 162(2), 1243–1248.
7. See text in Section 2.5.
8. See text in Section 2.4.
9. Grotius (1609), *Mare Librum*, Lodewijk Elzevir: Dutch Republic and also see Seldon (1618), *Mare Clausum*.
10. General Assembly Declaration of the Principles on the Seabed and Ocean Floor, GA Resolution 2749(XXV), 1970.
11. Legislation of FRG (1980) 19 *ILM* 1330; France (1982) 21 *ILM* 808; Japan (1983) 22 *ILM* 102; United Kingdom (1981) 20 *ILM* 1271; USA (1980) 19 *ILM* 1003.
12. Part XI of the 1982 Convention, UN Doc. A/CONF. 62/122; (1982) 21 *ILM* 1261.
13. 1958 Convention on the Continental Shelf, Article 2.
14. Daintith & Willoughby (1984), *Manual of United Kingdom Oil and Gas Law*, Oxford, UK, pp. 1–1107.
15. See Note 14.
16. Jennings (1961), 'State contracts in international law', 37BYIL 156.
17. *German Interests in Polish Upper Silesia Case*, PCIJ Series A no. 17 (1926); *Norwegian Shipowners Claims*, 1UNRIAA 307 (1922).
18. *Amoco International Finance Corp v Islamic Republic of Iran*, 27 *ILM* 1314 (1988).
19. 66 *ILR* 518, pp. 71–92.
20. There are numerous excellent books on International Law, but see generally Brownlie (2011), *Principles of International Public Law*, Oxford, UK; Shaw (2008), *International Law*, Oxford, UK.
21. Rosalyn Higgins (1994), *Problems & Process: International Law and How We Use It*, Oxford University Press: Oxford, UK.
22. John Ernst (1994), *Whose Utility? The Social Impact of Public Utility Privatisation and Regulation in Britain*. Open University: Buckingham, UK, p. 49.
23. The use of 'environmental adders' by US State Public Utilities Commissions is an example of an attempt at internalising the environmental costs, reflected then in the price of utilities. D. Foster (1992), *Privatisation, Public Ownership and the Regulation of Natural Monopoly*, p. 209. This, in turn, ensures further regulation to mitigate the effects of cost pass-through.
24. This is also consistent with the principle of 'shared responsibility' in the Fifth Environmental Action Programme of the EC 'Towards Sustainability' (1993–2000), which views government, producers, and consumers as sharing responsibility for environmental problems.
25. 14 JERL 1, 1996.
26. See Patricia Park (1998), 'Some challenges for science in the environmental regulation of industry', Michael Freeman & Helen Reece, Editors, *Science in Court*. Ashgate: Guildford, UK, p. 191.
27. See Sections 3.9 and 3.10.
28. Louka, Note 5.
29. Holdgate (1979), *A Perspective of Environmental Pollution*.
30. McEldowney & McEldowney (1996), *Environment and the Law*, Chapters 8 and 10, Edward Elgar Publishing: UK.
31. Royal Commission on Environmental Pollution, *Setting Environmental Standards*, Cm. 1200, HMSO, 1990.
32. Ibid, p. 130.
33. *This Common Inheritance, Britain's Environmental Strategy*, Cm. 1200, HMSO, 1990.
34. Ibid, paras 1.15–1.17.
35. *The Environment Agency and Sustainable Development. Statutory Guidance to the Environment Agency*, Department of the Environment, 1996.
36. These came into effect in 1996–1997.

37. An authorisation will require the operator to use techniques that ensure emissions are always within specified limits.
38. UK Environmental Protection Act 1990, Section 7(7).
39. Framework Directive on Combating Air Pollution from Industrial Plants, 84/360/EEC/L188.
40. Achieving the lowest possible emissions from a process may not, in reality, represent the BPEO if that requires large amounts of energy with resulting increase in emissions from a generating plant.
41. Council Directive 86/61/EC concerning integrated pollution prevention and control, O.J. L257.
42. See Note 40.
43. For further discussion, see Van den Burg (1995), *Beginning LCA: A Guide into Environmental Life Cycle Assessment*, United States E.P.A, www.epa.gov.
44. For example, from the extraction and purification of raw materials.
45. For example, from the use of the product or from its disposal after use.
46. Douben (1998), *Pollution Risk Assessment and Management*. Wiley: Chichester, UK.
47. UN Doc. A/Res/1803 (XVII), 14 December 1962.
48. Resolution 36/193, 17 December 1981.
49. White (1996), *The Law of International Organizations*, Manchester University Press: UK.
50. Anon, http://www.uccee.org. Accessed December 2010.
51. Article XII C of the Statute.
52. See Chapter 8 on European Energy Law.
53. Article II of the Statute.
54. NPT of 1 July 1969, 729 UNTS 161. Reviewed in 1995 when the Parties agreed to extend the Treaty indefinitely.
55. 888 *UNTS* 180 (1948).
56. Redgwell (2001), 'International law and the energy sector', Roggenkamp, Editor, *Energy Law in Europe*. Oxford University Press: Oxford, UK.
57. This was the first national standard to be developed and the author was part of the development team.
58. The author was also involved with a research group based at Imperial College, London, during the development of the European Eco-Management and Audit Scheme (EMAS), which fed into the EC's deliberations.
59. Muchilinski (2009), *Multinational Enterprises & the Law*. Oxford University Press: Oxford, UK.
60. World Commission on Environment and Development (1987), *Our Common Future*. Oxford University Press: Oxford, UK.
61. UNCTAD (2001), *Environment Series on Issues in International Investment Agreements*. New York, Geneva, p. 5.
62. Sands (2008), *Principles of Environmental Law*, Oxford, UK.
63. See Note 59.
64. Ibid.
65. Harvey, 'Time to clean up? The climate is looking healthy for investment in green technology', *Financial Times*, 22 June 2005, p. 15.
66. Directive 2004/35/CE of European Parliament and Council on environmental liability with regard to the prevention and remedying of environmental damage. (2004) O.J. L143/56.
67. IEA (2010), *Energy Technology Perspectives*, www.iea.org/.
68. For example, under civil law, for damage to the environment, human health, or third-party property.
69. See CCS regulations in different jurisdictions in Chapters 8, 9, 10.
70. See Chapter 10 on Norway.

71. Anon, http://yosemite.epa.gov/opel/RuleGate.nsf/byRIN/2050-AG60. Accessed January 2010.
72. This is a 'stopgap' federal indemnity programme.
73. While the risk of unintended migration and leakage should decrease after CO_2 injection ceases, injected CO_2 will continue to be mobile until it is eventually trapped through physical and chemical processes (IEA (2011), *Regulatory Review*, 2nd edition), www.iea.org/.
74. 1992 Convention for the Protection of the Marine Environment of the North East Atlantic.
75. 16th Conference of the Parties and 6th Conference of the Parties serving as the meeting of the Parties to the Kyoto Protocol to the United Nations Framework Convention on Climate Change.
76. IEA (2011), *Carbon Capture and Storage Legal and Regulatory Review*, p. 17.
77. Kulichenko & Ereira (2011), *Carbon Capture and Storage in Developing Countries: A Perspective of Barriers to Deployment*. World Bank: Washington, DC.
78. http://web.worldbank.org. Accessed January 2010.

2 Regulation, Energy Resources, and the Environment

This chapter deals with various theories of regulation and how they are used within the energy sector to protect the environment.

The expression 'regulation' is found in both legal and non-legal contexts. Essentially, it is a form of behavioural control by a number of methods but in this context, it is a politico-economic concept best understood by reference to different systems of economic organisations and the legal forms that maintain them. In most industrialised societies, there is a tension between the two main systems of economic organisation: the market system and the collectivist system.

Under the collectivist system, the state seeks to direct behaviour with the aim of directing the individuals and companies to achieve goals as specified in the public good. The legal system underpins these goals through public law. The state attempts to specify collectivist goals and identify, in the public interest, the circumstances in which these goals are unlikely to be met by the unregulated market. The public interest theory of regulation attributes to legislators and others responsible for the design of regulation, a desire to pursue collective goals.[1]

The market system relies on the fact that individuals must cooperate if they are to satisfy their fundamental human needs. These needs are met by different skills and expertise, whereby one expert sells his/her skills or produce to another who requires them in exchange for something that the first expert needs, whether it be money or some other requirement. Prices play an important role in this exchange because if the demand is high, then the supplier will be able to charge a higher price for his/her commodity. By extending this reasoning, Ogus[2] attempts to explain the private law, which underpins the market system as '. . . a series of utility-maximising but surrogate contracts between relevant individuals'. The expressed role of the market system is to preserve the economic liberty within the concept of 'allocative efficiency' using the *Kalder–Hicks* criterion.[3] The only coercive effect is under the criminal law to maintain law and order and defend the person and property, and under the constitutional law to secure administrative arrangements for the formulation and enforcement of private law.

Whatever the merits of the system, no industrialised country has relied exclusively on the market and private law system. Rationally, both individuals and companies will only enforce their private rights if the benefits will exceed the costs, and, given the general principle that compensation for the infringement of private rights should not exceed the losses of the plaintiff, many meritorious claims are not pursued. Given this possible 'private law' failure to enforce 'market failure', there is a good case for regulatory intervention based on the public interest. This is reflected not only in a mixed economy but also in the use of command and control, and market-based regulatory systems.

2.1 REGULATORY DESIGN AND ECONOMICS

In the past two or three decades, debates have focused on what role, if any, economic theory and analysis should play in the politics of expertise and environmental protection. Those who advocate the move towards a more economically orientated environmental protection regime claim that this approach would make policy and administration more effective and efficient.[4] It is argued that economics, combined with a greater reliance on market-based instruments, could provide a technical solution to many of the unintended consequences that have emerged under the more traditional command-and-control regimes, such as excessive costs, a certain amount of rigidity, and a lack of incentives to go beyond regulatory compliance. It is further argued by advocates of market-based instruments that if bureaucrats employ economic analysis early in the regulatory process and carry out a marginal cost–benefit analysis of additional incremental pollution reduction, they could then establish enforcement priorities *ex ante* and thus maximise the impact of scarce budgetary resources. This would mean replacing traditional command-and-control instruments with market-based instruments to reduce compliance costs and change the behaviour of businesses to promote environmental technology and achieve higher levels of environmental quality.

However, those who oppose excessive reliance on economics claim that this technical discourse of economics merely attempts to conceal a deeper desire to dismantle regulation and rely on market outcomes. They argue that environmental policy should seek to promote the quality of the environment and preserve that which may be difficult or impossible to place a price upon, such as ecosystems, endangered species, biodiversity, and human life.[5]

2.2 REGULATORY DEVELOPMENT

To further set these theories of regulation into context, it is necessary to consider the broader aspects of regulatory theory and trace the route towards the use of market mechanisms. The growth in regulation as an institutional phenomenon and as an academic area of scholarship does not mean that regulation is restricted to the two main systems as identified earlier. At its simplest, regulation refers to the promulgation of an authoritative set of rules, accompanied by some mechanism, typically a public agency, for monitoring and promoting compliance with these rules. A second, broader concept of regulation, commonly found in the political economic literature, takes in all the efforts of state agencies to steer the economy.[6] Thus, while rule making and application through enforcement systems would come within such a definition, a wider range of other government instruments based on government authority such as taxation and disclosure requirements might also be included. Even those tools that rely on government expenditure or direct organisation such as contracting and public ownership might be considered as alternative tools of regulation.[7] Such an approach has the merit that a variety of tools are considered as possible alternatives to the traditional 'command-and-control' type of regulation. Where rule making seems inappropriate as a means of achieving policy objectives, other tools may be used with enablement being the important aspect of regulation.

The development of economic analysis and economic instruments (the market system) for environmental regulation in the United States was essentially politically driven on the understanding that such instruments could contribute to efficiency and effectiveness. Environmental advocates and their congressional supporters clearly understood the challenge posed by economic analysis. The Congress provided the Environmental Protection Agency (EPA) with some discretion in setting many goals and selecting policy instruments.[8]

Regulatory theory includes the move from command and control by the use of economic instruments to self-regulation and emission inventories.

A great majority of regulatory forms adopted throughout the developed world, in relation to the energy sector, are coercive and therefore are usually referred to as *command-and-control regulations*. Failure to comply with the regulations or conditions applied to a licence to operate a certain process may lead to the imposition of penal sanctions, if only as a last resort. The sanction may be inherent in the form[9] or ancillary to it.[10] Most regulation is, therefore, underpinned by criminal law. Nevertheless, regulatory offences are not easily reconciled with the popular notion of crime and the concept has proved to be problematic.[11] There is perceived to be a distinction between conduct that is wrongful in itself, and which is thus treated as 'criminal' and activities that are subject to proscription.[12] The main means of regulating the energy sector is by the application of rules and procedures by a public regulatory body in order to achieve a desired level of control over the activities of firms operating within the sector or, in other words, administrative regulation. Having said this, administrative regulation involves more than the mere setting of rules on what can and cannot be done. It sets out a coherent system of control in which the regulatory body may easily monitor the ongoing activities within the sector. The main aim is to provide uniformity and fairness between those who are the subject of the regulations. The British system can be characterised as being both flexible and pragmatic. The regulator may discuss levels of emissions with the plant operators when conditions are applied to licences to operate. As long as the emissions are acceptable under general principles and standards and come within the overall accepted levels for the area, the specific levels can be negotiable for each plant.

A simple explanation of administrative regulation regarding the environmental effects of any activity in the energy sector is that any activity involved is prohibited by statute, unless the operator has been granted a licence by the regulatory body to carry out such defined activity. To this end, administrative regulation in the energy sector involves measures that are imposed prior to the commencement of any regulated activity as well as the continual monitoring of the activity. Measures that are imposed prior to an activity commencing are done so in order to forestall potential environmental problems. The objective is to prohibit the activity unless certain conditions are met. The licensing system is complemented by a combination of both criminal and administrative sanctions if the activity starts without permission or if the conditions attached to the permission are not met. There is an ongoing duty on the part of the licensee to comply with the terms of the licence once it is granted. There is also a range of other regulatory controls related to the monitoring of the activity and the enforcement of any conditions imposed. These may include the issuing of an 'enforcement notice' or even the revocation of the licence in some instances. Although these controls are mutually supportive, the licensing system relates to whether an activity should be carried out in the first place, whereas monitoring relates to how the activity is carried out once it has commenced.

Consents or licences obtained from a regulatory body involving the control of emissions to any environmental medium will normally combine the initial need for a consent with an ability, on the part of the regulatory body, to vary the requirements of the consent as situations change. This is reinforced by the fact that many activities in the energy sector may be subject to the requirements of more than one regulatory body.

2.3 MARKET MECHANISMS

Economic instruments, also known as *market mechanisms* or in some cases *fiscal measures*, can be defined as mechanisms designed to act via an economic signal in the market to which the polluter responds. They involve a financial transfer between polluters and the community or directly affect relative prices. Market mechanisms leave the polluter free to respond in the way that they choose. They have been defined as providing 'monetary incentives for voluntary, non-coerced action by the polluter'.[13]

These properties of market mechanisms are in contrast to traditional regulations and standards, which directly influence the actions of polluters by regulating the processes or products used, prohibiting or limiting the discharge of pollutants or placing geographical or time limits on activities. Polluters have no choice but to comply with the regulations or face penalties, although there may be a choice in the manner in which they achieve the compliance. Regulations have an impact on the market because they impose a cost on polluters, but they do not seek to use the market as a tool in environmental protection.

2.3.1 Types of Economic Instruments

These fall into five main categories:

1. Subsidies

   This is a general term for various forms of financial assistance, which must act as an incentive for polluters to alter their behaviour or which are given to firms facing problems in complying with imposed standards. They can involve support for capital investment in less polluting processes or pollution control equipment or assistance with research, development, and demonstration of new technologies or processes.

   a. Grants are non-repayable forms of financial assistance, which are provided by a central authority if certain measures are taken by polluters to reduce their levels of pollution.

b. Soft loans are conditional loans, which are provided to polluters in return for certain anti-pollution measures being taken. The interest rates of soft loans are preferential inasmuch as they are set below the market rate.
c. Tax allowances, which favour polluters by means of allowing accelerated depreciation or other forms of tax or charge exemptions or rebates if specified anti-pollution measures are taken. Investment in improving environmental performance may be allowed for environment-related investment.

2. Charges

Pollution taxes or charges are based on Pigou's concept of increasing the costs of polluters' activities so that they reflect the true cost to society (with respect to environmental damage) of those activities. The level of tax should therefore be based on an assessment of the costs to society of that pollution. In practice, this can be very difficult to determine, and several different types of taxes and charges have been developed:

a. Effluent or emission charges are based on the quantity or quality of pollution released into the environment, for example, tonnes of sulphur dioxide per year or pollutants into water.
b. User charges are direct payments for the use of collective or public pollution treatment facilities, such as sewage treatment works or solid waste disposal facilities.
c. Product taxes are charges levied on particular products to reflect the environmental damage caused by their production or use, for example, charges on fuels with high sulphur content, lubricating oils or non-returnable beverage containers.
d. Administrative charges are small-scale charges levied on polluters to contribute to the costs of the regulatory bodies, for example, for registration of certain chemicals or for implementation and enforcement regulations.
e. Tax differentials are a form of charging for a product or process already subject to taxation. The level of taxation is varied according to the polluting potential of the product or process, for example, lower taxation on unleaded petrol.

3. Deposit refund systems

In deposit refund systems, a surcharge is laid on the price of potentially polluting products. When pollution is avoided by returning these products or their residuals to a collection system, a refund of the surcharge follows.

4. Market creation

Most market creation mechanisms allow for the allocation of 'pollution rights' to polluters, based either on their current levels of emissions or on a level judged to be environmentally acceptable. Polluters may be required to pay initially for these 'rights' or they may be given free by the regulatory authority. After the initial allocation, if a discharger releases less pollution than its limit allows, the owners of these 'rights' are free to trade the difference between the actual discharges and its allowable discharges to another polluter who then has the right to release more than its initial limit allows. Under different regimes, these trades can take place in a number of different forms:

a. **Bubbles:** In this approach, a group of pollution sources is treated as a single source, as if a giant bubble with a single outlet surrounded them. If the total level of pollution from the bubble does not increase, emissions from different sources within the bubble can be changed as much and as often as polluters wish.
b. **Netting:** This is a similar concept to bubbles, but applies when there are special pollution control requirements for new or modified plants or processes above a certain size. By reducing pollution from other sources, polluters can avoid the new or modified source triggering the special requirements.
c. **Offsets:** This allows new pollution sources to be introduced in areas with environmental quality problems provided at least an equal reduction in pollution is made from an existing source.
d. **Banking:** This allows the storage of unused pollution rights, either for use in the future or for sale to other polluters.[14]

5. Financial enforcement incentives

This category may be considered as a legal rather than an economic instrument as non-compliance is punished either by requiring a payment returnable upon compliance or by charging a fine when non-compliance is a considered decision as an alternative:

a. Non-compliance fees or fines are charged when polluters do not comply with certain regulations. The amount charged is in direct correlation to the profits made because of non-compliance.
b. Performance bonds are payments to authorities in expectation of compliance with the regulations. A refund is then paid when compliance is achieved.

2.4 THE ECONOMIC ARGUMENT

Until as recently as a decade ago, both the regulators and environmental groups considered the market place and market forces as being fundamentally detrimental to the environment. That those forces acted to degrade the environment has been well documented and widely lamented. It is therefore not without difficulty that economists have gradually come to show that the power of the market can be harnessed and channelled towards the achievement of environmental goals.

Industrial polluters whose activities came within the ambit of regulators under the command-and-control system saw market mechanisms as a way of achieving regulators' goals in a manner that was less onerous and allowed more flexibility in how the environmental goal was achieved. Rather than relying on the regulatory authority identifying the best course of action, the individual company could use their own specific information to decide on the best means for meeting the regulators' goals. This flexibility achieves, it is claimed, environmental goals at lower cost, which in turn, makes the goals easier to achieve.[15]

A number of reasons have been given for the recent increase in interest in market mechanisms, including the perceived failure of traditional regulatory approaches to stimulate technical innovation in the control of pollution,[16] and the acceptance that the enforcement of regulation has been difficult, costly, and, in some cases, insufficient.[17]

Pearce has estimated that environmental policy is likely to become increasingly more expensive in the future,[18] and is likely to more than double, making the development of more cost-effective pollution control imperative if environmental improvement is not to be hampered by the high costs. A further factor promoting interest in market mechanisms is the changing scale of environmental problems. Such large-scale issues as global warming and climate change involve both large numbers of players (both the polluters and those affected by pollution) and large sums of money. It is likely that a transfer of resources will need to take place in order to solve the problem. It is claimed that this cannot take place by traditional regulation.

2.5 AN OVERVIEW OF TWO MARKET MECHANISMS THAT HAVE PARTICULAR RELEVANCE TO THE ENERGY SECTOR

Proposals and suggestions for new measures have been made by a range of governments, international organisations, and economists,[19] and arise mainly in response to two environmental concerns: firstly, carbon dioxide emissions and their contribution to global climate change; secondly, 'acid rain' and the problem of high emissions from old coal-fired power plants without emission controls.

They have, however, a number of common features inasmuch as they are transboundary in cause and effect[20] and the potential impacts are both significant and long lasting. There are also large numbers involved in both the emission sources and the individuals affected.

For these reasons, both commentators and legislators have suggested that regulation alone may not be sufficient and have turned to market mechanisms as an alternative to command and control. The two main mechanisms that have been considered and tried are taxes/charges and tradeable emission credits/permits.

2.5.1 Taxes/Charges

Energy taxes, even large ones, are not new.[21] However, most of such taxes are justified on specific microeconomic grounds, that is, to retain some of the large profits made from exploiting the state's natural resources when costs are below the international price. With regard to carbon taxes for pollution control, however, the implication is that taxes additional to the traditional ones are to be considered, and the economic principles would be different.[22] In other words, taxes designed simply to raise revenue seek to minimise impacts on behaviour, whereas taxes designed for pollution control seek to do exactly the opposite and influence the behaviour of the polluter.

Current proposals for carbon taxes share two features, namely, a differential tax on fuels (depending on their carbon content and/or carbon dioxide emissions resulting from their use), and the use of the revenue raised either to reduce other taxes or for redistribution. In this way, the proposed carbon tax would be somewhere between an emission tax and a product tax as defined earlier.

2.5.2 National Taxes

The main advantage of a national carbon tax is that it can be introduced unilaterally without the need for international agreement and could therefore be activated with little delay. The disadvantages are that it would lead to market distortion. Economies with a tax would reduce production of high-energy input goods, but imports of such goods from countries without a tax might increase. This could have an adverse effect on the economies of countries with carbon taxes and might result in a net reduction of emissions.[23]

2.5.3 International Carbon Taxes

These have been proposed as a way to overcome the problem of national taxes and ensure a global approach to carbon dioxide reduction. However, although a direct international carbon tax on fuels would in theory ensure its equal application in all countries, it would probably be too bureaucratic and cumbersome to operate,[24] apart from the complex and sensitive political issues involved.

The carbon tax models also vary in whether their main objective is to raise revenue or to provide an incentive to change behaviour.

A carbon tax scheme will determine both the way in which taxes are levied and the tax rates, the simplest way being to base tax on the carbon content of the fuel. This results in higher taxes for coal than for oil, and higher taxes for oil than for natural gas. Alternatively, carbon taxes could take account of carbon dioxide emissions per unit of useful energy generated, thus becoming a true emission tax.[25]

For international carbon taxes on countries, the charge basis would be the total carbon dioxide emissions from that country, in relation to agreed targets for reducing emissions. The level of taxes would be determined either by the need for revenue (raising schemes) or the marginal cost of reducing carbon dioxide emissions (incentive schemes).

2.6 EMISSION INVENTORIES

The current state of play with regard to the previous regulatory mechanisms is still not considered by environmentalists to be at a satisfactory level, but has set the scene for the latest regulatory system to come into being. This is the mandatory environmental information disclosure inventories. Information is widely acknowledged as a prerequisite to local, regional, and international environmental management. The availability of, and access to, that information thus ensures the participation of the public in many decision-making processes.

Knowledge and information can also influence the purchasing habits of the individual consumer in addition to corporate and government behaviour in capital and labour markets. Information allows the regulators to determine whether the regulated market participants are complying with their legal obligations.[26]

This relatively new regulatory instrument is used mainly in the United States and the European Union where over the past few years the change from request-driven information statutes towards mandatory collection and dissemination of standardised environmental information has been a major development. Secondly, the technological development in the area of information storage from physically decentralised public registers towards high-speed information technology and internet-based outline gateways have had a massive influence on the ability to handle the volume of available information.

It is expected that the public armed with good and accessible information will act in a potential pollution reduction capacity by responding to environmental information disclosure in three ways:

1. Community actions, which are frequently described as 'informal regulation';
2. Economic markets, such as capital markets, labour markets, product markets, brands and reputation;
3. The use of the judicial system by inclusion of citizen oversight against the competent authority as well as citizen enforcement actions against the polluter.

The public does not take isolated action, but normally acts in a variety of different ways as 'informal regulation':

- Protests and strikes
- Social proscription of the enterprise
- Adverse publicity
- Political pressure on regulators and community officials to enforce current regulatory standards, enact new regulations or exercise discretionary governmental authority against polluters.

The economic markets react to adverse news media reports and overall stock markets reactions are influential in motivating polluters to improve their environmental performance as measured in the pollutant release inventories.

Pollutant release data can sometimes be an important source of information concerning hazards faced by employees.[27] Therefore, adverse information disclosed in a pollutant release inventory could make it more difficult for a firm to recruit and retain workers.

One of the more recent developments in consumer behaviour is an increasing move towards seeking out environmentally friendly products. Consumers prefer to do business with companies demonstrating superior environmental credentials. Pollutant release inventories reveal both high and low rankings of companies and thus lead to changes in consumer behaviour.

Inasmuch as pollutant release inventory data allow comparisons to be made between companies and their environmental performance, this can result in either the enhancement or diminution of a specific company's reputation. If this happens, it can affect relations with customers, suppliers, employees, and investors and how they regard a particular brand. The brand value is an expression for the monetary identification of self-developed brands of a particular company.[28]

Effects on the judicial system can be through tort law actions and oversight enforcement actions against regulatory bodies. Within the judicial system, parties who are harmed directly by pollution can recover compensatory damages by suing polluters under tort law, but the basis for such an action depends on accurate, comprehensive, and timely data. It would be expensive and almost impossible for the claimants to collect this themselves but such information from the pollutant release inventories may well support the outcome of such actions.

The public may also bring enforcement actions against the polluter by bringing a citizen suit. These actions are not to gain compensation for the applicant but rather to enforce the polluter to come into compliance with the conditions of his licence to operate by reducing the environmentally damaging emissions from their plant. The company would also have to pay a fine for breaking the conditions of the company's licence to operate.[29]

ENDNOTES

1. Friedrich, Editor (1962), *The Public Interest*, Chapter 17, 'Prolegomena to a theory of the public interest'. Questia Publishers: New York, pp. 205–218.
2. Ogus (1994), *Regulation, Legal Form and Economic Theory*, Chapter 2, Clarendon Law Series, Oxford University Press: UK.
3. The criterion stipulates as efficient a policy that results in sufficient benefits for those who gain such that *potentially* they can fully compensate all the losers and still remain better off. However, under this test, the gainers are not *required* to compensate the loser, which means that, in practice, the test is satisfied when the gains exceed the losses and thus provides a theoretical basis for a standard cost–benefit analysis.
4. Marc Eisner (2004), 'Protecting the environment at the margins: The role of economic analysis in regulatory design and decision-making', Gary Edmond, Editor, *Expertise in Regulation and Law*. Ashgate Publishers: UK.
5. A.M. Freeman (2003), 'Economics, incentives, and environmental policy', Vig & Kraft, Editors, *Environmental Policy*, 5th edition. CQ Press: Washington, DC.
6. The French and German 'Regulationist' schools place state activity generally at the centre of their approaches. Jessop (1990), 'Regulation theories in retrospect and prospect', *Economy and Society* 19, 153–216.
7. Hood (1983), *The Tools of Government*. Macmillan; Dantith (1989), 'A regulatory space agency', *Oxford Journal of Legal Studies*. Oxford University Press: UK, 9, 534–546.
8. Freeman, Note 5.
9. For instance, a person who contravenes a regulatory standard commits a punishable offence.
10. For a majority of activities within the energy sector, a licence must be obtained; therefore, engaging in the activity without a licence is a punishable offence.
11. Williams (1955), 'The definitions of crime', *Current Legal Problems*. Oxford University Press: UK, 107; Ogus & Burrows (1983), *Policing Pollution: A Study of Regulation and Enforcement*, AbeBooks Co.: UK, 14–18.
12. For further discussion, see Wells (1993), *Corporations and Criminal Responsibility*.
13. Joseph Wheeler, OECD Report October 1989 C-(89) 146/Final.
14. This terminology has been developed in the United States and is specific to the legislation there and may not be directly transferable to other states.
15. See Tietenburg (1990), 'Economic instruments for environmental regulation', *Oxford Review of Economic Policy* 6(1), 17–33.

16. Kneese & Spofford (1986), 'The economics of integrated environmental management', *Proceedings of the Third Symposium on Integrated Environmental Control for Fossil Fuel Power Plants*. Pittsburgh, PA, USA. Oxford University Press: Oxford, UK.
17. See Note 13.
18. David Pearce (1990), *Economics and the Global Environmental Challenge*, Henry Sidgwick Memorial Lecture. Cambridge University Press: Cambridge, UK
19. Carbon taxes have been proposed in Sweden and discussed within the European Union. Finland's new fuel tax is related to carbon content.
20. The effects can occur outside the country of origin of the pollution.
21. The UK government has raised approximately 5% of its total revenues from North Sea oil over the past two decades. (*The Brown Book: UK Energy Statistics*, DUKES.)
22. For example, lower taxes for unleaded petrol encouraged motorists to use it. The landfill tax in the United Kingdom, on the other hand, is designed to encourage alternative disposal of waste.
23. The European Union made it a condition that their main industrial competitors introduce a similar tax when the Draft Directive on an Energy Taxation was published.
24. Michael Hoel (1990), *Efficient International Agreements for Reducing Emissions of* CO_2. Conference paper. Bergen, May 8-16, 1990. Proceedings published in *The Energy Journal*, Volume 12, number 2 (1991).
25. Cleg (October 1989), 'Can a carbon tax really stem the global warming problem?' Petroleum Times Report.
26. Sands (2009), *Principles of International Environmental Law*. Cambridge University Press: UK.
27. OECD (1996), *Pollutant Release and Transfer Registers: A Tool for Environmental Policy* (n 20), 94–95.
28. The environmental disaster at the blowout of the BP Macondo oil well was particularly damaging to the BP brand.
29. For an in-depth discussion on Pollution Release Inventories, see Bunger (2011), *Deficits in EU and US Mandatory Environmental Disclosure*. Springer-Verlag Publishers: Heidelburg, Germany.

3 International Law and State Sovereignty

This chapter considers the interrelationship between international regulation and state rules that companies must comply with. The chapter mainly considers the duties and liabilities of those multinational companies operating in the energy sector who have subsidiaries operating in different jurisdictions.

3.1 ENERGY, INTERNATIONAL, AND ENVIRONMENTAL LAWS

Phillippe Sands, in his introduction to the Earthscan publication *Greening International Law,*[1] states that '. . . the international community's recognition that environmental problems transcend national boundaries has resulted in the development of an important new field of public international law'.

Historically, international environmental law used to be exclusively concerned with conservation of flora and fauna, mainly under the auspices of the International Union for the Conservation of Nature (IUCN). It has now developed so as to address pollution of rivers and seas, the trans-boundary transportation and management of waste, and industrial and transport emissions into the atmosphere, all of which have particular relevance to the energy sector.

Thus, in energy law, international and environmental laws now work together in various directions in the regulation of the energy sector. International law secures the right of the state over its national resources and its right to explore and exploit its resources on the continental shelf, whereas international environmental agreements deal with the type of energy policy adopted by states, either collectively or individually.

3.2 STATE SOVEREIGNTY OVER ITS NATURAL RESOURCES

Through the initiative and expertise of the multinational oil companies, oilfields were discovered and developed in the states situated in the Middle East. These companies were granted concession agreements,[2] which gave them the exclusive right to search for oil in a large area and the right of ownership of the oil at the point of extraction. The producing country was not entitled to any share of its natural resources nor was it entitled to take part in the development or exploitation of those resources. It was in 1958 that the Organisation of Petroleum Exporting Countries (OPEC) passed a resolution on the right of producing countries to renegotiate these contracts with the multinational oil companies on the basis of inequity.[3] Following this, in 1962, the UN General Assembly adopted a resolution on 'Permanent Sovereignty over Natural Resources'.[4] Under the 1973 Riyadh agreement, the oil-producing states were given the right to participate in the production of the oil. This was initially at a level of 25%

but eventually rose incrementally to 100%. By the end of 1973, the price of oil had been raised to $5.119 a barrel, and since then the OPEC has periodically reviewed the price of oil and raised or frozen it from conference to conference.

It was the 1973 crisis, when the OPEC raised the price to above a level that the developed countries considered acceptable, which heralded the new legal framework on natural resources. The efforts of the United Nations to address the crisis culminated in the convening of a Special Session of the General Assembly on Raw Materials in 1974. The purpose of this session was to create a climate for future economic cooperation and this was, in part, met by the adoption by the United Nations of the Charter of Economic Rights and Duties of States (CERDS). This was accepted as part of the New International Economic Order,[5] which was intended to correct inequalities and redress existing injustices. It was also intended to make it possible to eliminate the widening gap between the developed and developing countries, and to ensure steadily accelerating economic and social development, peace, and justice for present and future generations.

The declaration was adopted by the General Assembly by consensus and it stated that 'In order to safeguard these natural resources, each State is entitled to exercise effective control over them and their exploitation with means suitable to its nationals, this right being an expression of the full permanent sovereignty of the State'.[6]

The principles of the New Economic Order include full permanent sovereignty of every state over its natural resources and all economic activities, including the right to nationalisation or transfer of ownership to its nationals. 'No State may be subjected to economic, political, or any other type of coercion to prevent the free and full exercise of this inalienable right' (para. 4(e)).

The principles also include the right of all states and peoples to restitution and full compensation for the exploitation and depletion of, and damage to, their natural resources and all other resources under foreign occupation, alien or colonial domination, or apartheid (para. 4(f)).

This concept of permanent sovereignty over natural resources is now vigorously asserted by most producing countries as a *sine qua non* of national independence and many have sought to translate the principle into action by nationalising the mining and petroleum interests within their jurisdiction, whereas others have achieved the same objective by taking a majority share interest in the local operating companies.[7]

Although it is not, in fact, legally binding, the resolution has had considerable effect on the developments in international law. It has been cited and relied on as evidence of the existing law on the subject and referred to as such by international arbitral tribunals.[8] However, despite these categorical pronouncements, they have not constrained the development of treaties and rules of customary international law concerning conservation and environmental protection that qualify this sovereignty.

3.3 STATE RIGHTS OVER MINERAL RESOURCES AND COMMON SPACE

Even prior to the UN Resolution, it was assumed that natural resources were allocated to sovereign states according to the boundaries established to deliminate their respective land territory and territorial seas. It was, therefore, the question of boundaries and jurisdiction over the natural resources of the continental shelf, including the subsoil and seabed, which led to the Truman Proclamations in 1945.[9] These proclamations are now regarded as the starting point of the positive law on the subject.[10] The first Proclamation stated '. . . the Government of the United States regards the natural resources of the subsoil and seabed of the continental shelf beneath the high seas but contiguous to the coasts of the United States as appertaining to the United States, subject to its jurisdiction and control'.

The second Proclamation claimed that the United States regarded it as '. . . proper to establish conservation zones in those areas of the high seas contiguous to the coasts of the United States, wherein fishing activities have been, or in the future may be developed and maintained on a substantial scale'. Following the Truman Proclamations, many maritime states followed suit and made various claims of differing extents in relation to their continental shelves. A number of disputes concerning these resources, therefore, centred around boundary delimitations, as in the *Norwegian Fisheries Case*.[11] In an attempt to regularise these claims, the United Nations held the Convention on the Territorial Sea and Contiguous Zone in Geneva in 1958.

The principle of the coastal state's sovereignty over the 'maritime belt' was not new, and had been well established by the beginning of the eighteenth century, but only extended as far as the power of arms carried to ensure the safety of the state concerned. At a later stage, the 'cannon shot' range was expressed as a definite figure of 3 nautical miles. By the nineteenth century, the 3-mi limit had received widespread recognition by jurists as well as the courts[12] and was adopted by many major maritime states, including the United States and also Great Britain. Some states adopted a wider 'maritime belt', even up to 12 mi. The evolution as well as the general acceptance of the doctrine of the continental shelf was recognised by the International Court of Justice (ICJ) in the *Anglo-Norwegian Fisheries Case*.[13]

3.4 THE 1958 CONVENTION ON THE CONTINENTAL SHELF

The 1958, Geneva Convention described the rights of the coastal states in accordance with the continental shelf as 'sovereign rights for the purpose of exploring it and exploiting its natural resources'. Under the Convention, these rights are exclusive. No other state can undertake these activities without the consent of the coastal state. In addition, these rights do not depend on 'occupation, effective or notional, or on any express proclamation'. The ICJ asserted the inherent right of the coastal state to exploit its natural resources of the continental shelf as it is '. . . the natural prolongation of their land territory into and under the sea'.[14] The Court described the existence of those rights as *ab initio and ipso jure* and accordingly they need not be proclaimed or established by treaty.

In Article 1 of the Convention, the term *continental shelf* is defined by a legal rather than a geomorphological definition. It refers to the seabed and the subsoil of the submarine areas adjacent to the coast but outside the area of the territorial sea, to a depth of 200 m or, beyond that limit, to where the depth of the superjacent waters admits the exploitation of the natural resources of the said areas, and the seabed and subsoil of similar submarine areas adjacent to the coast of islands.

The assumption made is that the 200-m depth and exploitability were virtually identical, but rapidly evolving technology soon made it apparent that it was possible to exploit the resources at deeper levels.

From the given description of the limit, it appears that, although a depth of 200 m forms in many places the geomorphological limit of the continental shelf, it was not intended that the coastal state should be restricted to this water depth if it would prove technically possible to develop and produce natural resources beyond that depth. However, even if it were technically possible to search and explore for natural resources in waters deeper than 200 m, this would not in itself, allow the coastal state to claim an expanded continental shelf.

The definition of the outer limits of the continental shelf, and thus of lawful exploitability, speedily became problematic.

The question remains: whether the coastal state must itself have demonstrated the ability to exploit natural resources at depths greater than 200 m before it can claim a continental shelf extending to such depths, or whether it is sufficient for such a coastal state to demonstrate that at some other place on the continental shelf exploitation is actually carried out at that water depth or even at greater water depths. If the latter interpretation were correct, it would mean that if natural resources were actually exploited at a water depth greater than 200 m by another coastal state on the same continental shelf, then any coastal state on that continental shelf could automatically claim a continental shelf extending to the greater water limit, without actually being sure that operations at that depth would also be technically feasible on its own continental shelf.[15] It may be assumed that this was never intended by the First Law of the Sea Conference.[16]

It cannot, however, be concluded that petroleum reserves beneath the continental shelf, from the legal perspective, are exactly the same as reserves beneath an area of land in the interior of a state. There is clearly sovereignty over all of a state's land mass. No other state has an entitlement and it is a matter for the discretion of the state concerned as to whether it takes title itself over all mineral reserves, or allows them to be owned by the owners of the superjacent soil. However, the Continental Shelf Convention provides that the coastal state has sovereign jurisdiction for purposes of the exploration and exploitation of the resource. It is a *functional* sovereignty.[17] This distinction is one of legal importance. Firstly, states have had to take care that any legislation they pass, which purports to have application on the continental shelf is limited to matters relating to the exploration and exploitation of the resources thereon. The issues of criminal jurisdiction and civil liability and the question of taxation of companies operating on the continental shelf, which are not engaged in exploring and exploiting the resources, have raised particular difficulties in the North Sea. Service companies, with vessels providing food or entertainment on platforms, have resisted their liability to tax at the demands of the coastal state.[18]

More importantly, if onshore mineral resources have been vested in a state, it is clear that it owns them. However, the state has no right of ownership in the resources *in situ* in the continental shelf, merely sovereignty for the purposes of exploring and exploiting those resources. This means that no one else may explore and exploit without permission; that the coastal state may grant licences for that purpose, but it does not itself own the petroleum.[19] The holder of an offshore licence does not, therefore, get title to a resource over which the licensing government does not itself have title. What it does get is the entitlement to explore and exploit and to reduce into possession. It is the actual reduction into possession, which then gives the licensee title to the recovered petroleum.

Although the 1958 Convention recognises the sovereign rights enjoyed by the coastal state for the purpose of exploring the continental shelf and exploiting its natural resources, it must be remembered that prior to the Truman Proclamations of 1945, the maritime zone, which became the legal continental shelf, was simply a part of the high seas and, as such, subject to the fundamental principle of the freedom of the high seas. Given the continuing importance in terms of economics and security of such freedoms as those of fishing, navigation, and cable laying, it was only to be expected that the new continental shelf rights would be carefully circumscribed. The need to not cause unjustifiable interference with existing users is reflected in the provisions of Articles 3–5 of the Convention, the general point being that under the 1958 Convention, the sovereign rights of the coastal state constitute an exception to the predominant principle of the freedom of the high seas and, like all legal exceptions, must be restrictively interpreted in the event of conflict with one of the preexisting freedoms. Article 3 aims to preserve these freedoms by providing for a legal presumption in favour of the various freedoms of the high seas in cases of conflict.

The Convention was meant to regulate the exploration and exploitation of natural resources on the continental shelf. The subject of deep sea mining fell outside the scope of the Convention and also the sister Convention on the High Seas 1958 under the rights and freedoms of the high seas. However, during the 1960s, further developments in drilling technology and bottom sampling made it clear that in the near future, deep sea mining would become technically feasible. In the light of these developments, many states became anxious that the issue of the seabed should be considered. The absence of any legal regime, it was considered, would have an inhibiting effect on the mining industry as it embarked on deep sea mining.

Consideration was given to an adaptation of the Continental Shelf Convention by which the Convention would become applicable to natural resources beyond the continental shelf. This would mean the removal of the 200-m water depth criterion, which would leave the technological water depth[20] as the only test for determining the limits of the jurisdiction of the coastal state. The practical obstacle to application of this solution lies with the application of the test and the differing results for differing coastal states as outlined previously. In addition, the presence of mid-oceanic ridges and plateaus, separated from the continental shelf by greater water depths, cannot be accommodated within the Convention.

Alternatively, if water depth were not the limiting factor, application of the Convention would have the result that the oceans and other sea areas of the world would be divided by and between oceanic coastal states. This would not be acceptable to landlocked states nor would it be consistent with the ICJ reasoning that the continental shelf was 'the natural prolongation of their land territory into and under the sea'.

Another solution that presented itself was to treat deep sea mining as one of the freedoms of the high seas. This approach, however, would have the consequence that any person would be entitled to search for and exploit the deep sea mineral resources, as long as they did not interfere with the legitimate rights of others. The result would be that a chaotic situation would develop, with no means for a prospector to protect his/her discovery and the mining site from other prospectors. Without such legal protection, there would be little incentive for the industry to make the huge investment needed to search for the resources of deep sea areas.

Meanwhile, developing countries feared that the deep sea resources would come under the control of the developed countries whose enterprises in deep seabed mining technology had made considerable progress. A proposal was put forward to the General Assembly of the United Nations by Malta in 1967 to the effect that the deep seabed and its mineral resources should be reserved for peaceful purposes in the interest of mankind.

In 1968, the 'Permanent Committee on the Peaceful Uses of the Seabed and Ocean Floor beyond the Limits of National Jurisdiction' was set up, which secured the adoption of the General Assembly Moratorium Resolution,[21] which declared that pending the establishment of an international regime

a. States and persons, physical or judicial, are bound to refrain from all activities of exploitation of the resources of the area of the seabed and ocean floor, and the subsoil thereof, beyond the limits of national jurisdiction,
b. No claim to any part of that area or its resources shall be recognised.

The United Nations continued its efforts, despite disagreement among Member States, and in 1970, by Resolution, the UN General Assembly laid down the following principles with respect to the international area (The Area):[22]

a. The Area and its natural resources are the common heritage of mankind,
b. No occupation by States or other natural or legal persons is permitted,
c. No rights may be granted or claims be made, which contradict the future international mining regime and the principles enunciated in this declaration,
d. All mining operations have to fall under the scope of the aforesaid international mining regime,
e. The Area must be open for peaceful use by all States without discrimination in accordance with the international mining regime,
f. All mining operations must be carried out in the interest and to the benefit of mankind, taking into account the interests and needs of the developing countries,
g. The international regime must be incorporated in an universal Law of the Sea Convention to which every State may accede,
h. The status of the waters of the Area and the air space above it may not be affected.

It was subsequently left to the Third United Nations Conference on the Law of the Sea (UNCLOS III) to reach a satisfactory conclusion.

3.5 THE UN CONVENTION ON THE LAW OF THE SEA, 1982

The 1982 Convention was the result of the UNCLOS III, which had been convened by the UN General Assembly for the purpose of giving effect to the principles contained in the 1970 Resolution. The Conference held its first session in 1973. The text of the Convention was adopted in 1982 and the Convention entered into force on November 16, 1994. The 1982 Convention now takes precedence over the 1958 Geneva Conventions, but will not alter any interstate agreements that have been made under the Geneva Conventions.

Pursuant to the UN General Assembly Resolution of 1970 regarding the Area, the Law of the Sea Convention created and defined the international Area and declared the resources of the Area to be the common heritage of mankind. All rights in the resources of the Area are vested in mankind as a whole, on whose behalf the International Seabed Authority shall act. The International Seabed Authority was established by the Convention as the organisation through which States Parties must organise and control activities in the Area.

The conference had considerable difficulty in reconciling the interests of the major groups of countries. The states that had continental margins extending beyond the 200-mi limit were not prepared to give up any claims to the resources in the outer limits. Further, the recognition of the Exclusive Economic Zone (EEZ) only served to make the situation more complex with Article 76 of UNCLOS III bringing a new definition to the continental shelf. It states that if the outer edge of the continental margin does not exceed 200 nautical miles, then the rights of the coastal states shall end either by the outer edge of the continental margin or at 200 nautical miles from the baselines. Alternatively, if the continental shelf extends beyond 200 nautical miles, then the boundary may extend to another given line, which cannot exceed 350 nautical miles from the baselines.

The state has, according to international customary law, the exclusive right to the use of its continental shelf. Additionally, the ICJ held that the state has the right to an EEZ as 'part of modern international law'.[23] The ICJ did not, however, deal with the question whether claims to the margin beyond 200 nautical miles are part of customary international law.

Although some states combine their claim for a continental shelf with a declaration of jurisdiction of the EEZ, it must be noted that although these two may coexist and relate to the same geographic area, they are different concepts. The continental shelf is the physical bed of the sea area,[24] whereas the EEZ is an area of maritime jurisdiction, which gives the state the right to exploit not only the mineral and other nonliving resources of the seabed and subsoil and sedentary species[25] but also commercial fishery in the Zone.

3.6 PROTECTION OF THE MARINE ENVIRONMENT

It had been the intention from the outset, that the 1982 Convention should cover extensively the duties of states with regard to the protection and preservation of the marine environment.[26] So, states have the sovereign right to exploit their natural resources but within the constraints imposed by their duty to preserve and protect the marine environment under Articles 192 and 193. It should also be noted that the Convention imposes these duties on *all* states not only coastal states.

Under Article 194(1), states must take, individually or jointly as appropriate, all measures that are necessary to prevent, reduce, and control pollution of the marine environment from any source and must endeavour to harmonise their policies. Although Part XII imposes a duty on states to deal with *all* sources of pollution of the marine environment, the Convention lays specific obligations as to the minimisation to the fullest possible extent of the following:

i. The release of toxic, harmful, or noxious substances from land-based sources, from or through the atmosphere or by dumping.

This is because about 80% of the pollution of the marine environment is caused by land-based sources.[27]

ii. Pollution from vessels.
iii. Pollution from installations and devices used in natural resources operations and from other offshore installations and devices.

Although states are obliged to cooperate on a global basis and, as appropriate, on a regional basis directly or through competent international organisations[28] and states must jointly develop and promote contingency plans for responding to pollution incidents in the maritime environment,[29] it is the coastal state and not states generally who must adopt laws and regulations in addition to 'the taking of measures' in relation to pollution from seabed activities subject to national jurisdiction.[30] These laws and regulations are to prevent, reduce, and control pollution of the marine environment arising from or in connection with seabed activities subject to the state jurisdiction and from artificial islands, installations, and structures under their jurisdiction and further, the state must take other measures as may be necessary to prevent, reduce, and control such pollution.

3.6.1 Article 208

> Coastal States shall adopt laws and regulations to prevent, reduce and control pollution of the marine environment arising from or in connection with sea-bed activities subject to their jurisdiction and from artificial islands, installations and structures under their jurisdiction, pursuant to articles 60 and 80.
>
> States shall take other measures as may be necessary to prevent, reduce and control such pollution.
>
> Such laws, regulations and measures shall be no less effective than international rules, standards and recommended practices and procedures.
>
> States shall endeavour to harmonise their policies in this connection at the appropriate regional level.
>
> State, acting especially through competent international organisations or diplomatic conference, shall establish global and regional rules, standards and recommended practices and procedures to prevent, reduce and control pollution of the marine environment referred to in paragraph 1. Such rules, standards and recommended practices and procedures shall be re-examine from time to time as necessary.

3.7 OIL POLLUTION

It was the *Torry Canyon* disaster in 1967, involving the contamination of large areas of coastline by oil, which exemplified the risks posed by the transport of large quantities of toxic and hazardous substances at sea. Since then, scientific studies carried out by GESAMP and others have shown a steady increase in the pollution of the seas by oil, chemicals, nuclear waste, and the effluents of urban, developed countries.[31]

The control, reduction, and elimination of marine pollution has now created a substantial body of new law.

It was the 1954 London Convention that marked the first successful attempt at international regulation of oil pollution from tankers. This was later replaced in 1973 by the MARPOL Convention, although some of the minor tanker-operating states are still bound only by the 1954 Treaty. The London Convention aimed to control the location of oil tankers by defining prohibited areas and excluding coastal zones. It also controlled the need for discharges by setting construction and equipment standards for vessels, which were intended to reduce the volume of waste oil, or to separate oil from ballast water. Governments were required to provide port discharge facilities, which included in a 1969 amendment, the 'load on top system', which enabled the tankers to discharge oily residues to land-based reception facilities.

The MARPOL Convention was first adopted in 1973 and revised in 1978 to facilitate its entry into force. Over 85% of the gross registered tonnage of the world's merchant fleet belong to the present participating States Parties, and is thus included in the 'generally accepted international rules and standards' prescribed by Article 211 of the 1982 UNCLOS. It could also be argued that the MARPOL Convention is now a customary international standard that must be complied with by vessels of all states, whether or not they have chosen to ratify the Convention.

The MARPOL Convention employed the same approach to the regulation of oil pollution as did the 1954 London Convention inasmuch as it relied mainly on technical measures to limit oil discharges. New and more stringent construction standards are set for new vessels. All enclosed or semi-enclosed seas are designated as 'special areas' as regards oil pollution where more stringent standards are necessary. These provisions are aimed at taking advantage of modern technology and operating methods and so eliminate, as far as is possible, all but minimal levels of oil discharge. The MARPOL is not, however, confined to oil pollution alone, but it also regulates other types of ship-based pollution and thus, provides for internationally agreed standards of environmentally sound management for the transport of chemicals and hazardous wastes. The MARPOL also involves the cooperation of coastal states, port states, and flag states in a system of certification, inspection, and reporting whose purposes are to make the operation of defective vessels difficult.

Unfortunately, the MARPOL Convention is only of limited relevance to the prevention of pollution from continental shelf operations and there is no comparable international convention that proceeds from the general principles of the Geneva Conventions. The feasibility of establishing uniform standards for construction and use of offshore installations and incorporating them in a new treaty was considered by a Working Group set up by the Conference on Safety and Pollution Safeguards in the Development of North West European Offshore Mineral Resources, which met in 1973. Its conclusions were endorsed by the United Nations Environment Programme (UNEP) Governing Council and by the UN General Assembly in 1982, but so far very few governments have made use of the guidelines produced.

The precise regulation of the construction and operation of offshore installations and devices has mainly been achieved through national legislation,[32] which has, to a minor extent, been coordinated by bilateral treaties relating oil pipelines and the desirability of harmonising national safety requirements.[33]

There is a large and increasing number of regional conventions that have been negotiated under the UNEP Regional Seas Programme. The obligations embodied in these conventions are normally formulated in very broad terms but are gradually being developed in more detail through protocols specifically addressing pollution from offshore exploitation.

Under the International Convention on Oil Pollution Preparedness, Response and Cooperation, 1990 (OPRC Convention),[34] States Parties undertake to take all appropriate measures in accordance with the Convention to prepare for and respond to an oil pollution incident. Under Article 2, the undertaking covers, *inter alia*, oil pollution from an 'offshore unit' defined as 'any fixed or floating offshore installation or structure engaged in gas or oil exploration, exploitation or production activities, or loading or unloading of oil'. Oil pollution emergency plans are required and the persons in charge of such units are required to report any 'event' involving a discharge of oil.[35] The Convention also provides for actions to be taken by a State Party on receipt of such an oil pollution report and the duty to establish national and regional systems for preparedness and response.

The primary responsibility for responding effectively will thus fall in most cases on the relevant coastal states; however, flag states also have a responsibility for ensuring that their vessels are adequately prepared to deal with emergencies.[36] Nevertheless, it remains unrealistic to expect flag states themselves to maintain the capacity to respond to accidents involving their vessels wherever they occur, and despite the provisions of Article 3, the 1990 Convention does not attempt to do so.

The safeguard for the coastal states in protecting themselves from risks posed by oil tankers and other vessels carrying toxic substances in passages near their coastlines remains the right of coastal states to intervene on the high seas in cases of maritime casualties involving foreign vessels as provided for under the International Convention Relating to Intervention on the High Seas in Cases of Oil Pollution Casualties, 1969 and its 1973 Protocol.

The 1969 Convention permits parties to take

> Such measures on the high seas as may be necessary to prevent, mitigate, or eliminate grave and imminent danger to their coastline or related interests from pollution or threat of pollution of the sea by oil, following upon a maritime casualty or acts related to such a casualty, which may reasonably be expected to result in major harmful consequences.[37]

Note that this definition does not cover operational pollution, however serious, or dumping at sea. In addition, to avoid the danger of precipitous action by coastal states causing undue interference with shipping beyond the territorial sea, the references to 'grave and imminent, and major harmful consequences' were inserted.[38]

State Jurisdiction over Coastal State Adjacent Seas

With the ever-increasing exploration and exploitation of mineral resources and wind offshore, it becomes important for the energy sector to be aware of the delimitations and what rights and duties a state has over its territorial sea, continental shelf, and EEZ. This section also includes the transport of hazardous waste.

Similarly, the energy sector is, potentially, one of the most polluting of the atmosphere with regard to CO_2 and greenhouse gas emissions. International conventions regulate emissions into the ozone layer, trans-boundary pollution, and issues regarding climate change.

3.8 STANDARD SETTING WITHIN INTERNATIONAL TREATIES AND THEIR PROTOCOLS

The main sources of trans-boundary air pollution are sulphur dioxide (SO_2) and nitrogen oxide (NO_x); these are mainly produced anthropocentrically by the combustion of fossil fuels for power generation. Once emitted into the atmosphere, they are distributed by the prevailing winds and deposited back onto the terrestrial environment. Scientific monitoring has shown that this deposition can be thousands of miles away and in another jurisdiction from that within which the emissions arose. It has also been noted that sulphur and nitrogen can be deposited in dry form or as acid rain.[39] This acid deposition has been blamed for increased acidity of soil and freshwater, which in turn has led to reduced crop growth and degradation of forests, and the disappearance of both fish and wildlife.[40] In the *Trail Smelter* arbitration, the panel had tried to address the problem of trans-boundary pollution when it held that

> no state has the right to use or permit the use of its territory in such a manner as to cause injury by fumes in or to the territory of another or the properties or persons therein, when the case is of serious consequence and the injury is established by clear and convincing evidence.[41]

The evidential questions were resolved by scientific evidence and the tribunal laid down a regime regulating the liability of states in international law for harm caused in breach of obligations not to cause harm by trans-boundary pollution. In other words, Canada was ordered to adopt a regime for regulating the future operation of the smelter, but the right to continue to operate was maintained. It also emphasises that the *compromise* expressly empowered the tribunal to prescribe measures.[42] This created a balance of interests between the two parties, which was achieved through the order of the tribunal and led to the acceptance of negotiated settlements to arrive at an equitable solution, without weakening the underlying rules of international law that structured their negotiations.[43] This was a landmark case in which the principle was established that states may be held accountable in interstate claims, and are required to take adequate steps to control and regulate sources of serious global environmental pollution or trans-boundary harm within their territory or subject to their jurisdiction. It was an arbitral decision then that established the need for a negotiated regime to protect one state from pollution arising in another jurisdiction. This establishes a basic precept of an obligation to prevent future harm rather than merely a basis for reparation after the event. This obligation has been carried forward in a number of later judicial decisions,[44] a wide range of global and regional treaties,[45] and in the Stockholm and Rio Declarations.

Inasmuch as the problem was most marked in Europe, by 1976, a European monitoring programme had been put in place. This, in turn, led to the adoption of the 1979 Geneva Convention on Long-Range Transboundary Air Pollution, which remains the only major regional multilateral agreement devoted to the regulation of transboundary air pollution. It also enables the European states to treat their air mass as a shared resource and so impose common pollution control methods and emission standards. As we shall see, Europe is considered in the same manner under the Kyoto Protocol.

3.9 THE DEPLETION OF THE OZONE LAYER

Increasing scientific certainty that particular substances,[46] generally known as *ozone-depleting substances (ODS)*, were damaging the ozone layer provided the international community with the impetus to take more positive action. However, the 1985 Convention signed at Vienna[47] was merely a framework convention as many parties remained uncertain about the true effects of certain chemicals on the ozone layer. The Convention itself is largely an empty framework that requires States Parties to agree to take further action rather than establishing any targets or timetables for the phasing out of the ODS. It encourages parties to cooperate and exchange information in order to better understand and assess the effects of human activities on the ozone layer.[48] Although a Secretariat was set up to receive all the information gathered, it was just that—an information-gathering exercise to assess the causes and effects of ozone depletion. However, these provisions laid the basis for ensuring adequate monitoring and research, and for making substitute technologies and substances available, in particular, to developing countries.

By 1987, the scientific evidence was much more convincing with the hole in the ozone layer having become larger at an alarming rate. The scientists involved planned to hold their conference in Montreal the week before the States Parties to the Convention were due to meet. They stayed on to demonstrate their scientific models to the politicians who were then sufficiently convinced that they agreed to the Montreal Protocol at the Convention.

3.10 THE 1987 MONTREAL PROTOCOL ON SUBSTANCES THAT DEPLETE THE OZONE LAYER

This was a much more significant agreement than the Convention itself inasmuch as it sets out firm targets for reducing and eventually eliminating both production and usage of a range of ozone-depleting substances. This was an important step as far as policymakers and legislators were concerned because it was an innovative use of command and control at the international level. It was decided that it would be easier to control the production of CFCs first because not many industries actually produce the substances but a large number use them. However, to control production without restricting consumption could create an illegal market,[49] and so recycled and reused substances were not to be considered as production.[50] The Montreal Protocol then regulated both CFCs and halons[51] with further substances put under regulatory control, and a schedule was put in place for the acceleration of the elimination of these substances.

The Protocol has been amended on numerous occasions, which have had the effect of tightening the procedures for trade in recycled substances with this licensing system being incorporated into the 1997 amendments to the Protocol.[52] Because there is also a non-compliance procedure, which incorporates both a 'carrot' and a 'stick',[53] the control of the production and use of ODS had a beneficial effect on the hole in the ozone layer, which began to heal itself. However, the current issue is how the Protocol will be implemented by developing countries that must begin phasing out their CFCs, halons, and carbon tetrachloride. Asian countries have actually increased their CFC consumption as a result of a high rate of economic growth and their dependence on CFC products will increase future demand.[54] If these developing countries do not comply, then the regime based on the Montreal Protocol will collapse and the migration of CFC-intensive industries to less-regulated countries will reduce the benefits of the Protocol.[55]

3.11 CLIMATE CHANGE

When addressing climate change, the challenge is to *control* the production of a substance or group of substances from one section of industrial production under command and control, which is entirely different from legislating to *stabilise* various emissions known as *green house gases* (*GHGs*) from a large number of different sources, which contribute to climate change.

Although the scientific issues surrounding climate change had been known for over a 100 years, it was not until the late twentieth century that climate change emerged onto the international political agenda when, in 1988, the UN General Assembly took up the issue for the first time and adopted Resolution 43/53. The Resolution declared that climate change was a 'common concern of mankind'. The Intergovernmental Panel on Climate Change (IPPC) had been established that year by the World Meteorological Organisation (WMO) and the UNEP who jointly set up the panel with a mandate to assess the emerging science of climate change and subject it to intergovernmental scrutiny. The latest set of principles governing its work[56] state that it is to.

> . . . assess on a comprehensive, objective, open and transparent basis the scientific, technical and socio-economic information relevant to understanding the risk of human-induced climate change, its potential impacts and options for adaptation and mitigation.

Although the IPPC does not carry out the scientific research itself, it does conduct a massive review of climate change research that has been published in peer-reviewed journals by government bodies, universities, intergovernmental organisations, and individual researchers from around the world. Therefore, what the IPCC provides is an objective analysis of all the scientific research in order that policymakers can make informed decisions.

3.11.1 Framework Convention on Climate Change, 1992

Although the need for a Framework Convention on Climate Change was agreed to in 1990 at the World Summit in Rio de Janeiro, it was not until the third Conference of the Parties (COP) when they met in Kyoto that a new regulatory structure was devised, which included a number of flexible market mechanisms.

The Convention itself defines an ultimate objective and principles—which also apply to the Protocol—and divides countries into Annex I, which are Organisation for Economic Co-operation and Development (OECD) countries and economies in transition, Annex II which are OECD countries only and Non-Annex I countries, which are mostly developing ones. All parties have general commitments including reporting obligations, but Annex I Parties have a specific 'aim' to return emissions to 1990 levels by 2000. Annex II Parties must provide financial assistance to the developing countries and promote technology transfer.[57]

The objective of the Convention was the 'stabilisation of greenhouse gas concentrations in the atmosphere at a level that would prevent dangerous anthropogenic interference with the climate system'. This objective is framed in terms of an environmental quality standard inasmuch as it establishes an environmental threshold that Parties must not exceed. However, the threshold that is established (dangerous anthropogenic interference with the climate system) does allow activities that cause such interference up to this point. Article 2 of the Convention goes on to provide additional guidance concerning the timing of any actions to stay within the threshold.[58] The objective, therefore, has a precautionary emphasis. This preventative focus of the objective also applies to the Kyoto Protocol as the Convention states that 'any related instrument' shall share the ultimate objective set out in Article 2. This is also affirmed in para. 2 of the Preamble of the Protocol.[59]

3.11.2 The Kyoto Protocol 1997

Under the Protocol, all Parties have general commitments but Annex I Parties have individual emission targets, which require a total reduction of emissions by 5%. The targets range from –8% for most countries to +10%. These targets are set out in Annex B to the Protocol. The emission targets cover CO_2, CH_4, N_2O, HFCs, perfluorcarbons (PFCs), and sulphur hexafluoride (SFs), which are all counted together as a 'basket'. The targets also cover certain carbon sequestration activities in the land use, land use change, and forestry (LULUCF) sectors, which have their own specific rules. These targets must be met by the 'commitment period' 2008–2012. All of the flexible mechanisms as set out in the Protocol may be used to help meet the targets and groups of countries may also meet their targets on a regional basis. So far, this provision has been taken up only by the European Union.

The Kyoto protocol recognised that the costs of climate change and any adaptation to mitigate its impacts are unknown. In fact, the biggest factor influencing Annex I Parties' compliance cost is the geographic availability of mitigation measures. However, the principle that groups of Parties with differential compliance costs could cooperate in implementing mitigation measures was accepted in the Convention.[60] The Protocol provided further clarity on this by including the three innovative mechanisms allowing Annex I Parties to achieve their Article 3.1 mitigation commitments by undertaking, financing, or purchasing emissions reductions generated overseas.[61] The inclusion of these flexible mechanisms enabled countries to commit to more environmentally stringent targets at Kyoto than they would normally have been inclined to do. However, their implementation in an international context on so large a scale was unprecedented and raised novel moral, equity, and environmental considerations.[62] Many European and developing countries were, and remain, morally concerned by the concept that the most polluting countries can 'buy' their way out of taking any kind of domestic action. Others are concerned that emissions trading may entrench existing inequalities by endorsing the 'right to emit' of those who were historically responsible for creating the problem by emitting the greatest share of GHGs while shifting the burden of pollution control to those who have contributed little. Environmental concerns were expressed at the lack of more stringent Kyoto targets for the Russian Federation and Ukraine, whose most polluting industries had collapsed, and the fact that these countries might sell their surplus allowances.[63] These surplus allowances, which were based on historic emission levels and bore no relation to climate mitigation policies, were known as *hot air* and could be used by richer countries to avoid making politically unpopular domestic reductions.

However, in response to these moral and equity concerns, the Marrakesh Accords[64] state that the 'Kyoto Protocol has not created or bestowed any right, title, or entitlement to emissions of any kind on Parties included in Annex I'.[65] Further, prioritising domestic action, which has both moral and environmental dimensions, was one of the most divisive elements of post-Kyoto negotiations. However, the Marrakesh Accords provide that 'use of the mechanisms shall be supplemental to domestic actions and domestic action shall thus constitute a significant element of the effort made'. However, there is no definition of the word 'significant' but Annex I Parties must submit information in accordance with Article 7, which will be reviewed under Article 8. A further report must be submitted on how each Annex I Party is making 'demonstrable progress' under Article 3.2 of the Protocol. In addition, the rules limiting banking constrains the use of the Kyoto mechanisms and could serve to create more incentives for domestic action.

3.11.2.1 Use of the Flexible Mechanisms under the Kyoto Protocol

The heart of the climate change regime is the mitigation commitments by certain parties. The mitigation commitments, which are applicable to all parties, are provided in Article 4.1 of the Convention, and these are known as *general commitments* as they cover a broad range of issues relevant to addressing climate change. These include planning, research, and adaptation. Annex I Parties have more stringent mitigation commitments, which are identified in Article 4.2 of the Convention and are often known as *specific commitments*. Articles 4.3–4.10 contain commitments relating to specific situations with regard to particular groups of countries or Parties. However, it is the legally binding mitigation commitments that are linked to almost every aspect of the climate change regime and both the Convention and the Protocol allow groups of Parties to fulfil these mitigation commitments jointly. In addition, Annex I Parties may make use of Kyoto Protocol flexible mechanisms to meet their Article 3 commitments. The eligibility of Annex I Parties to use the flexible mechanisms is overseen by the Enforcement Branch of the Compliance Committee[66] and any determination by the Enforcement Branch of non-eligibility would lead automatically to the suspension of eligibility to use the flexible mechanisms by the Party concerned.[67] The consequence of ineligibility is that the relevant Annex I Party, and any legal entities authorised by them, cannot undertake transactions dealing with Kyoto units. Decision 24/CP.7 provides that the suspension 'is to be in accordance with the relevant provisions' under Articles 6, 12, and 17. Therefore, if an Annex I Party is not eligible to deal with units under one flexible mechanism, they may fulfil the criteria for eligibility under one of the other mechanisms. However, if an Annex I Party is suspended from eligibility for emissions trading, it would appear that no transactions relating to *any* of the flexible mechanisms units can be completed until the eligibility criteria are fulfilled. The effects of being barred from the use of any of the flexible mechanisms would be very significant for any Annex I Party, which placed a high reliance on the use of these mechanisms to achieve their Article I commitments.

In Decision 15/CP, 7, Annex, para. 8, all Parties agreed that

> environmental integrity is to be achieved through sound modalities, rules and guidelines for the mechanisms, sound and strong principles and rules governing land use, land use change and forestry activities and a strong compliance regime.

However, unresolved differences concerning the binding nature of the compliance procedures meant that a number of countries were unwilling to agree that acceptance of the compliance procedures was to be a condition of eligibility to use the flexible mechanisms. The final wording agreed at Marrakesh is as follows:

> The eligibility to participate in the mechanisms by a Party included in Annex I shall be dependent on its compliance with methodological and reporting requirements under Articles 5.1 and 5.2 and articles 7.1 and 7.4 of the protocol. Oversight of these provisions will be provided by the enforcement branch of the compliance committee, in accordance with the procedures and mechanisms relating to compliance as contained in decision 24/CP.7, assuming approval of such procedures and mechanisms by the [COP/MOP] in decision form in addition to any amendment entailing legally binding consequences, noting that it is the prerogative of the [COP/MOP] to decide on the legal form of the procedures and mechanisms to compliance.

However, this final wording was not intended to undermine or prejudice the oversight of eligibility conditions by the Enforcement Branch as all Parties agreed that such oversight is necessary.

ENDNOTES

1. Phillippe Sands, Editor (1993), *Greening International Law.* Earthscan: UK.
2. Schanze (1978), 'Mining agreements in developing countries', *Journal of World Trade Law* 12, 135–173.
3. Under a further Resolution XV/90 of 1968, the OPEC prescribed the rules to determine excessive profits and the criteria to be used for renegotiation.
4. Resolution 1803 (xvii); UN Doc. A/5217 (1962).
5. UN Doc. A/RES/3201 (S-VI) (1974); UN Doc. A/RES/3281 (XXIX) (1974).
6. Ibid., para. 4.
7. P. Ellis Jones (1988), *Oil: A Practical Guide to the Economics of World Petroleum.* Woodhead-Faulkener: Oxford, UK.
8. *Texaco v Libya, 53 ILR* (1977), 389.
9. Proclamation no. 2667, 3 C.F.R. 67, White House Press Release (28 September 1945) and 13 *Department of State Bulletin* 484 (1945).
10. *The North Sea Continental Shelf Cases*, ICJ Rep., p. 32 (1969).

11. *Norwegian Fisheries Case*, ICJ Rep., 116 (1951).
12. *The Anna*, 5 Ch Rob 373 (1805).
13. See Note 11.
14. *The North Sea Continental Shelf Cases*, ICJ Rep., 1, p. 22 (1969).
15. Taverne (1994), *An Introduction to the Regulation of the Petroleum Industry*. Graham & Trotman Ltd.: Oxford, UK.
16. The Convention on the Continental Shelf was part of the outcome of the First Law of the Sea Conference, 1958.
17. Rosalyn Higgins (1998), *Problems & Process: International Law and How We Use It.* Oxford University Press: Oxford, UK.
18. T. Daintith & G. Willoughby (1984), *Manual of United Kingdom, Oil and Gas Law*, Oxford University Press: Oxford, UK.
19. Rosalyn Higgins, p. 518.
20. The water depth where exploitation of minerals was possible.
21. Resolution 2574 (XXIV), 1970.
22. Resolution 2749 (XXV), 17 December 1970.
23. *Case Concerning Delimitation of the Maritime Boundary in the Gulf of Maine Area* ICJ Rep., p. 52 (1984).
24. As defined in Article 3 of the 1958 Convention on the Continental Shelf.
25. These are organisms which, at the harvestable stage, either are immobile on or under the seabed or are unable to move except in constant physical contact with the seabed or the subsoil.
26. Part XII, Articles 192–212.
27. See Bernard Taverne, p. 64 & 65.
28. Article 197.
29. Article 199.
30. Article 208.
31. Group of Experts on the Scientific Aspects of Marine Pollution (GESAMP) (1990), *The State of the Marine Environment*. UNEP, DEFRA: London, UK. See also UK Department of the Environment (1987), *Quality Status of the North Sea*; Clark, Editor (1986), *Marine Pollution*. Oxford University Press: Oxford, UK.
32. See Chapter 9.
33. For example, the Anglo-Norwegian Agreements.
34. OPRC Convention, 30 November 1990, IMO-55OE (1991st edition); 30 *ILM* (1991), p. 733.
35. Articles 3 and 4.
36. Article 3 of the Convention requires the parties to ensure that vessels flying their flag have on board an oil pollution emergency plan in accordance with International Maritime Organization (IMO) provisions.
37. Article 1(1).
38. Birnie & Boyle, *International Law and the Environment*, 2nd edition, Chapter 7, Oxford University Press: Oxford, UK.
39. GESAMP (1990), *The state of the marine environment*, 2nd International Conference on the Protection of the North Sea; *Quality Status of the North Sea*, UNEP, IMO: London, UK, 1987.
40. UN/ECE, *Air Pollution Studies*, nos 1–12 (1984–1996).
41. 35 *AJIL* (1941), 716.
42. 35 *AJIL* (1941), 712ff. Gray, *Judicial Remedies in International Law.*
43. Birnie & Boyle, *International Law and the Environment*, Oxford University Press: Oxford, UK.
44. *Lac Lanoux Arbitration*, 24, *ILR* (1957); *Icelandic Fisheries Case (Germany v Iceland)*, ICJ Rep. (1974).

45. Convention on Environmental Impact Assessment in a Transboundary Context (Espoo) (1991).
46. Chlorofluorocarbons (CFCs), hydrochlorofluorocarbons (HFCs), carbon tetrachloride, and methyl chloroform.
47. Convention for the Protection of the Ozone Layer 1985, 26 *ILM* 1529 (1985).
48. Article 2(2).
49. Weiss (1989), 'The five international treaties: A living history', *Strengthening Compliance with International Environmental Accords*, ISSD Library, International Institute for Sustainable Development: Winnipeg, Canada.
50. Article 1(6).
51. Articles 2A and B.
52. Decision VII/9: Basic Domestic Needs, Seventh Meeting of the Parties, 5–7 December, 1995.
53. Annex V to the Montreal Protocol on substances that deplete the ozone layer.
54. UNEP, Ozone Secretariat, http://www.unep.org/ozone. Accessed January 2011.
55. Bial et al. 'Public choice issues in international collective action: Global warming regulation', http://www.ssrn.com. Accessed January 2011.
56. IPPC, 1998, para. 2.
57. At the forefront of the discussions for a New International Economic Order was the assumption that the economic and social advancement of nations would be achieved through the access of those nations to the world's technological resources on the basis of equitable entitlement. (Michael Blakeney (1989), *Legal Aspects of the Transfer of Technology to Developing Countries*. ESC Publishing Ltd: Oxford.)
58. Article 2, '. . . such a level should be achieved within a timeframe sufficient to allow ecosystems to adapt naturally to climate change. . .'.
59. Sands (2009). *Principles of International Environmental Law*, 2nd edition, Cambridge University Press: Cambridge, UK.
60. FCCC Articles 3.3 and 4.2(a) and (d).
61. Yamin & Depledge (2004), *The International Climate Change Regime*, Cambridge University Press: Cambridge, UK.
62. Yamin (2000), 'Joint implementation', *Global Environmental Change* 10(1), 87–91.
63. Yamin & Depledge, Note 51.
64. Cop 7. fccc/CP/2001/13 Marrakesh.
65. Decision 15/CP.7, Preamble, para. 5.
66. Decision 24/CP.7.
67. Decision 24/CP.7, Annex, section XV, para. 4.

4 Trade, Competition, and the Environment

This chapter completes the general section of the book by considering trade, competition, and environmental law with reference to the energy sector.

4.1 INTERNATIONAL COMPETITION LAW AND THE GENERAL AGREEMENT ON TARIFFS AND TRADE

The General Agreement on Tariffs and Trade (GATT) was signed in 1947 as a reaction to the protectionism of the 1920s and 1930s. It was founded within the framework of the establishment of the new world order to re-establish an international marketplace and promote free world trade. At the same time, the International Monetary Fund (IMF), the International Bank for Reconstruction and Development (IBRD), and the International Trade Organisation (ITO) were set up. Although they did not come into existence then, the GATT was applied from January 1948 with the high Contracting Parties creating an institutional structure by a series of decisions similar to those in an international treaty system rather than an organisation. In the absence of an executive decision-making organ, a mechanism of 'rounds' of multinational trade negotiations was developed with a number of supplementary treaties or side agreements that did not have to be signed by all parties. As the GATT only affects international trade in goods, during the Uruguay Round in 1994, it was decided to establish the World Trade Organisation (WTO), which encompasses the GATT 1994 as well as the General Agreement on Trade in Services (GATS), the General Agreement on Trade-Related Aspects of Intellectual Property Rights (TRIPS),[1] and the Agreement on Trade-Related Investment Measures (TRIM).[2] The WTO recognises the importance of environmental measures and sustainable development and is an actor within the energy field as energy is a 'good' within the framework of GATT.

The main purpose of the GATT was to provide a framework of rules for the orderly conduct of world trade and the reduction of trade barriers. In addition to that, the goals of the WTO itself recognise the need to protect the environment and to promote sustainable development. The WTO administers all the trade agreements, provides a forum for multilateral trade negotiations, administers the trade dispute-settlement procedures, and reviews national trade policies. At Marrakech, the trade ministers agreed to establish a WTO Committee on Trade and Environment (CTE) with analytical and prescriptive functions covering all areas of the multilateral trading system including goods, services, and intellectual property (IP).

4.2 INTELLECTUAL PROPERTY RIGHTS AND ENVIRONMENTAL TECHNOLOGY

Effective and timely development of new environmental technology is crucial for a concerted global action towards the reduction of greenhouse gas emissions. Environmental technologies provide solutions to decrease material inputs, reduce energy consumption and emissions, recover valuable by-products, and minimise waste disposal problems.[3] They also support the application of environmental management systems and make production processes cleaner by collecting information about the environment and monitoring and gathering data to identify the presence of pollutants. According to the European Environment Agency, environmental technologies have the potential to contribute to the reduction of greenhouse gas emissions by 25%–80%; to the reduction of ozone depletion by 50%; and to acidification and eutrophication by up to 50%.

It is also commonly understood that in order to supply the increasing global demand for energy there will need to be the use of a more diverse collection of technologies different from the current established technologies used for accessing conventional oil and gas reserves. The technologies required will include technologies for unconventional oil and gas reserves such as shale gas and heavy oil; for renewable energy sources; and for achieving cleaner coal energy production.

4.2.1 Transfer of Environmental Technology and Corporate Social Responsibility

Under a number of international environmental conventions, there is a general requirement for the transfer of environmental technology to developing countries.

Given the numerous forms that technology can take, it is not surprising that there are as many methods and devices for effecting the transfer of technology; however, the principal mode of technology transfer to developing countries is on commercial terms as a business transaction between enterprises or institutions in different countries. Not only will the complexity of the technology to be acquired have an important influence on the method of transfer but also the nature of the domestic legislation and legal practices in the recipient country. Financial considerations will inevitably play a part as foreign exchange considerations will invariably have an important role. There will also be the issue of the technological sophistication of the acquirer. For the less-experienced transferee, technical assistance agreements will need to be in place. However, all the methods by which technology is transferred will need to be backed by a legal agreement.

4.2.2 Theory, Intellectual Property Rights, and Technology Transfer

A striking correlation has been noted between industrialisation and the protection of industrial property rights[4]; the express objective for the first grants of patent protection in Europe in the Middle Ages was to encourage industrial development.[5] The subsequent industrialisation of Europe and the United States is now taken to be evidence of the causal nexus between industrial property protection and industrialisation. However, this assumption cannot be as easily made because of the strong movement in Europe for the abolition of patent protection, influenced by free trade theories.[6] At that time, patent protectionism was ideologically linked with tariff protectionism and both were anathema to the free trade spirit of the time. In the patent debate of the nineteenth century, four lines of argument were advanced against the free trade opponents of the industrial property systems. Blakeney[7] claims that it was asserted that man had a natural property right to his own ideas; that patent protection was a just reward for the useful service performed by inventors; that the prospect of reward was an incentive for invention; and, finally, that in the absence of protection against the immediate imitation of novel technological ideas, it was asserted that an inventor would keep his invention secret, thereby at least delaying technological progress. The contemporary relevance of these arguments is present in the debate on the transfer of technology to developing countries.

In the contemporary debate on the transfer of technology to developing countries, the status of the natural rights justification of industrial property protection is very much subordinate to the various economic arguments. In fact, the concept of private natural property rights in the creations of the intellect exists in uneasy tension with countervailing public rights such as the 'universal human right to share in scientific advancement' and the 'right to development' as the 'right' is only a limited one.[8] In the current way in which property rights are defined as 'socially recognised economic rights to act, to dispose and use', Lehmann[9] asserts the ascendancy of the economic touchstone of private rights and compels an evaluation of the alleged economic justifications for the protection of industrial property.

The assumption of almost all the proponents of the transfer of technology is that such transfer is a prerequisite, even an imperative, for the desirable economic and social development. Solow[10] attributed 87.5% of the growth of per capita income in the United States in the first half of the twentieth century to technological progress and the remainder to the use of capital. In respect of developing countries, the deprivation and poverty suffered by the indigenous peoples has been attributed almost entirely to their technological dependence. To try to address the question of the technological transformation of developing countries, an UNCTAD[11] report in 1980 noted that industrialised countries spent 17 times more of their gross national product on research and development than developing countries. The WIPO (World Intellectual Property Organisation) Licensing Guide for Developing Countries (1977) commences with the assertion that 'Industrialisation is a major objective of developing countries as a means to the attainment of higher levels of well being of the peoples of such countries. . .'. These statements reflect the philosophy animating the Declaration on the Establishment of the New World Order adopted by the General Assembly of the United Nations in December 1974. Interpreting the Declaration, Fikentscher[12] asserted the right of access for all nations to 'the universal heritage' of technology. This right was to be secured through the institution of an appropriate legal regime to facilitate technology transfer to developing countries on an 'equitable' basis. This was an article of faith that the transfer of technology to developing countries would improve their material circumstances to levels approaching those of the industrialised countries. The theory itself proceeds on the assumption that the transfer of technology can facilitate the more productive use of resources and provide a technology base from which the development of indigenous technology can proceed. This is undeniable, but it is questionable whether it is possible or prudent to use technology that has been developed for markets in industrialised countries. Hansen[13] explains, 'the realisation of developmental objectives is crucially dependent on whether the mechanism of transfer is rightly adapted to the absorptive capacity of the economy'. So, where does this leave the international corporation with a strong corporate social responsibility (CSR) policy and a desire to assist the development of its suppliers from the developing countries?

It is now a given that global warming is one of the most pressing problems faced by the international community. It has also been accepted by the Intergovernmental Panel on Climate Change (IPPC) that a large proportion of these greenhouse gas releases are due to anthropocentric activity. Should patent law help cool the planet? This is the question asked by Estelle Derclaye.[14] The main goal of patent law is to create incentives for industry to develop new technology, which in turn may create pollution including greenhouse gases. However, despite the perceived neutrality of patent laws, they in fact already address the issue of protection of the environment through Art. 53(a) of the European Patent Convention (EPC) and the corresponding national provisions. In addition, under the European Community Treaty, patent laws must take account of environmental laws. Therefore, any new technology developed by corporations with a view to protecting the ideas through patent, must take account of environmental principles and legal instruments.

The transfer of technology that will help mitigate climate change has always been an item on the agenda at the United Nations Framework Convention on Climate Change (UNFCCC) Conferences of the Parties (COP) meetings. Technology transfer is a major feature of the UNFCCC, which requires developed countries to promote technology transfer to developing countries to enable them to implement the provisions of the Convention (Article 4.5). In 2001, UNFCCC COP 7 established a framework to enhance the implementation of Article 4.5 of the Convention. The five key themes identified were technology needs and needs assessments, technology information, enabling environments, capacity building, and mechanisms for technology transfer. The different views on the role of IP in the transfer of technology were particularly highlighted at the UNFCCC 2007 meeting in Bali. Many developing countries proposed further options and expansions on the existing WTO TRIPS Agreement, including development of compulsory licensing. Technology transfer from developed countries to the developing countries was discussed at the Beijing International Conference on Carbon Abatement Technology Transfers in 2008. Countries such as China and India suggested that IP rights could be a potential barrier to technology transfer and proposed that the TRIPS compulsory licensing flexibilities provided for medicines should be extended to cover carbon abatement technologies. The reasoning was that, similar to public health, the climate is a public good and there should be greater access to the technology used for priorities such as carbon abatement. However, many developed countries, non-governmental organisations (NGOs) such as the WIPO, Japan, and the European Union argued that the climate situation is not the same as the health situation because in the pharmaceutical industry generally one company owns the patents for a key technology, whereas in the carbon abatement industry the technologies and patent ownership are spread over a large number of companies. The argument from the developed countries is that IP rights are not a barrier to technology transfer and that only a strong IP regime, including patent laws, will encourage the large investment in resources needed for the necessary technology innovation, transfer, and diffusion. During the debate on whether or not IP rights are a barrier to technology transfer, there were various investigations into the patenting activity and patent ownership for carbon abatement technologies. A large study on the patenting and licensing activities for carbon abatement technologies was carried out by the United Nations Environment Programme (UNEP), the European Patent Office (EPO), and the International Centre for Trade and Sustainable Development (ICTSD), and the final report 'Patents and Clean Energy: Bridging the Gap between Evidence and Policy' was published in 2010. The results of the study found that patenting activity had increased at the rate of 20% per annum since 1997 and that, for certain technologies, patenting was dominated by developed countries. The six leading countries innovating and patenting were Japan, the United States, Germany, the Republic of Korea, the United Kingdom, and France. However, there were some emerging countries as well, such as India, which was among the top five for solar technologies, while Brazil and Mexico were the top two countries in hydro and marine technologies. The conclusion of this report was the patenting activity indicated that, for many of the technology sectors, there seemed to be sufficient competition in the marketplace, taking account of the market share of the major players. The report suggested further analysis and monitoring of patent assignments between companies to indicate technology transfer or a building of a dominant position by one company.

Other reports reviewing the patenting activity for climate change technology include a report by the WIPO in 2009 'Patent-Based Technology Analysis Report: Alternative Energy Technology'. This review on the patent-filing activity finds a similar situation where the developed countries dominate the patenting activity.

The debate on whether IP rights are a barrier to transfer of climate change technology is reviewed in a study commissioned in 2009 by the European Commission (DG Trade) authored by Copenhagen Economics and The IPR Company, 'Are IPR a Barrier to the Transfer of Climate Change Technology?'. The conclusion of this study was that IP rights do not in themselves constitute a barrier to the transfer of carbon abatement technology from developed countries to low-income developing countries or to emerging market economies. The report suggests the reason for an alleged lack of transfer of technologies to developing countries could be insufficient technical knowledge and absorption capacity to produce locally; insufficient market size to justify local production; and insufficient purchasing power and financial resources to acquire the technology or the technology products.

Another paper that reviews the generation and diffusion of climate change mitigation technologies and their links to key policies is the one by the Organisation for Economic Co-operation and Development (OECD) Environment Directorate under the title *Climate Policy and Technological Innovation and Transfer: An Overview of Trends and Recent Empirical Results* Working Paper no. 30. The conclusions of this paper are similar to the other studies in this area, in that the most important determination of innovation in the area of climate change technologies is innovation capacity. Mostly technologies were created and advanced in developed countries and were not being diffused to the developing countries with sufficient speed.

4.3 THE MANAGEMENT OF RISK IN ENERGY INFRASTRUCTURE

The International Energy Agency has estimated that global requirements for primary energy will increase by 1.6% per annum or 45% in total between 2007 and 2030. This does not include the replacement of the ageing infrastructure. The Agency also projects that the total investment in energy supply infrastructure between 2007 and 2030 will require approximately $26.3 trillion.[15] Whatever the actual demand growth in energy, the investment required to satisfy it will rest upon an array of complex international trade, macroeconomic, political, technical, environmental, and social challenges. These will present areas of risk that warrant careful analysis by policymakers and private sector investors.

The mismatch in the location of supply and demand for oil and gas is increasingly marked with the majority of the lower costs hydrocarbon reserves concentrated in the Middle East, North Africa, and the former Soviet Union, while the high-consuming markets are situated in North America, Europe, and, more recently, the Asia Pacific. Following the 2008 financial crisis, economic growth in the emerging markets has driven demand for oil higher and thus increased commodity prices. This upward pressure on oil prices has been accompanied by instability across the Middle East and North Africa; it was further compounded by the Fukushima tragedy in Japan.[16]

This supply/demand dislocation has driven disparate and competing policy responses with the distance between the conventional producing fields to markets increasing as governments and international oil and gas companies are forced to seek out new reserves further afield. Diversification away from cheaper oil in the Middle East, North Africa, and the former Soviet Union also forces governments and international oil companies to develop and produce reserves in new and more politically unstable environments such as Iraq, or more environmentally sensitive areas such as the Arctic.

The pragmatic reality is that the projected rise in primary energy demand requires large amounts of investment to satisfy such demands. Policymakers, industrial players, and financiers will need a clear understanding of the nature and range of potential risks involved.[17] Coincidental with the financial crisis, two recent and spectacular industrial accidents occurred in the United States.[18] Both were in part blamed on inadequate regulatory oversight by agencies otherwise charged with protecting health, safety, and the environment. With regulators being less than vigilant, the idea that respect for the rule of law is greater among OECD Member States than in emerging markets may now have less of a deterrent effect on investment in developing economies.[19]

4.3.1 Bilateral and Multilateral Investment Agreements

As noted earlier, the extending distances between the source and use of energy products is becoming a pressing geopolitical issue, especially in Western Europe. Trans-boundary issues can arise in the energy sector in any circumstances where infrastructure and transit routes cross several state borders. There may be the added problem of resources subject to the claims to sovereignty by more than one state. The nature of trans-boundary energy transportation and transmission gives rise to concerns that do not occur in other energy infrastructure projects. In this section, consideration is given to what protection is available to investors against governmental interference in trans-boundary energy infrastructure and the network of bilateral and multilateral investment treaties that exist.

4.3.2 Bilateral Investment Treaties

As global trade increased and became more sophisticated, treaties developed to protect investments rather than simply to encourage trade. Coincidentally, arbitration gained in popularity as an alternative dispute settlement in a neutral forum. By 1966, the International Centre for Settlement of Investment Dispute (ICSID) Convention[20] came into force. The Convention is the culmination of efforts by investor states in the wake of the highly publicised nationalisation of the investments of British Petroleum (BP) in Iran during the 1950s. The Convention established the International Commission on Irrigation and Drainage (ICID), which is dedicated to the conciliation and arbitration of international investment disputes; this was followed by a number of bilateral investment treaties providing for arbitration directly between investors and host states, which are now the norm. There are now over 2600 bilateral investment treaties and more than 270 bilateral investment agreements, such as economic partnership agreements. All countries are party to at least one bilateral investment treaty or agreement.[21]

Many bilateral investment treaties are usually short and contain 10 key provisions that offer similar protections. Some states have model bilateral investment treaties such as the 2004 US Model Bilateral Investment Treaty. Claims arising under such treaties have produced a significant body of international arbitration decisions, key trends emerging from which are the varying approaches of tribunals, possibly reflecting the political nature of the issues that they are dealing with.

4.4 THE ENERGY CHARTER TREATY

The Energy Charter Treaty is the only multilateral investment treaty devoted to the energy sector, and has its origins in the early 1990s, following the collapse of the Soviet Union.

The treaty came into force in April 1998 and covers the following:

- The promotion and protection of investments in energy;
- Trade in energy, energy products and energy related equipment;
- Freedom of energy transit;
- Improvement of energy efficiency; and
- Provisions for the resolution of disputes through international arbitration.

By 2011, 51 countries were signatories to the treaty, in addition to the treaty having been signed collectively by the European Union and the European Atomic Energy Community. Of those who have signed the treaty only Australia, Iceland, and Norway have yet to complete ratification. Although it signed the treaty in the first instance in October 2009, Russia terminated its provisional application of the treaty and gave notification that it did not intend to become a full contracting party. A number of states, including the United States, are 'observer states', which means that they are entitled to participate in certain aspects of the treaty without having any binding obligations under the treaty nor are there any requirements for their having to contribute to the treaty's budget. These observer states also include the Association of Southeast Asian Nations (ASEAN), which itself enjoys observer status.

4.4.1 Investment Protection under the Energy Charter Treaty

The Secretariat of the Charter Treaty states that 'the fundamental aim of the Energy Charter Treaty is to strengthen the rule of law on energy issues, by creating a level playing field of rules to be observed by all participating governments, thereby mitigating risks associated with energy related investment and trade'.[22]

The investment protection regime is provided for in Part III of the Treaty and seeks to ensure the protection of investments based on the principle of non-discrimination. The drafters of the Treaty adopted many of the protections contained in the bilateral investment treaties with the outcome containing provisions for the following:

- Fair and equitable treatment
- Protection from unreasonable or discriminatory treatment
- National treatment and most-favoured nation treatment
- Provision for compensation in the event of expropriation
- Transfer of capital
- Full protection and security

Once investments have been made, the obligation of Contracting Parties to avoid unreasonable and discriminatory treatment is triggered but not until then, although the treaty does draw a distinction between 'investments', which are afforded full protection, and the 'making of investments',[23] in relation to which Contracting Parties are obliged to endeavour to accord non-discriminatory treatment to investors. Thus, there is an ongoing debate as to what rights/expenditure should be protected under an investment treaty if protection is not extended to the pre-investment phase.

Definitions are equally important as they tend to be rather broad under the Energy Charter Treaty. The treaty sets out a non-exhaustive list of investment categories that are mostly found in the wording of the bilateral investment treaties:

> Investment means every kind of asset, owned or controlled directly or indirectly by an investor and includes:
>
> a. tangible and intangible, and moveable and immovable, property, and any property rights such as leases, mortgages, liens, and pledges;
> b. a company or business enterprise, or shares, stock, or other forms of equity participation in a company or business enterprise, and bonds and other debt of a company or business enterprise;
> c. claims to money and claims to performance pursuant to contract having an economic value and associated with an investment;

d. intellectual property;
e. returns;
f. any right conferred by law or contract or by virtue of any licences and permits granted pursuant to law to undertake any Economic Activity in the Energy Sector.

Basically, investments that are protected are those associated with economic activity in the energy sector, which relates to certain energy materials and products.

The definition of 'economic activity in the energy sector' includes exploration, extraction, refining, production, storage, land transport, transmission, distribution, trade marketing, or sale of 'energy materials and products' in the nuclear, coal, natural gas, petroleum, and electrical industries. Most importantly, the definition of 'investment' expressly extends to investments made prior to the entry into force of the Energy Charter Treaty in 1998. However, protection is only extended to investments after that date.

Investors under the Energy Charter Treaty can be both natural and legal persons such as companies, and investors can incorporate companies in Energy Charter Treaty Contracting Parties and invest through them and so gain access to the protection of the treaty. However, there is a 'denial of benefit' provision in the treaty, which denies protection to a company that has no substantial business activities in the treaty contracting party. The treaty also requires Contracting Parties to ensure that any state enterprises through which they may conduct their activities in relation to the treaty observe the investment protection provisions.

4.4.2 Transit under the Energy Charter Treaty

The Energy Charter Treaty provides for a special regime for the transit of energy goods through Contracting Parties with Article 7 establishing a legal framework for relationships between governments in relation to transit.[24]

Article 7 sets out the rules on the following:

- Non-discriminatory passage with no distinction allowed as to origin, destination, or ownership of products or materials
- Non-discriminatory pricing
- Absence of unreasonable delays, restrictions, or charges
- Modernisation of infrastructure
- Non-interruption of transit in case of dispute, and clear dispute and conciliation procedures

Article 7 also requires each contracting state to take

> . . . necessary measures to facilitate the transit of Energy Materials and Products consistent with the principle of freedom of transit and without distinction as to the origin, destination or ownership of such Energy Materials or Products or discrimination as to pricing on the basis of such Energy Materials or Products or discrimination as to pricing on the basis of such distinctions, and without imposing any unreasonable delays, restrictions or charges.

The principle of 'freedom of transit' is referred to without any legal basis for such a principle,[25] but the obligation of non-discrimination in transit relates to both terms of access to the energy transport facilities and to the terms of conditions of carriage. Since 1998, a draft Protocol has been contemplated and discussed, which considers the uninterrupted energy flow between Contracting Parties even in circumstances where such transit would endanger the security of supply. Should this draft Protocol on transit be agreed on, it would greatly clarify the rights and obligations of transit states.

4.4.3 Dispute Resolution

The Energy Charter Treaty provides for international arbitration. It offers investors the choice of ICSID arbitration, ad hoc arbitration under the United Nations Commission on International Trade Law (UNCITRAL) rules, and arbitration under the rules of the Arbitration Institute of the Stockholm Chamber of Commerce. However, not all of the provisions of the Energy Charter Treaty entitle investors to institute arbitration proceedings against defaulting states. For example, breach of the transit and trade provisions are to be resolved at the contracting party level, and so an investor who is adversely affected by a transit-related issue that does not breach the investment protection provisions will be unable to commence arbitration under the Energy Charter Treaty and so enforce its rights.

4.5 CASE STUDIES

Hulley Enterprises Limited v The Russian Federation (Permanent Court of Arbitration [PCA] Case AA 226), *Yukos Universal Limited v The Russian Federation* (PCA Case AA 227), and *Veteran Petroleum Trust v The Russian Federation* (PCA Case AA 228).

These parallel cases arose from the collapse of the Russian oil company OJSC Yukos Oil Corporation in 2006.

Russia disputed the jurisdiction of the arbitral tribunal in these parallel cases on the grounds that inter alia, as it was only subject to the provisional application of the Energy Charter Treaty, Russia was not liable for a failure to provide investment protection to the claimants. The Russian Federation claimed that provisional application under the treaty only has effect where such application is not inconsistent with the constitution, laws, or regulations of the contracting party. Russia argued that submission to international arbitration was inconsistent with its domestic laws, which required sovereign acts to be referred to the Russian courts.

The tribunal disagreed and found that there was nothing in Russian law that prevented the provisional application of treaties. This was the fundamental issue and not whether arbitration under the treaty was contrary to Russian law. In addition, the tribunal found that Russia's termination of provisional application of the treaty did not preclude the investor's claims. Parties that had made investments prior to the termination date would continue to benefit from investment protection for a further 20 years from termination. The tribunal, therefore, ruled that it had jurisdiction to consider the merits of the investors' cases.

The Yukos decision is a landmark one and is highly significant to investors in Russia under the Energy Charter Treaty prior to October 18, 2009 (when provisional application was terminated); they will benefit from protection under the treaty until 2029. Given the importance, in particular to the European energy markets, this ruling strengthens the force of the Energy Charter Treaty as a mechanism for energy sector investment protection.

4.6 MODEL AGREEMENTS

The Energy Charter Secretariat has developed model forms of agreements for trans-boundary pipeline developments. They are intended to assist as the starting point for negotiated agreements between potential participants in infrastructure.

They are intended to be a neutral and non-prescriptive starting point and thus facilitate project-specific talks and can be downloaded from the Energy Charter web site.[26] Currently, there are Model Agreements for Cross-Border Pipelines and Cross-Border Electricity Projects. The Secretariat is working on Investment Model Agreements (IMAs).

4.6.1 Structuring Investment Protection

In order to enhance the likelihood of benefiting from investment protection under the Charter, investors in trans-boundary energy infrastructure should seriously consider the options available to them as follows:

1. Are there bilateral investment treaties that may offer protection for the specific investment?
2. Does the Energy Charter Treaty or any other multilateral investment treaty apply?
3. If so, could the investor adopt a contractual structure that increases the available protection?
4. Check the substantive provisions in each treaty and the ones that are best suited to the investment.
5. Consider other mechanisms to mitigate the risk of investing.[27]

These extensive bilateral and multilateral investment treaties are sometimes overlooked by investors and yet they can afford valuable protection. Intergovernmental risk, however, remains and despite some criticism that the current regimes are too investor friendly, they remain the best means available to investors to mitigate the risks of investing in energy infrastructure.

ENDNOTES

1. (1994) O.J. L336/213.
2. (1994) O.J. L336/100.
3. http://www.eea.europa.eu/themes/technology. Accessed December 2010.
4. Beier (1980), 'The significance of the patent system for technical, economic and social progress', *IIC* 11, 663.

5. Machlup & Penrose (1950), 'The patent controversy in the nineteenth century', *Journal of Economic History* 10, 1–29.
6. Ibid.
7. Blakeney (1989), *Legal Aspects of the Transfer of Technology to Developing Countries.* ESC Publishing Ltd: Oxford.
8. See Note 7 above.
9. Lehmann (1985), 'The theory of property rights and the protection of intellectual and industrial property', *IIC* 16, 525–526.
10. Solow (1957), 'Technological change and the aggregate production', *Review of Economics and Statistics* 39, 312–320; quoted in Ewing (1976), 'UNCTAD and the transfer of technology', *Journal of World Trade Law* 10, 197–214.
11. UNCTAD (1980), 'Formulation of a strategy for the technological dependence of developing countries', UNCTAD Doc. TD/B/779.
12. Fikentscher (1980), *The Draft International Code of Conduct on the Transfer of Technology.* Max Planck Institute: Munich.
13. Hansen (1980), 'Economic aspects of technology transfer to developing countries', *IIC* 11, 429–430.
14. Derclaye (2009), 'Should patent law help cool the planet? An inquiry from the point of view of environmental law: Part 1', *European Intellectual Property Review* 31(4), 168–184.
15. Tansaka, www.iea.org/speech/2009/Tansaka/4th_OPEC_Seminar_speech.pdf. Accessed July 2010.
16. The tragedy in Fukushima has renewed the cloud over the use of nuclear energy as a potential substitution for hydrocarbons.
17. Dimitroff (2011), *Risk and Energy Infrastructure.* Globe Business Publishing: London.
18. Texas City oil refinery and the blowout at the Macando oilfield in the Gulf of Mexico.
19. Ibid.
20. This is a multilateral treaty promoted by the World Bank.
21. UN Commission on Trade and Development (2009), 'The role of international investment agreements in attracting foreign direct investment to developing countries'. www.unctad.org/Templates/Download.asp?docid=1254&dang. Accessed November 2011.
22. The Secretariat, www.encharter/org/index.php?id=7. Accessed November 2011.
23. These are defined as 'establishing new investments, acquiring all or part of existing investments or moving into different fields of investment activity'.
24. Cameron (2007), *Competition in Energy Markets.* Oxford University Press: Oxford, UK.
25. Ibid.
26. http://www.encharter.org/index.php?id=330. Accessed January 2011.
27. A more in-depth discussion is provided by Cummins & King (2011), 'Multilateral and bilateral investment agreements', Dimitroff, Editor, *Risk and Energy Infrastructure: Cross Border Dimensions.* Globe Law and Business: London.

Section II

This part addresses the individual regulation of the various energy sectors under international law.

5 International Law on Oil and Gas

This chapter looks at the oil and gas sector, both onshore and offshore.

5.1 LICENSING FOR ONSHORE OIL AND GAS EXPLORATION AND EXPLOITATION

All hydrocarbon resources in the soil and subsoil, in internal waters and in the territorial sea, on the continental shelf and in the exclusive economic zone come under the jurisdiction of the state. However, not all states have easy access to development investment or the technical expertise in the task of finding and extracting hydrocarbons. In these cases, such states delegate these operations to an international oil company (IOC) or, in some cases, to a domestic one. The relationship between the state and the international oil company must then be regulated by international law or domestic legal instruments and, in some cases, by contract. The most important parts of such a relationship are the licence, the concession, the production-sharing agreement (PSA), and the service contract.

Any tensions that may occur between the state and the international oil company would normally occur in the allocation of risk between the parties, and the provision of incentives to the international oil company to accomplish the state's objectives with regard to the development of the resource.

As far as the international oil company is concerned, it is the upstream petroleum sector that is faced with physical, commercial, and political risks at every stage of the exploration and production process. This is exacerbated by the significant investment required for the exploration and production of oil and gas and the large amount of uncertainty surrounding the ability to profit from the investment. Even if a commercial discovery is established, history is witness to the dramatic fluctuations in the international price of hydrocarbons and the high costs of extraction. Such fluctuations in both price and costs translate into risk with respect to the profitability of the venture.

Normally, under current upstream licence or contractual regimes, it is the international oil company that takes the greatest risk in respect of exploration. However, different states have different regimes particularly in respect of definitions of such issues as ownership of the hydrocarbons and the legal nature of the instrument granting the rights to the international oil company. Upstream regimes can also be quite different in how quickly the exploration costs can be recovered by the oil company at the production stage, as this would depend on the way the state allocates the other risks between the state and the oil company.

With regard to ownership under both licences and concessions, all hydrocarbons produced belong to the oil company. Under both production sharing agreements and service contracts, the state owns the hydrocarbons produced with a portion of them being allocated by the state to the oil company as a payment for the risks taken and the service performed. Further, under the production sharing agreements and service contract arrangements, the state retains the larger portion of the risk associated with the commercialisation of any discovery.

The type of regime chosen by the state will depend on the experience and capabilities of the state. The state will normally be keener to bear the exploration risks arising from those fields that show more certainty, rather than the significant risks of the more speculative fields. Some states will even argue to change from one system to another as the risks become less and the certainty greater.[1]

Under all upstream regimes, the international oil company cannot ignore the fact that by entering into a relationship with a State Party, that State Party is a sovereign state and can use a number of legal and regulatory instruments to expropriate *ex post facto* blocks, and even entire fields, or reduce the revenue received by the company. For example, the state can change the laws retrospectively, increase taxation on the product, or refuse to issue export licences. Nevertheless, the sovereign risk has historically been the greater because of favourable conditions negotiated by the oil company in the contractual arrangements.

Although a certain amount of sovereign risk is unavoidable, the state can include a number of contractual and legal provisions aimed at limiting its extent so the oil company will want to ensure that stabilisation provisions provide a certain amount of protection against future opportunistic behaviour by its sovereign partner.

A further objective of the licence or contractual arrangement between a state and the oil company is adequate incentives for both parties, but in particular for the oil company. No oil company will be motivated to explore an oilfield unless it is entitled to an acceptable level of benefit from any discovery made. Different regimes deal with this in different ways but fiscal terms will be at the heart and represent the revenues that the oil company will be able to earn from any production. If a field is a technical success, the level of royalties, taxation, and profit sharing imposed by the state will determine whether it is a commercial success for the oil company.

Sometimes, the State Party is interested in accomplishing other objectives from the relationship. These could include transfer of technology, the earning of foreign exchange, or the promotion of direct and indirect employment. They may also need to retain a percentage of the hydrocarbons for local consumption. These secondary objectives may come at a cost and are discussed in each of the regimes outlined in the following sections.

5.2 THE LICENCE REGIME

In some states, licences are referred to as leases but they are essentially the grant of a permission by the state to an international oil company to explore for and exploit a particular geographical area for a fee and/or a royalty. Ownership of the hydrocarbons remains with the state, if found on land or in the territorial sea, but if beyond the territorial sea on the continental shelf, then the state merely has the right under the United Nations Convention on the Law of the Sea 1982 to grant a licence to the international oil company. Once the hydrocarbons have been taken into possession by the oil company at the well head, the ownership gets transferred to the oil company. Therefore, the licence grants a proprietary right over the hydrocarbons and any profits made by the oil company will be taxed by the state. Generally, the licensing regime is a free market under which the oil company bears the majority risk but also enjoys the majority profits. The United Kingdom, Canada, Norway, and Russia are the main states that use the licensing regime to develop their hydrocarbon resources, and the specific regimes in the United Kingdom and Norway are discussed in dedicated chapters.

5.3 THE CONCESSION REGIME

The concession regime was most prevalent in the early twentieth century, whereby IOCs acquired the right to explore for and produce hydrocarbons on large tracts of the host state's territory. A concession is defined as a relationship between a state and an IOC under which the state concedes sovereign control over its hydrocarbon resources. These concessions were normally very favourable to the international oil company, who enjoyed absolute authority over the territory and ownership of all hydrocarbons *in situ*. These concessions were also granted for long periods of time and the state had minimal control over how the territory was developed or the hydrocarbons recovered and disposed of.

Under these concessions, the IOCs would pay an initial consideration and an annual rental fee, which was independent of any exploration or production results. On the other hand, all of the risks were with the IOC as were the rewards. These old concession regimes were to provide incentives to the IOCs who were developing technology and 'know-how' and seemed an efficient and cost-effective way to the state to exploit quickly what was in some cases the state's only economic resource. The problem was that the state had no control over other possible objectives such as environmental considerations, or over any wider developmental objectives.

Following the UN Resolution on ownership of minerals and hydrocarbons, these concessions changed fundamentally, with the state gaining permanent sovereignty over mineral and hydrocarbons within its territory. The IOC then only gained ownership of the hydrocarbons once they had been taken into possession at the well head. Concessionaires have control over the territory of the concession, but post-1950s these areas are limited to smaller geographical areas and may be subject to revocation depending upon the work programme and budget.[2] As with the earlier concessions, the IOC takes most of the risk but the state gains revenues from royalties, income taxes, and other payments.

As with licences, concession agreements contain terms and conditions. This forms a legal system that is applicable only to the concessionaire and covers a range of regulatory requirements including work permits, foreign exchange, health and safety, taxation, planning and environmental issues, and the right to import and export goods. Although historically, concessions would be by direct negotiation with the oil company, currently, most states have a standard form laid down in legislation, which the oil companies can take or leave. There is little room for negotiation and the IOCs will need to look carefully at any risks posed, in particular, the political stability of the state. These risks would then be reflected in the bidding price.

Is a concession agreement a contractual or regulatory instrument? This was the question considered by the court in the *Amaco*[3] and *Texaco*[4] oil arbitrations. In both cases, the arbitration panel held that concession agreements were not mere administrative instruments capable of unilateral amendments by the state. Interference with these contractual rights would amount to expropriation and therefore subject to compensation payable by the state.

Concessions are similar to licences but are more commonly used in developing countries that have a less than developed legal and political system. The concessions will also often require the IOC to provide local employment and training in addition to technology transfer. On the other hand, the IOC would require some sovereign assurance as to state support arrangements.

5.4 CASE STUDY OF THE CONCESSION REGIME IN BRAZIL

In 1995, the monopoly of Petrobras[5] was dismantled, and in 1997, the Brazilian Petroleum Law was enacted. The National Petroleum Agency became the regulator responsible for issuing tenders to grant concessions to both domestic and foreign companies. These tenders are open to any company meeting the technical, economic, and legal requirements. The state requires a signature bonus that is established by a public notice tender; royalties of 10% over gross production; and a special participation fee. There are variable percentages applied to the net revenue of highly productive fields and an annual rent determined by the square kilometre. Onshore production is also subject to an extra 5% royalty payable directly to the landowner. Concessionaires are entitled to all of the hydrocarbons produced.

In 2007, deepwater offshore discoveries were made, and discussions took place regarding the move from a concession regime to a product sharing agreement regime. On December 10, 2010, the Brazilian House of Representatives gave its final approval to the creation of the new PSA regime under which Petrobras will own at least 30% participating interest in every new venture within the pre-salt area, which is estimated to hold recoverable reserves of roughly 50 billion barrels of oil.

Petrobras will have the right, either solely or in consortium with other partners, to conduct all the exploration and production operations required within the pre-salt blocks, at its cost and risk, and in the event of a commercial discovery, be entitled to reimbursement of the costs incurred (cost oil) and a share of the surplus production (profit oil). The oil produced becomes government property after both the cost oil and profit oil are deducted.[6]

5.5 PRODUCTION-SHARING AGREEMENTS

Generally a production-sharing agreement is a contractual relationship between an IOC and the state. The agreement authorises the oil company to explore for and exploit oil and gas in a defined area and for a defined period. The state actually hires the oil company as a contractor to conduct the exploration and production work in exchange for the entitlement to a defined share of revenues from the produced hydrocarbons. This is in payment for the services rendered and the risks taken by the oil company.

The basic difference between a production-sharing agreement and a licence is that the relationship between the IOC and the state under a production-sharing agreement is that the oil company is a contractor or service provider. Under the production-sharing agreement the state remains the owner of the resources. The agreement establishes the amount of compensation that the oil company will receive for services rendered. This compensation will, either in part or in total, be in hydrocarbon revenues. Therefore, the oil company will have a contractual right to a portion of the hydrocarbon production post treatment and processing. Essentially, the production-sharing agreement regime is a service contract with payment in kind. As far as the state is concerned, it is the retention of ownership of the hydrocarbons that has its attractions as this is an important factor in states that are keen to protect their sovereignty over their natural resources. Production-sharing agreements are to be found in the postcolonial host states where there is a substantial nationalist culture.

Production-sharing agreements can easily be tailored to the specific circumstances of a particular field, as well as the national business environment and other local factors. The terms may be confined to exploration and development or may cover other areas such as the right to import and export goods, work permits, and foreign exchange. As with the concession regime, the production-sharing agreement can cover many areas not covered by a licensing regime, such as planning and environmental issues, and so can be particularly useful in a state with a less than developed legal system.

The key is the production-sharing agreement document itself. The key parties to the agreement will be the state,[7] the state oil company,[8] and the IOC. The rights of the oil company under a production-sharing agreement are meant to be contractual rather than proprietary and the issues typically dealt with under a production-sharing agreement include the following[9]:

1. Definition of the geographical area that the agreement covers and the length of the agreement
   The length of the agreement is usually split into two phases. The exploration phase, normally about 10 years, followed by a production phase if a commercial discovery is made, usually about 20 years. If any part of the designated area is not explored during the exploration phase, then that part will be expected to be relinquished and surrendered to the host state.
2. Specification of the operator who is in charge of the operations
   The operator would normally be the IOC who has the technical knowledge and experience. Sometimes, the operator under the production-sharing agreement is a consortium of oil companies and/or their backers; then one of them is chosen and designated as the operator. In China, once a commercial discovery is made, the State Party has the right to take over the role of the operator. In Egypt, the state has the right to operate jointly with the IOC through a joint venture company once a commercial discovery has been made.
3. Prescription of minimum work and expenditure commitments
   These are crucial negotiated points because of the small number of exploration activities that lead to a commercial discovery. The work commitments

and financial obligations define the extent of the exploration risk. In addition, who determines whether a field is economically viable and what is the definition of commerciality? Some states party to a production-sharing agreement will allow the IOC to make these determinations as to whether a development is commercially viable but it is more common for the state to have a greater say in the determination of commerciality.

4. The state's option to participate in the venture
   This does not imply that the state will share in the exploration costs and risks, but usually the IOC carries the state. This means that the oil company bears all the exploration costs and, if the field is designated as commercial, then the state has the right to participate in the production revenues.
5. The division of the hydrocarbons produced
   Obviously, this is a critical issue that depends on the relative bargaining power of the State Party and the IOC. The division essentially is determined by such issues as the strength of the competition among the IOCs, the risks of exploration and production in the specific field, and the characteristics of the field. Generally, however, the process of splitting the hydrocarbon produced will follow the outline given in the following[10]:
   a. An initial share of approximately 10%–15% will be set aside as royalty hydrocarbons.
   b. A share of the remaining hydrocarbons will be allocated to the IOC as a cost hydrocarbon.[11] It is important to determine how soon the oil company can recover its costs as the quicker the costs recovery, lesser is the chance that the hydrocarbons will run out before the company has recovered all its exploration costs.
   c. Taxation will be defined because sometimes the taxes are discharged by the state, but at other times the oil company will have to pay income tax to the host state.
   d. The remaining hydrocarbons are considered as 'profit hydrocarbons' and these are divided among the parties to the production-sharing agreement. These percentages may change over time with the greater the amount of hydrocarbons produced the greater the percentage share of the state in the 'profit hydrocarbons'. From the state's perspective, once the IOC has achieved a reasonable return on its investment, then the majority of the remainder should benefit the state.
6. Any obligation on the IOC by the state to sell hydrocarbons domestically
   Some states specify that a certain percentage of the production must be made available for domestic consumption, whereas others require that there is a general option that up to 100% should be made available to the domestic market should it be required.
7. A stabilisation clause
   This offers some protection to the IOC in case the state might seek to alter laws or taxes and thereby change the oil company's return from the development. The problem is that even where such clauses exist, they are not always enforceable against the host state.

8. The mechanisms by which the state will monitor the accomplishment of both its primary and secondary objectives
   Oversight by the state of the operations is usually achieved through management committees, which will approve investments, annual work programmes, and budgets. They will also confirm any declaration of commercial discovery.
9. Ownership of assets
   Normally, all moveable and fixed assets acquired by the IOC for the operations will immediately become the property of the state. The IOC will only retain the use of these assets for operational reasons, but the oil company does have the right to recover the costs out of the cost hydrocarbons. Therefore, the oil companies would be sensible to use subcontractors for their drilling and other operations and, where possible, lease rather than purchase any equipment needed.

5.6 CASE STUDY OF PRODUCTION SHARING IN RUSSIA

A good example of production-sharing agreements is the one operated in Russia as it highlights many of the issues surrounding the practicalities of such a production-sharing agreement regime. Russia embraced production-sharing agreements enthusiastically in order to encourage investment. However, this was followed by the state's dissatisfaction with the regime because of the nature of profit sharing during a period of unprecedented high energy prices. Finally, the state took steps to recapture control over their hydrocarbons.

Originally, a production-sharing agreement regime was established in Russia in parallel with the licensing regime with the intention of encouraging foreign investment in geographically isolated areas where there were technically complex hydrocarbon projects. Three large production-sharing agreements were reached in Sakhalin I, Sakhalin II, and Kharyaga. These agreements were signed at a time when the IOCs held a very strong position in the negotiations. Consequently, the conditions included were very favourable to the oil companies. Subsequently, with the increase in energy prices, the Russian Government became dissatisfied with the existing agreements and a series of amendments to the Production Sharing Agreement Law were enacted in 2003. These amendments effectively forced the Production Sharing Agreement operators to either switch to the licensing system under the Subsoil Law or abandon the projects altogether. Sakhalin I and II, in addition to Kharyaga, were initially exempt from the new law, but in 2006, when Shell, who were the operators of Sakhalin II, announced that their project costs would increase, the Russian government withheld its approval and renegotiated the Sakhalin II agreement. The Russian government ultimately took a 51% stake in the field.

5.7 SERVICE CONTRACTS

This is a straightforward service contract under which the host state retains full ownership of all hydrocarbons being produced within its jurisdiction and the IOC performs the exploration and production operations as a service to the host state. These regimes have normally been adopted in states where the constitution actually prohibits foreign control or ownership of natural resources. Examples of these are Saudi Arabia, Kuwait, and Iran. These states have substantial capital at their disposal but lack the technical expertise of the IOCs to carry out the exploration and production activities themselves.

There are two types of service contract, the first being a *risk service contract*, when the IOC is responsible for the capital expenditure and management of the whole operation of exploration and development. All the exploration is undertaken at the oil company's risk, and unless hydrocarbons are found in commercially viable quantities, the IOC will not be compensated. However, if oil is produced in commercial quantities, then the capital expenditure and operating costs incurred by the oil company are treated as a loan by the IOC to the state, but this may be recovered with interest from the state by cash payments, usually out of the proceeds from the oil sales. These cash payments will cover the operating costs and a fixed sum per unit of production and so the risks in oil price fluctuations will be borne by the state. Alternatively, the IOC may be entitled to a share of the market value of the production for a fixed period, with an option to convert the sum into actual hydrocarbons. Finally, the IOC may be appointed as agent for the state for marketing and sales of the petroleum. In that case, the oil company will be entitled to retain a percentage of sales proceeds as reimbursement of the loan plus a commission.

In a *pure services contract*, all exploration and production costs and risks will be met by the state, but then so will the rewards be retained by the state. The IOC is contracted to perform a service, which may include such things as consulting, engineering, construction, operational, and managerial services as defined in the contract. The IOC is a contractor and has no legal interest in the enterprise itself. The oil company will be responsible only for the construction and commissioning of the facility as described in the agreement.

5.8 CASE STUDY OF A SERVICE CONTRACT IN IRAN

The Constitution of Iran prohibits the granting of concessions to foreign companies involved in the extraction of hydrocarbons, and so all exploration and extraction in Iran is conducted by the Ministry of Petroleum and the National Iranian Oil Company (NIOC). However, some indirect participation by foreign companies is permitted under a buy-back scheme, which is, in fact, a risk service contract.

Under the Iranian buy-back contract, the IOC funds all investment costs on behalf of the NIOC in return for remuneration through NIOC's allocation of production. This can be up to 60% of production under a long-term export oil sales agreement, which will continue until the contractor has fully recouped his investment costs. This will normally be for a development period of 2 or 3 years and a remuneration period of 5–8 years.

5.9 THE REGULATION OF OFFSHORE INSTALLATIONS UNDER INTERNATIONAL LAW

The legal status of offshore oil rigs and/or production platforms has been a matter of discussion from a number of practical points of view. Should we consider them to be ships, artificial islands, offshore installations and structures, or give them a category of their own?

If they were to be considered as 'ships', under international law they would have the right of innocent passage and must fly under a state flag. The flag state would have rights of jurisdiction over the oil rigs and those working on board. The rig would be subject to shipping regulations concerning marine pollution, arrest, collision, and salvage. This issue came before the International Court of Justice as *The Case Concerning Passage Through the Great Belt.*[12] European lawyers spent long months arguing whether certain kinds of oil rigs and mobile drilling units are ships for the purpose of innocent passage or not. In fact, the case was settled out of court in 1992.[13] Alternatively, oil rigs may be included in other categories such as artificial islands, which we will consider under the United Nations Convention on the Law of the Sea 1982 (Section 5.11).

5.10 SHIPS AND OIL RIGS IN INTERNATIONAL LAW

Owing to the different definitions in the various texts of the international conventions in respect of the word 'ship' or 'vessel', there is no clear definition of the word 'ship' in international law. It is also not clear if all or some types of drilling ships are considered ships. In some cases, some types of oil rigs have been treated as ships or vessels, whereas in others, oil rigs are treated as artificial islands or as separate entities altogether. Even from a technical point of view, there are no common standards for the description of a ship for juridical purposes.[14]

Various international conventions may clearly be applicable to offshore oil rigs and may affect the legal situation of drilling rigs. In a number of situations, an offshore oil rig may be treated as a ship for certain purposes. Unfortunately, there are no uniform rules or common set of standards as to what objects may qualify for the legal status of a ship in either international or domestic state law. However, the actual practice of states in certain situations, such as registry and innocent passage, indicates that mobile oil rigs are treated like ships for legal purposes.[15] In both international and state laws, there are a number of characteristics of ships that may be applicable to oil rigs, such as seagoing and navigation, and in some situations of flag registry and collisions, an oil rig may be considered as a ship for certain legal purposes.

Drilling ships are considered by many states as ships. They have almost all the characteristics of a 'ship', in particular the capability of navigation. But are they 'ships' when they are wholly engaged in the act of drilling? Other types of oil rigs, such as fixed oil rigs, would not qualify as a 'ship'; nevertheless, certain state legislation and international treaties have designated them to be so.

5.11 OIL RIGS UNDER THE LAW OF THE SEA CONVENTION, 1982

The 1982 Convention provides certain rules and regulations in respect of artificial islands and offshore installations and structures for the purpose of exploration and exploitation of the natural resources of the sea and other economic purposes.[16] The Convention refers in many articles to 'artificial islands, installations and structures' and also to 'installations' and 'installations and devices', but nowhere does the

Convention actually give a definition for these structures. Much discussion took place during the negotiations prior to the 1982 Convention. In fact, the United States prepared a draft article that included provisions on offshore installations with the intention to define the term *installations*.[17] However, the proposal was unacceptable to the conference and so led to different interpretations.

Articles 60 and 80 of the Convention include the main provisions with regard to oil installations and these articles do make a distinction between offshore installations for the purpose of exploration and exploration of the natural resources of the sea and other economic purposes, primarily oil rigs, and artificial islands such as 'floating hotels'.

Article 60 is based on the provisions of Article 5 of the 1958 Geneva Convention on the Continental Shelf, which refers to 'installations and other devices' for the purpose of the exploration of the continental shelf and its natural resources.[18] However, the 1958 Convention makes no difference between oil rigs and artificial islands, but artificial island and oil rigs have variously been created for different purposes. Floating hotels and sea cities such as established in Dubai have a very different reason for existence as compared to oil rigs.

It would appear that both the 1958 Convention on the Continental Shelf and the 1982 Law of the Sea Convention (LOSC) have created a separate legal category for offshore installations and structures for the purpose of exploration and exploitation of natural resources of the sea, which are neither ships nor islands. As oil rigs are the main category of installations and structures for the purpose of exploration and exploitation of natural resources of the sea, the following state legislations and international treaties have distinguished oil rigs from ships in this regard.

5.12 JURISDICTION OF STATES IN RELATION TO OFFSHORE INSTALLATIONS

Questions of jurisdiction are a recurrent theme throughout this book, and in this section, the rights of states over offshore installations are examined. The rights of states to construct installations in the various maritime zones under the 1982 LOSC, the general control of coastal states with regard to the criminal and civil jurisdiction on board from an international law perspective, and jurisdiction in respect of fiscal matters, health, and safety are considered.

5.12.1 A State's Right to Construct Offshore Installations

Although there are nine identifiable maritime zones, there are only five that are applicable to the development of offshore hydrocarbon production and they are addressed here. Those that have not been addressed are the exclusive fishing zone (EFZ), the contiguous zone, and the archipelagic waters.[19]

5.12.1.1 Internal Waters

The internal waters of the states are defined as 'waters on the landward side of the baseline of the territorial sea'.[20] This area includes ports, harbours, bays, lakes, canals and river mouths, and closely related waters.[21] The coastal state enjoys full sovereignty and jurisdiction over its internal waters as they are considered to be legally equivalent to a state's land. Therefore, the construction of offshore installations in internal waters is a matter for the coastal domestic law and regulation. Any access to internal waters requires the permission of the coastal state but there are some exceptions. In the *Anglo-Norwegian Fisheries Case* the International Court of Justice acted on the belief that for access to internal waters the permission of the coastal state is required.[22]

In exceptional cases, there may be a right of innocent passage under Art 5(2) of the Geneva Convention on the Territorial Sea and the Contiguous Zone and Art 8(2) of the 1982 Convention. According to Art 8(2), 'Where the establishment of a straight baseline in accordance with the method set forth in Art 7 has the effect of enclosing as internal waters areas which had not previously been considered as such, a right of innocent passage as provided in this convention shall exist in those waters'. In addition, the International Court of Justice considered this point in the *Land, Island and Maritime Frontier Dispute* in 1992. In this case, the court held that '. . . rights of innocent passage are not inconsistent with a regime of historic waters. . .'.[23] Because of these two exceptions, the question arises as to whether the rights of innocent passage will be hampered if an installation is being constructed within the internal waters of a state. Where the right of innocent passage exists under international law, the state will be liable for any interference with that right.[24]

5.12.1.2 In the Territorial Sea

Under Art 3 of the Law of the Sea 1982, 'every State has the right to establish the breadth of its territorial sea up to a limit not exceeding 12 nautical miles, measured from baselines determined in accordance with this Convention'. Under Article 2(1), the coastal state has sovereignty over its territorial sea. Article 2(2) states that this sovereignty extends to the air space over the waters as well as the seabed and subsoil under the water. By this sovereignty over its territorial sea, the coastal state has the legal authority to build offshore oil rigs and other installations within its territorial waters.[25] Again the right to construct and operate oil rigs or installations in territorial waters is restricted by the right of innocent passage as under Article 2(3) of the Convention 'subject to this Convention, ships of all States, whether coastal or land locked, enjoy the right of innocent passage through the territorial sea'. Article 19 explains the meaning of 'passage' as

> . . . navigation through the territorial sea for the purpose of
>
> (a) traversing that sea without entering internal waters or calling at a roadstead or port facility outside internal waters; or
> (b) proceeding to or from internal waters or a call at such roadside or port facility . . . passage includes stopping and anchoring . . . ,

Under Article 14(4), 'passage is innocent so long as it is not prejudicial to the peace, good order or security of the coastal State. . .'.

Article 24 describes the duties of the coastal state in respect to this right as

> 1. The coastal State shall not hamper the innocent passage of foreign ships through the territorial sea except in accordance with this Convention.

In particular, in the application of this Convention or of any laws or regulations adopted in conformity with this Convention, the coastal State shall not:

a) Impose requirements on foreign ships which have the practical effect of denying or impairing the right of innocent passage; or
b) Discriminate in form or in fact against the ships of any State or against ships carrying cargoes to, from or on behalf of any State.

2. The Coastal State shall give appropriate publicity to any danger to navigation, of which it has knowledge, within its territorial sea.

Therefore, the right of the coastal state to construct and operate offshore installations as well as undertake other activities in their territorial sea must be consistent with the rights of other states' innocent passage.

5.12.1.3 The Exclusive Economic Zone

The concept of the exclusive economic zone (EEZ) acquired clear legal definition in Part V of the 1982 Convention, although it had been foreshadowed by development in state practice before then. Iceland asserted an EFZ in 1948 and extended it periodically thereafter up to 50 nautical miles. In parallel to assertions of fishery jurisdictions, other states claimed jurisdiction over economic resources on appurtenant continental shelves, such as the United States through the Truman Proclamations of 1945.[26]

The EEZ, as established by the 1982 Convention, is a claimable maritime zone. This sets it apart from the continental shelf, which inherently belongs to the coastal state. Article 57 indicates that states may claim an EEZ up to 200 nautical miles from the baselines from which the breadth of the territorial sea is measured. When a state does claim an EEZ of 200 nautical miles, this means that the zone covers the band of water, seabed, and subsoil that is effectively 188 nautical miles in breadth beyond the 12 nautical mile territorial sea.

Most states, including non-parties to the 1982 Convention, have claimed an EEZ up to the 200 nautical mile limit, and many have passed legislation applicable to these waters giving effect to Part V of the Convention, although states are free to claim an EEZ of less than this limit because of overlap of the rights of states.

The United Kingdom only claimed an EFZ as it considered that in conjunction with continental shelf rights, it had sufficient rights and authority as the rights regarding offshore installations and structures were addressed in the 1958 Convention on the Continental Shelf. However, once the United Kingdom decided to grant licences for offshore wind farms on the continental shelf, they discovered that their international rights under the EFZ and the continental shelf rights did not give them the authority to do so. The United Kingdom has since claimed and registered an EEZ.[27]

Under Article 56(1)(a), the coastal state has sovereign rights within the EEZ for the purpose of exploring and exploiting, conserving, and managing the natural resources of the waters super adjacent to the seabed and of the seabed and its subsoil. The coastal state also has jurisdiction with regard to the establishment and use of artificial islands, installations, and structures.[28] Detailed rules with regard to the construction, operation, and use of offshore installations and artificial islands are provided in Article 60 of the 1982 Convention:

> 1. In the Exclusive Economic Zone, the coastal State shall have the exclusive right to construct and regulate the construction, operation and use of:
>   (a) Artificial islands;
>   (b) Installations and structures for the purposes provided for in Article 56 and other economic purposes;
>   (c) Installations and structures which may interfere with the exercise of the rights of the coastal State in the zone.
> 2. The coastal State shall have exclusive jurisdiction over such artificial islands, installations and structures, including jurisdiction with regard to customs, fiscal, health, safety and immigration laws and regulations . . .

The coastal state also has the right to establish safety zones, which would normally not exceed 500 m in breadth, around oil installations, which is discussed in the following.

5.12.1.4 The Continental Shelf

The new concept of the EEZ has not changed or weakened the position of the continental shelf provided for under the 1958 Convention on the Continental Shelf as restated in the 1982 Convention, although it has created a certain amount of duplication and confusion.[29] The coastal state has sovereign rights over its continental shelf for the purposes of exploring and exploiting the nonliving natural resources. Under the EEZ legislation, coastal states have rights over exploration and exploitation of the natural resources, whether living or nonliving, and sovereign rights with regard to other economic interests such as the production of energy from water under Article 56(a).

Substantively, the same legal regime will be applied on both the continental shelf and the EEZ except in cases where the geomorphologic continental margin extends beyond 200 nautical miles, and the coastal state has not registered an EEZ. In this case, it will be the 1958 Convention that prevails, which means that the coastal state is entitled to exercise jurisdiction over the adjacent continental margin beyond 200 nautical miles. However, the rights of the coastal states will be limited to the exploitation of nonliving resources of the seabed as in the adjacent continental margin beyond 200 nautical miles, the waters are legally considered to be the high seas.

The definition of the continental shelf under Article 76(1) of the 1982 Convention provides:

> the continental shelf of a coastal State comprises the sea bed and sub soil of the submarine areas that extend beyond its territorial sea throughout the natural prolongation of its land territory to the outer edge of the continental margin, or to a distance of 200 nautical miles from the baselines from which the breadth of the territorial sea is measured where the outer edge of the continental margin does not extend up to that distance.

Article 76(3) provides the legal definition of the geomorphic feature as

> The continental margin comprises the submerged prolongation of the land mass of the coastal State, and consists of the sea bed and sub soil of the shelf, the slope and the rise. It does not include the deep ocean floor with its ocean ridges or the sub soil thereof.

This gives the coastal states the authority to exercise jurisdiction over their adjacent continental margin beyond 200 nautical miles.

Within the continental shelf area, the coastal state only has sovereign rights for the exploration of the continental shelf and the exploitation of its natural resources of

> . . . the mineral and other non-living resources of the sea bed and subsoil together with living organisms belonging to sedentary species, that is to say organisms which, at the harvestable stage, either are immobile on or under the sea bed or are unable to move except in constant physical contact with the sea bed or the subsoil. (Article 77(4) 1982 LOSC.)

Under Article 81, the coastal state also has the exclusive right to authorise and regulate drilling on the continental shelf for all purposes, and no other state has the right to undertake such activities without the coastal state's consent. Under Article 77(3) these rights do not depend on occupation, effective or notional, or any express proclamation, unlike the EEZ.

Under Article 80, the provisions of Article 60 concerning artificial islands and installations and structures in the EEZ apply, *mutatis mutandis* to artificial islands and installations on the continental shelf.[30]

However, there is some confusion with regard to the establishment and operation of offshore oil installations and artificial islands in the geomorphologic continental margin beyond 200 nautical miles as according to Article 80 of the 1982 Convention, the provisions of Article 60 of the Convention only apply to installations on the continental shelf. However, the coastal state does have the exclusive right to construct, authorise, and to regulate the construction, operation, and use of those installations that may interfere with the exercise of the rights of the coastal state under Art 60(1)(b). This means that the coastal state has the right to establish installations and artificial islands for all economic purposes if the geomorphologic continental margin extends beyond 200 nautical miles, with the super adjacent waters in that area being part of the high seas.

5.12.1.5 On the High Seas

Article 1 of The Geneva Convention on the High Seas defines the term *high seas* as 'all parts of the sea that are not included in the territorial sea or in the internal waters of a state'. The 1982 Convention on the Law of the Sea does not define the term but provides the rules in Part VII, which apply to 'all parts of the seas that are not included in the exclusive economic zone, territorial sea or in the internal waters of a State, or in the archipelagic waters of an archipelagic State'.[31]

All states, whether coastal or not, have the right to the historic freedoms of the high seas, which are freedom of navigation; freedom of over flight; freedom of fishing; and freedom of scientific research. The freedom to lay submarine cables and pipelines and the freedom to construct artificial islands and other installations permitted under international law, subject to Parts VI and XIII of the 1982 Convention, were new freedoms conferred by virtue of the 1982 Convention, and these freedoms were not mentioned in the 1958 Convention on the High Seas because at that time, technology was not sufficiently developed to enable the exploration and exploitation of the seabed of the high seas. However, by the 1980s, the exploitation of mineral resources under the high seas was becoming more practicable and accordingly the Convention under Article 87(1)(d) provided that both coastal and land-locked states 'have the freedom to construct artificial islands and other installations permitted under international law, subject to Part VI'.

States that authorise the construction of installations in the high seas have a duty not to create unreasonable interference with fishing,[32] scientific research,[33] laying or maintaining submarine cables or pipelines,[34] and the conservation of living resources.[35] Given these freedoms, the corollary is that where, as a result of the establishment of offshore installations, any damage is caused to the rights or interests of another state, the normal rules of state responsibility will apply.

5.12.1.6 The International Deep Seabed Area

One of the most important, and yet the most contentious, developments with regard to offshore activities is in the regulation of the deep seabed. This is the submarine area of the seafloor beyond the continental shelf where there is known to be a wealth of mineral resources. These include polymetallic nodules containing concentrations of valuable metals. Prior to the 1982 Convention on the Law of the Sea, which set up the deep seabed regime, this area was subject to the freedom of the seas, and was, therefore, open to exploration and exploitation by all states. Although there was little prospect of exploiting deep seabed resources, this situation went largely unchallenged. With developments in mining technologies, and with the rise of prices for minerals, particularly, oil and gas, the emerging economies were keen to secure a fairer international economic order. This led to a shift in international opinion and thus to the designation of the seabed as the 'common heritage of mankind'.

This doctrine includes the non-appropriation of the seabed by states or private entities; internationalised management via the International Seabed Authority (ISBA), which was established by the 1982 Convention; the sharing of benefits for the common good of humankind; and the peaceful use of the area.[36] Twelve years after the LOSC was concluded, the seabed mining provisions were modified by the 1991 Agreement Relating to the Implementation of Part XI.[37] This Agreement went a long way to wards addressing some of the concerns of the more-developed countries in relation to the common heritage regime.

The ISBA is now discharging a number of the primary functions for which it was set up.[38] It has drafted and adopted a complete 'mining code' to address not only contractual arrangements for mining in the area but also the protection of unique and vulnerable seafloor habitats during the exploration and exploitation process. Prior to the 2008 economic downturn, it appeared that seabed mining would soon become economical.[39] However, the global financial crisis has meant that it is state-supported entities that are seeking licences from the ISBA rather than contractors from the commercial sector.

Article 147 provides that 'installations used for carrying out activities in the Area shall be erected, emplaced and removed solely in accordance with Part XI and subject to the rules, regulations and procedures of the Authority'. Article 1(1)(3) defines 'activities in the area' as all activities of exploration for, and exploitation of, the resources of the Area. Resources of the Area are defined as 'all solid, liquid or gaseous mineral resources *in situ* in the Area at or beneath the seabed, including polymetallic nodules'.[40] Therefore any erection of installations for the exploitation of liquid or gaseous minerals will be subject to the rules and regulations of the Authority. As these rules and regulations are only concerned with exploring and exploiting solid natural resources of the seabed, any artificial islands and installations for all other purposes are subject to the provisions of the regime covering the high seas. The LOSC also provides for restrictions in relation to activities that are not related to the exploration and exploitation of the natural resources of the seabed. According to Article 147(3), all activities in the marine environment that are unrelated to the exploration and exploitation of the natural resources of the seabed must be conducted having reasonable regard for activities in the Area. Thus, the Authority should also have the power to authorise any activities, including the erection of installations or artificial islands that are unrelated to the exploration or exploitation of the natural resources of the seabed, which may interfere with the exploration and exploitation of the natural resources. This was put to the UN Seabed Committee by Belgium but was rejected, and so any other installations or artificial islands constructed for purposes other than the exploitation of natural resources will be subject to the freedom of the high seas.

5.13 STATE CONTROL AND JURISDICTION OVER OIL RIGS

Once a state authorises the establishment of an offshore installation, the question of jurisdiction over the activities of those on board comes into being, as this is different in the various delimitations of the marine environment.

The question of jurisdiction over oil rigs whether mobile or fixed is different within territorial waters, on the continental shelf, within the EEZ, and on the high seas beyond the EEZ and the continental shelf.

It is obvious that under Article 89 of the LOSC, a territorial type of jurisdiction is not available over the high seas as no state may validly purport to subject any part of the high seas to its sovereignty. This means that a state cannot claim jurisdiction over oil installations it has authorised beyond the EEZ and continental shelf except through a flag state jurisdiction over ships. However, in respect of internal and territorial waters, the coastal state has full jurisdiction.

The jurisdiction of the coastal state in relation to activities and people on board oil rigs and installations on the continental shelf and the EEZ is not clear; however, Article 60(2) of the LOSC does indicate that the coastal state has exclusive jurisdiction over oil rigs and other installations with regard to customs, fiscal, health, safety, and immigration laws and regulations.

5.14 CRIMINAL JURISDICTION

Again, the application of criminal law on board or in relation to oil rigs and installations is not clear in international law. There are many workers who operate both on oil rigs and installations and service them. These workers may be involved in violence and/or criminal activities. Does the coastal state have jurisdiction over crimes committed and criminals on board offshore installations in the EEZ and on the continental shelf? Who is competent to try offenders on board installations on the high seas beyond national jurisdiction?

As *internal waters* are considered as an integral part of the coastal state, it is evident that the criminal law of the coastal state will apply and is enforceable on board fixed oil rigs and installations. As far as mobile oil rigs and drilling ships are concerned, it may differ among states. Some states treat mobile rigs as ships, and then the law of the flag state will apply, whereas under the Anglo-American position, the jurisdiction of a coastal state over foreign ships in its ports and internal waters is absolutely applicable. However, some coastal states may forgo the enforcement of jurisdiction as a matter of policy.[41] On the other hand, the French approach is that a coastal state does not have jurisdiction over the internal affairs of foreign ships when they are within its ports or internal waters.[42] In practice, coastal states will normally assert jurisdiction if any offence endangers the peace or security of the coastal state or when there is a request from the captain or consul of the flag state.

If the coastal state domestic law considers any type of oil rig to be a 'ship', then it is either the law of the flag or coastal state which will be applied over any offences committed on board. If, on the other hand, oil rigs are considered to be a separate category in the coastal state's domestic law, then it is the coastal state's jurisdiction over all criminal offences committed on board that will prevail.

The general sovereignty of the coastal state over its territorial seas extends from the outer limit of the internal waters to the delimitation of the territorial waters. This includes the air space above the sea as well as the seabed and subsoil, and is only restricted by the right of innocent passage.[43]

As most drilling vessels would be considered ships under international law if they are passing through territorial waters of another state, under Article 27 of LOSC, they would be subject to the same laws and regulations as merchant ships on a commercial voyage. The coastal state, with some exceptions, should not normally exercise criminal jurisdiction on board a foreign ship:

1. Where the consequences of the crime extend to the coastal state.
2. The crime is of a kind to disturb the peace of the country or the good order of the territorial sea.
3. The assistance of the local authorities has been requested by the master of the ship or by a diplomatic agent or consular officer of the flag state.
4. Such measures are necessary for the suppression of illicit traffic in narcotic drugs or psychotropic substances.

Categories (1), (2), and (3) are based on the traditional principle of jurisdiction in regard to the territorial sea. Category (4) indicates the recognition that the illegal trade in narcotics, hijacking, and terrorism need a universal international approach.[44]

Where a drilling ship is operating under license from the coastal state or the installation is fixed, then under the exclusive jurisdiction of the coastal state over its territorial sea under LOSC Article 2(1), but bearing in mind that the LOSC does not directly address the issue, by implication, international law would not prevent a coastal state from extending its domestic criminal jurisdiction to include criminal activity on board oil installations within its territorial sea.

It should be noted that the coastal state is not empowered to apply its criminal law over oil installations, in particular mobile rigs and drilling ships in its territorial waters, except to apply preventative and punishment actions in relation to mobile vessels.

The same applies to offshore installations on the continental shelf and in the EEZ as it would depend on the domestic law of the coastal state, and nothing within international law prevents coastal states from applying their criminal law on board oil installations on the continental shelf.

In the *North Sea Continental Shelf Case*, the International Court of Justice asserted that 'the right of the coastal State to its continental shelf areas is based on its sovereignty over land domain, of which the shelf area is the natural prolongation into and under the sea'.[45]

The *Convention for the Suppression of Unlawful Acts Against the Safety of Maritime Navigation 1988* is relevant here as under the Convention, States Parties are permitted to make certain acts on board a ship criminal, such as performing an act of violence against another, or destroying, or causing damage to a ship. Under Article 1 of the Convention, a ship is defined as 'a vessel of any type whatsoever not permanently attached to the seabed, including dynamically supported craft, submersibles, or any other floating craft'.

The original Convention applies only to ships, but under the 1988 Protocol for the Suppression of Unlawful Acts Against the Safety of Fixed Platforms located on the Continental Shelf, the provisions are extended and apply *mutatis mutandis* to fixed oil rigs.

Article 2 of the Convention sets forth the offences as follows:

1. Any person commits an offence if that person unlawfully and intentionally:
   (a) Seizes or exercises control over a fixed platform by force or threat of or any other form of intimidation; or
   (b) Performs an act of violence against a person on board a fixed platform if that act is likely to endanger its safety; or
   (c) Destroys a fixed platform or causes damage to it which is likely to endanger its safety; or
   (d) Places or causes to be placed on a fixed platform, by any means whatsoever, a device or substance which is likely to destroy that fixed platform or likely to endanger its safety; or
   (e) Injures or kills any person in connection with the commission or the attempted commission of any of the offences set forth in subparagraphs (a) to (b).

Under Article 2 of the Protocol, the coastal state is empowered to take measures necessary to establish its jurisdiction over offences provided for in Article 2 when the offence is committed:

(a) Against or on board a fixed platform while it is located on the continental shelf of that State; or
(b) By a national of that State.

A number of coastal states have enacted domestic legislation to apply their criminal law to activities in relation to oil rigs on their continental shelf and EEZ, and this is discussed within the relevant chapters on the states.

5.15 CIVIL JURISDICTION

Inasmuch as a large number of people work on board oil installations, and this work is arduous and physical involving much bending, stooping, climbing, and lifting, there are many possible hazards. Collisions that occur at sea can take place between ships and oil installations; some accidents may result from design problems.

International law does not provide for compensation suits and civil actions for damages, injuries, or death on an oil installation; therefore, such actions must be brought under private international law. Article 219 of LOSC provides for civil jurisdiction as applied to foreign ships on the territorial sea; however, the Convention does not provide for civil jurisdiction on oil installations erected in territorial waters or on the continental shelf. Nevertheless, many states have enacted domestic legislation in relation to civil jurisdiction over or in connection with offshore oil installations and these are discussed in the relevant state chapters that follow (Chapter 7 on USA and Chapter 9 on UK).

5.16 POLLUTION CONTROL AND ENVIRONMENTAL ISSUES

Any exploration of the seabed and exploitation of its natural resources will inevitably cause pollution of some kind. Although it is accepted that 44% of marine pollution is caused by land-based discharge, 33% from atmospheric inputs, 12% from maritime transport, 10% from dumping, and only 1% from oil exploration and production,[46] according to the Group of Experts on the Scientific Aspects of Marine Pollution (GESAMP), that 1% has been addressed by international treaties. When accidental pollution from offshore oil installations occurs, it is usually a major incident that attracts the attention of world media. Furthermore the chance of a catastrophic blowout always exists. The latest incident was the blowout at the BP Macondo field in the Gulf of Mexico, which is examined as a case study later in Chapter 7.

5.17 DEFINITION OF MARINE POLLUTION

The GESAMP defined marine pollution as 'the introduction by man, directly or indirectly, of substances or energy to the marine environment which results in deleterious effects on marine activities, such as fishing and other living resources, the impairment of the quality and the use of seawater, and the reduction of amenities'.[47] The 1982 Convention on the Law of the Sea defines marine pollution as

> Pollution of the marine environment means the introduction by man, directly or indirectly, of substances or energy into the marine environment, including estuaries, which results or is likely to result in such deleterious effects as harm to living resources and marine life, hazards to human health, hindrance to marine activities, including fishing and other legitimate uses of the sea, impairment of quality for use of sea water and reduction of amenities.[48]

Although the Convention adopts a version similar to the GESAMP one, the Convention expands it somewhat.

Oil, being the main pollutant, can affect the marine environment including the flora and fauna and can affect the ecosystem by the destruction of sensitive immature life forms or through the elimination of food sources, with a large oil spill having a substantial impact on the ecology of the sea.[49] Added to that, pollution from the physical act of seabed operations, such as drilling mud, drill cuttings, produced water, and dumping, can create significant threats to the marine environment.

The definition of dumping includes the deliberate disposal of oil platforms at sea and is addressed under the section on decommissioning and disposal of oil installations (Section 5.20.3).

The international community has responded to the increase in marine pollution by ratifying as many as 85 international conventions regarding pollution and liability for compensation and clean-up in the marine environment. However, there is no single international agreement that comprehensively covers pollution from offshore oil installations, and therefore only a few of the principal agreements are considered here.

The 1958 Geneva Convention on the High Seas and the Continental Shelf established a number of rules with regard to pollution from offshore operations. Under Article 5(1) and 5(7) of the Continental Shelf Convention and Article 24 of the High Seas Convention, the coastal state must take all appropriate measures to protect the living resources of the sea from harmful agents by establishing safety zones around offshore installations. Article 24 of the High Seas Convention also provides that 'every State shall draw up regulations to prevent pollution of the seas by the discharge of oil from ships or pipelines or resulting from the exploitation or exploration of the seabed and its subsoil, taking account of existing treaty provisions on the subject'.

The 1982 LOSC includes a separate part entirely for the protection of the marine environment, with the general obligation under Article 192 subject to other rights and duties contained within the Convention. According to Article 194, measures should be taken to minimise

> pollution from installations and devices used in exploration and exploitation of the natural resources of the seabed and subsoil, in particular measures for preventing accidents and dealing with emergencies, ensuring the safety of operation at sea, and regulating the design, construction, equipment, operation and manning of such installations or device.[50]

The Convention thus directs coastal states to establish global and regional rules, and national legislation to regulate pollution from seabed activities.[51] Under the 1994 New York Agreement on the implementation of Part XI of the Convention, a plan of work on the international seabed area must be 'accompanied by an assessment of the potential environmental impacts of the proposed activities and by a description of a programme for oceanographic and baseline environmental studies in accordance with the rules, regulations and procedures adopted by the Authority'.[52]

Although the 1982 Convention provided a much more comprehensive framework for the protection of the marine environment than the 1958 Convention, in particular with regard to offshore oil and gas activities, by providing a complete Part to pollution in the marine environment, it lacks in listing any particular remedies for the protection of coastal states whose coastline has been damaged as a result of offshore accidental pollution.

The 1972 Convention on the Prevention of Marine Pollution by the Dumping of Wastes and Other Matter (the London Dumping Convention, LDC) has been regularly amended and updated. The Convention provides that the High Contracting Parties, individually and collectively, promote the effective control of all sources of pollution of the marine environment, and pledge themselves to take all practicable steps to prevent the pollution of the sea by the dumping of waste and other matter that may harm marine life.[53] The definition of dumping under the Convention is 'a deliberate disposal at sea of wastes or matter from platforms or other man-made structures at sea, aircraft and vessels'.[54] This includes any deliberate dumping of oil platforms at sea.[55]

In 1979, a Code for the Construction and Equipment of Mobile Offshore Drilling Units (MODU) was issued by the International Maritime Organisation (IMO).[56] This Code provided international regulations in respect of technical matters of offshore installations, and was revised in October 1989. The revisions came into effect in May 1991. Although the Code is not mandatory, a number of states have applied it and its operation would go to due diligence should anything go wrong.

The 1990 International Convention on Oil Pollution Preparedness, Response and Cooperation (OPRC) requires that all of the High Contracting Parties take appropriate measures, and based on the provisions of each article, prepare for and respond to oil incidents. The Convention expressly identifies offshore oil installations by referring to pollution from offshore units. An 'offshore unit' is defined as 'any fixed or floating offshore installation or structure engaged in gas or oil exploration, exploitation, or production activities, or loading or unloading of oil'.[57] The Convention lays a duty on offshore operators to formulate oil pollution emergency plans,[58] and the duty holder is required to report any event involving the discharge of oil.[59]

The Convention also provides for the establishment of regional and national systems for the preparedness and response to accidental oil discharges and the cooperation between authorities. The IMO provides various functions with regard to information, education and training, and technical assistance.

The OPRC is the most important international treaty with regard to issues of efficient response to pollution from oil installations. It deals with oil pollution in far more detail than the 1982 LOSC, given the exponential increase in offshore oil and gas exploration and production. The OPRC also recognises that offshore production is a dangerous business and not all eventualities can be accounted for. Oil pollution and major or minor slicks are an occupational hazard and must be anticipated and prepared for.

In addition to the major international treaties with regard to oil pollution of the marine environment, there are also a number of relevant regional treaties in respect of offshore oil and gas exploration and production.

The 1992 Oslo/Paris (OSPAR) Convention was born out of the Oslo Convention for the Prevention of Marine Pollution by Dumping from Ships and Aircraft, which was adopted in 1972 and the Paris Convention for the Prevention of Marine Pollution from Land based Sources adopted in 1974, which both provided for the protection of the North Sea and the North-East Atlantic Approaches. They were brought together in 1992 as the OSPAR Convention for the Protection of the Marine Environment of the North-East Atlantic. Together with the adoption in 2008 of the European Commission Marine Strategy Framework Directive, the Convention plays a key role as one of the regional sea conventions with specific responsibilities for the protection of the marine environment.

5.18 CIVIL LIABILITY FOR ENVIRONMENTAL HARM CAUSED BY OIL POLLUTION

The main international convention to cover civil liability from offshore installations is the 1976 Convention on Civil Liability for Oil Pollution Damage resulting from Exploration for and Exploitation of Seabed Mineral Resources but this is not yet in force. Therefore, those damaged by oil pollution from offshore installations must rely on the 1982 LOSC Article 235, which provides the following:

> 1. State shall ensure that recourse is available in accordance with their legal systems for prompt and adequate compensation or other relief in respect of damage caused by pollution of the marine environment by natural or juridical persons under their jurisdiction.
> 2. With the objective of assuring prompt and adequate compensation in respect of all damage caused by pollution of the marine environment, States shall cooperate in the implementation of existing international law and the further development of international law relating to responsibility and liability for the assessment of and compensation for damage and the settlement of related disputes, as well as, where appropriate, development of criteria and procedures for payment of adequate compensation, such as compulsory insurance or compensation funds.

Article 235 is based on Principles adopted by the United Nations Conference on the Human Environment in 1972. They are Principle 22 and Principle 7, which basically laid a duty upon the states to provide a system of compensation for those damaged inter alia by oil and gas activities. It is, therefore, the states' responsibility to provide a system either at state or regional level for compensation. There are, in fact, a number of regional schemes that are discussed in the later relevant chapters.

One of the accepted principles of environmental law is 'the polluter pays', which was examined in Chapter 2. The principle is intended to ensure that the costs of cleaning up pollution are borne, as far as is possible, by the polluter. The Organisation for Economic Cooperation and Development (OECD) recommended that this principle should be taken into consideration when calculating the costs of the prevention of oil spills at sea. The liability for the costs of 'reasonable remedial action' should be borne by the polluter and these costs can also be recovered in a variety of different ways including civil actions for damages.[60]

The *Torrey Canyon* disaster in 1976 galvanised public opinion, to which the politicians responded, and prompted the IMO to call an international conference in 1969.[61] The resulting Convention on Civil Liability for Oil Pollution Damage, together with the 1971 Convention on the Establishment Fund for Compensation for Oil Pollution Damage established a regime for liability for oil pollution. The 1969 Convention channels liability to the shipowner, who is alone responsible for the safe and efficient operation and seaworthiness of the vessel. The liability is strict so that no fault or negligence need be shown. The usual defences of war, hostilities, insurrection, civil war, and unforeseen inevitable accidents are available as is the negligence of those responsible for navigational aids.[62] In recognition of the acceptance of strict liability, the owner is permitted to limit his liability, according to a formula relating to tonnage of the vessel and an overall total.[63] The overall total soon proved to be insufficient and was substantially increased under the 1984 Protocol. Additionally, the main purpose of the International Oil Pollution Compensation Fund (IOPC Fund) established in 1971 is to provide further compensation in order that victims are fully compensated.[64] However, not only does the Fund provide a defence to liability under the same circumstances as does the Civil Liability Convention (CLC) but additionally where the claimant cannot prove that the damage was caused from 'an incident involving one or more ships'.[65] The importance of this provision is that where the source of the oil is unidentified, no compensation is obtainable.[66] The innocent victim may then be left without effective recourse.

In parallel with the CLC and the Fund Conventions, two industry schemes existed until 1998. The Tanker Owners Voluntary Agreement on Liability for Oil Pollution (TOVALOP) applies to tanker owners, whereas Contract Regarding an Interim Supplement to Tanker Liability for Oil Pollution (CRISTAL) provides a fund comparable to the IOPC. These industry-based schemes overlapped the CLC and Fund Conventions and covered those vessels from non States Parties to the 1969 and 1971 Conventions.

5.19 THREE CASES BROUGHT UNDER THE CLC

5.19.1 The *Amoco Cadiz* Case

The *Amoco Cadiz*[67] was owned by Amoco Transport Co. (Transport) and had been designed and constructed in Cadiz, Spain, by Astillieros. Her steering gear had been approved by the American Bureau of Shipping.

In March 1978, *Amoco Cadiz*, with a cargo of 121,157 tons of Iranian light crude oil was approaching Western Europe when she sailed into a severe storm. The vessel rolled heavily on March 15 and rolled more heavily that night and into the morning of March 16 as she encountered rough seas and severe winds. The steering gear consequently failed and attempts to repair it were unsuccessful. The tug Pacific went to the assistance of *Amoco Cadiz*. While the Pacific was attempting to tow *Amoco Cadiz*, *Amoco Cadiz* grounded with the result that oil was seen in the water.

Various actions were brought. The cargo owners claimed the value of the cargo. Actions for oil pollution damage and clean-up costs were brought by the Republic of France, the French administrative departments of Finistere, and Conseil General des Cotes du Nord, in addition to numerous municipalities, and a number of French individuals. Bugsier, the operator of the salvage tug Pacific, was also sued for negligence in carrying out the salvage operation. Amoco International Oil Co. (AIOC) was sued for its failure as the manager and operator of the vessel. An action was brought against the shipbuilders Astilleros by the Cote du Nord, claiming that the shipyard was negligent in respect of the design and construction of the vessel.

Judge Frank McGarr[68] held that the CLC was the law in France and not in the United States, as the United States was not a party to the Convention. Even if French Law, including the CLC, were applicable to the issue, none of the provisions of the CLC would bar suits against AIOC. It therefore was not entitled to limitation of liability.

The design and construction faults contributed to the grounding of *Amoco Cadiz*, therefore Astillerios, the shipbuilders, were liable in part for the consequent damage. The French claimants and the cargo owners were entitled to the full extent of their incurred damages. Amoco was entitled to damages against Astilleros to the extent that its own liability was contributed to by the negligence and fault of the shipbuilder. All claims against Bugsier, the salvor, were rejected.

5.19.2 The *Sea Empress* Case

In February 1996, the Liberian Tanker *Sea Empress*[69] was carrying over 130,000 tonnes of crude oil when she ran aground at the entrance of Milford Haven, spilling 2000 tonnes of the oil. A pilot had joined the tanker outside the harbour and was on board at the time. After numerous attempts to refloat the tanker without success, she was towed out of Milford Haven. Over 72,000 tonnes of crude oil and 360 tonnes of heavy fuel oil escaped into the sea.

Later that month, the Welsh Office imposed an order, under the Food Environment Protection Act, prohibiting the landing of fishery and aquaculture products taken from a designated zone. A further ban was imposed on salmon and migratory trout in all freshwater rivers and streams that flow into a specified area of the sea. The bans were not lifted until some months later.

By December of the following year, 927 claims for compensation had been made and 621 were approved and paid out, to the total of £12.7 million. In addition the UK Government submitted a claim for approx. £11.5 million for clean-up costs.

Claims that were approved included those from the Wildlife Trust for cleaning birds, the National Trust for rescuing birds and monitoring the coastline, the French government in respect of assistance in offshore pollution response, and Youth Hostel clubs for loss of income. Among claims that were rejected was one submitted by a motorist alleging an accident and resultant damage to his car caused by the oil left on the road by vehicles carrying sludge from the contaminated beaches for disposal.

5.19.3 The *Braer* Case

In January 1993, the Liberian tanker *Braer*[70] was grounded south of the Shetland Islands. The tanker broke up and lost most of her cargo and bunkers into the sea. Owing to heavy weather, most of the oil was dispersed, but the wind caused the oil spray to be blown ashore, affecting farmland and houses close to the coast. Two thousand claims for compensation were made against the 1971 Fund, the shipowner and the P&I Club (Skuld). Some claims were settled, some withdrawn, and some came to the Court of Sessions in Edinburgh.

The dispute, which was the subject of the instant case, was brought by Landcatch Ltd, a business that supplied smolt to salmon farmers on Shetland from its hatcheries on mainland Scotland 500 km from Shetland. The Shetland salmon farmers were the main market for Landcatch smolt.

As a consequence of the discharge of oil from the *Braer*, the Secretary of State for Scotland issued an order designating an area that could be affected by the oil or other chemical substances. These substances were likely to cause a hazard to human health if fish or shellfish taken from within the designated area were consumed. Landcatch Ltd claimed for the economic loss suffered as a result of the oil pollution and the consequent exclusion zone designated.

Lord Gill held that the liability for pure economic loss could be satisfactorily interpreted to mean a liability for such loss where it was directly caused by the contamination in accordance with established principles of law. However, Landcatch had failed to establish the necessary proximity on which claims for economic loss depended, and that they were no more than a potential supplier to salmon farmers carrying out business in the contaminated area.

In a further case, *P&O Scottish Ferries Ltd. v The Braer Corporation*, it was held that losses suffered by the ferry company serving the Shetland Islands also failed the proximity test and were unable to recover.

5.20 DECOMMISSIONING

Originally, the major oil companies referred to 'abandonment' of offshore installations. This was because they considered that once the licence period was complete, they could 'abandon' the structures under the international maritime provisions for the abandonment of shipwrecks. This, of course, was never the case and so the alternative terms of removal, disposal, and decommissioning became terms of common usage to describe the process of managing and/or disposing of installations that had come to the end of their economic life. The most appropriate term is *decommissioning* and is now the term increasingly gaining currency within the sector.

Decommissioning describes the set of activities that are undertaken to manage and dispose of installations and platforms that are placed on the continental shelf and are subject to international law. There are approximately 6500 such installations in place around the world.[71] The various options for decommissioning may include leaving in place, dismantling, removing, or disposing of on the continental shelf. The option chosen would reflect the impact upon the environment, safety, and cost effectiveness after an environmental impact assessment for each installation.

Initially, the international law relating to decommissioning stated a 'clean bed' policy of absolute removal under Article 5(5) of the 1958 Geneva Convention on the Continental Shelf. However, lobbyists from the oil and gas commercial sector claimed that this would be too expensive, and during the discussions for the 1982 LOSC, a more permissive regime was provided for under Article 60(3). Article 5(5) simply provided that

> Due notice must be given of the construction of any such installations, and permanent means for giving warning of their presence must be maintained. Any installations which are abandoned or disused must be entirely removed.

On the other hand, the more liberal 1982 LOSC Article 60(3) provides that

> Any installations or structures which are abandoned or disused shall be removed to ensure safety of navigation, taking into account any generally accepted international standards established in this regard by the competent international organisation.

Such removal shall also have due regard to fishing, the protection of the marine environment and the rights and duties of other States. Appropriate publicity shall be given to the depth, position and dimensions of any installations or structures not entirely removed.

5.20.1 The IMO Guidelines

Before the 1982 Convention had even come into force, the IMO assumed the mantle of the unspecified 'competent international organisation' as the IMO authority covered 'safety of navigation'. Under the auspices of the Technical and Legal Committees, the IMO produced *'Guidelines and Standards for the Removal of Offshore Installations and Structures on the Continental Shelf and in the Exclusive Economic Zone 1989'.*[72] The general principle of the Guidelines is that all abandoned or disused offshore installations or structures on the continental shelf should be removed, so reflecting the Article 5(5) of the 1958 Convention. However, the Guidelines do go on to state 'except where non removal or partial removal is consistent with the detailed standards'.[73] The final decision remains with the coastal state as to whether partial or non-removal is concerned and this can be fraught with difficulties, as Shell UK discovered when they tried to dump the *Brent Spar* in the North-East Atlantic approaches.[74]

The factors that a coastal state must consider include the following:

- The potential effect on the safety of surface and subsurface navigation and other uses of the sea (with regard to, among other factors, the number, type, and draught of vessels expected to transit the area, their cargoes, the tide, the hydrographic and climatic conditions, proximity to shipping lanes and port access, and location of commercial fishing areas);
- The rate of deterioration of material and its present and future effect on the marine environment;
- The potential effect on the marine environment, including living resources and the presence of endangered species, local fishery resources and the potential for pollution by residual products;
- The risk that material may move from its present position;
- The costs, technical feasibility and risks of injury associated with any removal, and
- Any new use or reasonable justification for allowing the installation to remain.[75]

The Guidelines require that all structures in place prior to January 1, 1998, which are standing in less than 75 m of water and weighing less than 4000 tonnes in air (excluding the deck and superstructure), must be entirely removed. For those structures in place on or after January 1, 1998, the water depth increases to 100 m.[76] There is an exception if entire removal is not technically feasible or by way of extreme costs or unacceptable risk to personnel or the marine environment.[77]

It is the coastal state which determines whether an installation may be left wholly or partially in place, for example, to serve a new use, or it may be left without unjustified interference with other users of the sea.

Should the state permit partial removal, the decommissioning programme must provide for unobstructed water, sufficient to allow safety of navigation, with a minimum of 55 m of water above the highest point of the partially removed material. In addition, any part of the structure not removed must be maintained to prevent structural failure.[78] It is for the coastal state to ensure that partially removed structures are marked on nautical charts.[79]

5.20.2 The OSPAR Decision 98/3

Following the publication of the IMO Guidelines and the *Brent Spar* affair, the OSPAR Commission came to its Decision 98/3.[80] The Decision sets out highly detailed criteria for the removal or partial removal of offshore installations but an overall assessment must be made:

> The assessment shall be sufficient to enable the competent authority of the relevant Contracting Party to draw reasoned conclusions on whether or not to issue a permit under paragraph 3 of this Decision and, if such a permit is thought justified, on what conditions to attach to it. These conclusions shall be recorded in a summary of the assessment which shall also contain a concise summary of the facts which underpin the conclusions, including a description of any significant expected or potential impacts from the disposal at sea of the installation on the marine environment or its uses. The conclusions shall be based on scientific principles and the summary shall enable the conclusions to be linked back to the supporting evidence and arguments . . . .[81]

Table 5.1 indicates the options that may be considered for various categories of offshore installations.

Treatment of concrete infrastructure is different from that made of steel. This reflects the fact that steel structures left in place will be more vulnerable to corrosion and may collapse over time. As a result, they are considered to be more dangerous to shipping.

TABLE 5.1
Options for Various Categories of Offshore Installations

Installation (Excluding Topside)	Weight (tonnes)	Complete Removal to Land	Partial Removal to Land	Leave Wholly in Place	Reuse	Disposal at Sea
Fixed steel	<10,000	Yes	No	No	Yes	No
Fixed steel–concrete	>10,000	Yes	Yes[a,b]	No	Yes	No
Concrete gravity	Any	Yes	Yes[b]	Yes	Yes	Yes
Floating	Any	Yes	No	No	Yes	No
Subsea	Any	Yes	No	No	Yes	No

[a] Only the footings or part thereof may be left in place.

[b] Minimum water clearance of 55 m is required above any partially removed installation that does not project above the surface of the sea.

5.20.3 The 1972 London Dumping Convention

The LDC came into force in 1975. As with OSPAR, the Convention applies to activity wider than just decommissioning, and although the provisions of the Convention are wide enough to affect other aspects of petroleum operations, this section is limited to decommissioning. The word 'dumping' is defined as including 'any deliberate disposal at sea of vessel, aircraft, platforms or man-made structures at sea'. This definition would also include oil and gas offshore installations of any kind, although the Convention does cover industries other than oil and gas.

The 1996 Protocol to the LDC is intended to update the Convention and adopts the precautionary principle. The Protocol came into force in March 2006 with its purpose being to

> protect and preserve the marine environment from all sources of pollution and take effective measures, according to their scientific, technical and economic capabilities, to prevent, reduce and where practicable eliminate pollution caused by dumping or incineration at sea of wastes or other material.[82]

The Protocol reverses the burden of proof so that no item other than those specified may be dumped at sea.[83] Platforms and other man-made articles are included in the specified list that may be dumped if granted a permit. In respect of decommissioning, an offshore installation may be considered for dumping if

> material capable of creating floating debris or otherwise contributing to pollution of the marine environment has been removed to the maximum extent and provided that the material dumped poses no serious obstacle to fishing or navigation.[84]

Coastal states must take into consideration the following before granting a permit:

1. Alternatives to dumping should be considered (including a hierarchy of waste management options of reuse, off-site recycling, destruction of hazardous constituents, and treatment to reduce or remove the hazardous constituents and disposal on land, into air and in water).
2. Any appropriate opportunities to reuse, recycle, or treat the waste without disproportionate costs or undue risks to human health or the environment. The practical availability of other means of disposal should be considered in the light of a comparative risk assessment involving both dumping and the alternatives.[85]

Compared to the LDC, the Protocol takes a more restrictive approach towards dumping but does include in its definition of 'dumping' 'any abandonment or toppling at site of platforms or other man-made structures at sea, for the sole purpose of deliberate disposal'.[86] This means that, unlike the Convention, which arguably permitted the abandonment of installations without a permit, the Protocol requires permission of the coastal state.[87]

After the *Brent Spar* incident, coastal states have been reluctant to issue permits for disposal at sea notwithstanding the provisions under the Protocol for a monitoring system to verify that the permit conditions are being complied with,[88] and an arbitration procedure to resolve disputes relating to non-compliance with permit conditions.[89]

The main significance of the Protocol is that it directly identifies oil rigs and offshore installations as an issue when they come to the end of their economic life.

The increase of the number of offshore oil installations worldwide, and the fact that a high percentage of them are coming to the end of their economic life within the present and next decade, only goes to exacerbate the situation. The *Brent Spar* incident is not far from multinational oil companies' minds when contemplating decommissioning, and the requirement for ongoing monitoring would make a company think very hard before either partial removal or dumping on the seabed of installations that have come to the end of their economic life.

ENDNOTES

1. Brazil made a case for changing from a concession system to a production sharing agreement following large offshore discoveries in 2007.
2. Easo (2009), 'Licences, concessions, production agreements and service contracts', Pickton-Turbervil, Editor, *Oil & Gas a Practical Handbook.* Globe Law and Business: London.
3. (1963) 27 *ILR* 117.
4. (1977) 53 *ILR* 389.
5. Petrobras was the state-owned oil company set up in 1953 to take control of the hydrocarbon sector in Brazil.
6. Chequer & Costa (2010), *Brazilian Pre-Salt: PSA Regime Approved in the Congress.* Tauil & Chequer: Rio de Janeiro.
7. The state will be acting through the relevant ministry or government official.
8. The state oil company will take delivery of the state's share of the production.
9. See Easo, Note 2, pp. 27–41.
10. See Easo, Note 2, pp. 27–41.
11. This is to compensate the oil company for the costs it incurred during exploration and production.
12. 1991 ICJ, *ILR* (1994), vol. 94 at 446.
13. Finland had filed an application against Denmark in the International Court of Justice arguing that a Danish plan to build a high-level bridge over the main navigational channel of the Great Belt Strait would make it impossible for drilling ships and oil rigs, which had deep draughts and required a clearance of more than 65 m to enter and leave the Baltic. Finland further contended that the Great Belt was a strait used for international navigation and that there was a right of free passage through it for vessels, including mobile offshore drilling rigs. Denmark denied that the right of passage would apply to oil rigs, as it did not consider them to be ships.
14. Taggart (1980), *Ship Design Construction.* Society of Naval Architects and Marine Engineers, SNAME: New York, USA.
15. Esmaeili (2001), *The Legal Regime of Offshore Oil Rigs in International Law.* Ashgate: Guilford, UK.
16. Law of the Sea Convention (1982), Articles 11, 56, 60, 80, 87, 147, 194, 208, and 262.
17. Esmaeili, ibid.
18. 1958 Convention, Article 5, paras 2–6.
19. Rothwell & Stephens (2010), *The International Law of the Sea.* Hart Publishing: Oxford.
20. 1958 Geneva Convention on the Territorial Sea and Contiguous Zone, Article 5(1); 1982 Convention on the Law of the Sea, Article 8(1).

21. Rothwell & Stephens, ibid.
22. *Anglo-Norwegian Fisheries Case*, ICJ Rep. 116 (1951).
23. (1992) ICJ Rep. 351, p. 593.
24. Rothwell & Stephens, ibid.
25. Esmaeli (2001), *The Legal Regime of Offshore Oil Rigs in International Law.* Ashgate: Guilford, UK.
26. Proclamation 2667, 'Policy of the United States With Respect to the Natural Resources of the Subsoil and Seabed of the Continental Shelf', 28 September 1945, Basic Documents no. 5.
27. See Chapter 9 on United Kingdom.
28. Article 56(b)(i).
29. Esmaeili (2001), *The Legal Regime of Offshore Oil Rigs in International Law*, p. 77.
30. Esmaeili, ibid.
31. Article 86 of Law of the Sea Convention, 1982.
32. Article 116.
33. Article 87(f).
34. Article 112(1).
35. Article 117.
36. Rothwell & Stephens (2010), *International Law of the Sea.* Hart Publishing: Oxford.
37. Basic Documents no. 37.
38. Wood (1999), *International Seabed Authority: The First Four Years*, vol. 3. Max Planck Yearbook of United Nations Law, Leiden: The Netherlands, 173.
39. Secretary General (2008), 'Report of the ISBA Secretary General', 14th session, ISBA/14/A/2, European Commission: Kingston Jamaica.
40. Article 133.
41. Esmaeili, ibid.
42. Francioni (1975), *Criminal Jurisdiction Over Foreign Merchant Vessels in Territorial Waters: A New Analysis.* Italian Yearbook of International Law. Martinus Nijhoff: The Hague, Netherlands, 42.
43. LOSC Articles 17, 18, and 19.
44. Esmaeili, ibid.
45. (1969) ICJ Rep. 29.
46. Gold (1997), *Gard Handbook on Marine Pollution.* Gard P&I Club: Norway.
47. UN Doc. A/7750, Part 1, p. 3, 10 November 1969.
48. LOSC, Article 1(1)(4).
49. Esmaeili (2001), *The Legal Regime of Offshore Oil Rigs in International Law.* Ashgate: Guildford, UK.
50. LOSC Article 194 3(c).
51. LOSC Articles 208 and 209.
52. The 1994 Agreement, Annex, Section 7.
53. Article 1 of the 1972 Convention on the Prevention of Marine Pollution by the Dumping of Wastes and Other Matter.
54. LDC Article 3(1)(a)(i).
55. LDC Article 3(1)(a)(ii).
56. IMO Assembly Resolution A.414 (XI), 15 November 1979.
57. OPRC Article 2.
58. OPRC Article 3.
59. OPRC Article 4.
60. OECD (1982), *Combating Oil Spills.* OECD: Paris.
61. Keaton, 21 *CLP* (1968), 94.
62. Articles 3(2) and 3(3).
63. Article 3.

64. Article 14.
65. Article 4(2).
66. Birnie & Boyle, *Official Journal of the European Union*, 347–404.
67. (1984) 2 Lloyd's Rep. 304.
68. United States District Court, Northern District of Illinois.
69. The Annual Report of the International Oil Pollution Compensation Funds, 1997, 86–94.
70. *Landcatch Ltd v The International Oil Pollution Compensation Fund (IOPCF) (The 'Braer')*, 2 Lloyd's Rep. 552 (1998).
71. www.info.ogp.org.uk/decommissioning. Accessed July 2010.
72. www.imo.org/home.asp. Accessed July 2010.
73. IMO Guidelines, Clause 1.1.
74. For further discussion, see Chapter 9 and Igiehon and Park (2001), 'Evolution of international law on the decommissioning of oil and gas installations', *IELTR* 198.
75. IMO Guidelines, Article 2.1.
76. IMO Guidelines, Articles 3.1 and 3.2.
77. IMO Guidelines, Article 3.5.
78. IMO Guidelines, Article 3.6.
79. IMO Guidelines, Article 3.8.
80. This Decision does not cover pipelines and there are no other international guidelines on the decommissioning of disused pipelines.
81. OSPAR Decision 98/3, Annex 2, 12.
82. LDC Article 2.
83. LDC Article 4.
84. Protocol to LDC, Annex 1, Article 2.
85. Protocol to LDC, Annex 2, Articles 5 and 6.
86. Protocol to LDC, Article 1(4)(1).
87. Protocol to LDC, Article 4(1)(2), Annex 1(1)(4).
88. Protocol to LDC, Annex 2, Article 16.
89. Protocol to LDC, Annex 3.

6 International Regulation of the Nuclear Industry

6.1 NUCLEAR ENERGY AND THE ENVIRONMENT

After World War II, there dawned an age of optimism as far as nuclear energy was concerned. It was widely believed that the benefits of cheap energy would contribute to world peace.[1] In 1956, the International Atomic Energy Agency (IAEA) was created with the objective of encouraging and facilitating the spread of nuclear power.[2] The belief was that health and environmental risks could be managed successfully by governments and the IAEA through cooperation on the regulation of safety matters.

Successive declarations by international bodies maintained the belief in the benefits of nuclear energy and supported the transfer of nuclear technology to developing countries. In 1977, the UN General Assembly reaffirmed the importance of nuclear energy by proclaiming that every state had the right to use it and have access to the technology.[3]

From these earliest days of the use of nuclear energy for peaceful purposes, a notable spirit of international cooperation had helped to lay the foundations of a new body of law consisting of agreements, rules, and regulations, both international and national, known collectively as *nuclear law*. At the time, this body of law was considered to accurately reflect the international community's determination to protect workers, the public, and the environment from any dangers of ionising radiation and also to provide equitable compensation in the event of an accident. The primary objective of these regulations was to create a legal regime that could take account of the specific problems relating to the development of nuclear energy. The exchange of information and scientific knowledge also enhanced the possibility of the consensus reached internationally, to filter down to national level and so develop a harmonised national legislation in the field of nuclear energy production.

In February 1955, the Minister of Fuel and Power in the United Kingdom presented the Parliament with a proposal for a programme of nuclear power.[4] Nuclear energy was to be the energy of the future and the coming of nuclear power promised the beginning of a new era. Although the risks were known to be high, it was considered that the final reward would be 'immeasurable'.[5] The White Paper heralding the UK government's commitment to a programme of nuclear power stations pronounced:

> This formidable task must be tackled with vigour and imagination. The stakes are high but the final reward will be immeasurable. We must keep ourselves in the forefront of the development of nuclear power so that we can play our proper part in harnessing this new form of energy for the benefit of mankind.

Following this announcement of government commitment to nuclear energy, the United Kingdom established the world's first nuclear power programme.

6.2 THE EMERGENCE OF ENVIRONMENTAL CONCERNS

The 1973 crisis, when the Organization of Petroleum Exporting Countries (OPEC) raised the price of oil to above the level that the developed countries considered acceptable, heralded a marked expansion in nuclear energy production. This, in turn, brought the long-term health and environmental consequences to the attention of the international community. In 1972, the Stockholm Conference had considered the problems of radioactive emissions and the disposal of radioactive waste in addition to the problems of reprocessing. The Conference called for a register of emissions, and international cooperation on the disposal of radioactive waste.[6] Although the Conference recognised that reprocessing was a growing problem, it was unable to develop a clear policy on how to address it. The dumping of radioactive waste in the marine environment was partially banned in 1972 under the London Dumping Convention, and was also the subject of a moratorium pending further study in 1983.[7]

However, the nuclear accidents at Three Mile Island in the United States, the conflagration at Chernobyl in the then Soviet Union, and, more recently, in Fukushima in Japan dramatically demonstrated the risks for health, agriculture, and the environment associated with nuclear energy. Like the sinking of the *Torry Canyon*,[8] the accident at Chernobyl in 1986 revealed the limitations of the international policy on the containment of catastrophic risks, and identified unrealised costs of nuclear power for the environment. The adequacy of national and international controls on nuclear facilities was questioned as the issues of liability and state responsibility were brought into sharp relief.

Adjacent states realised the importance of the siting of nuclear power plants and the right to be consulted on issues of safety in addition to prompt notification of potentially harmful accidents. Further, it demonstrated that the benign view of nuclear power as held in the 1950s was fundamentally flawed and required modification.[9] In addition, for the first time, an international body, in the shape of the European Community, was prepared to describe nuclear energy as potentially dangerous, and to recommend a moratorium on the construction of new facilities.[10] Nevertheless, most states continue to believe that through stronger international regulation and cooperation, the risks could be contained in order to reduce reliance on fossil fuels and so help reduce the prospect of global warming.

6.3 THE INTERNATIONAL REGULATION OF NUCLEAR ENERGY

The international community has realised that nuclear installations are potentially hazardous undertakings; that the risks to health, safety, and the environment are best met by strong, enforceable regulation; and that the failure to respond to the need for the setting of common standards and their enforcement by international institutions may result in injury or pollution damage to other states and the global environment. Such regulation would benefit the international community as a whole, but the burden falls on national governments, which lose the freedom to develop their own energy policies and assess their own risks.[11]

However, this idea is not new. The choice of strong international regulation has already been made in the case of oil tankers, with minimum duties imposed on flag states in matters of environmental protection having been laid down in international conventions.[12] The majority of major maritime states have incorporated the MARPOL requirements, which provide for a strong system of enforcement, into their national law. Unfortunately, no similar regime exists for nuclear installations, with national sovereignty prevailing and the consequent freedom to set national standards of health and safety regulations.[13] In the case of nuclear installations, although the international bodies have the responsibility for formulating standards on health and safety, these regulations are not binding on states in most instances. The result is that without binding international standards, no comparable level of international enforcement is possible.

6.4 THE IAEA AND THE REGULATION OF NUCLEAR POWER

The primary responsibility for developing the scientific basis for health and safety standards and for promoting the development of regulations in the field involved three major international organisations: the IAEA, the Organisation for Economic Cooperation and Development (OECD), Nuclear Energy Agency (NEA), and the European Atomic Energy Community (EURATOM). All three agencies were created within a year, 1957/1958, and had complementary roles. The IAEA represents the largest and most diverse group of countries, including both developed and developing states. The NEA represents the industrialised market economies and EURATOM represents those European states that are part of an integrated nuclear community.[14]

The member countries of the IAEA have entrusted the Agency with the responsibility of establishing health and safety standards, the purpose of which is to protect health and minimise any risks to life and property. The Agency also has the right to make arrangements for the application of these standards to particular operations.

The Agency's Statute provides that the Agency must consult or collaborate with other international organisations that are concerned with the establishing or adoption of safety standards, whenever appropriate. This involves both the exchange of scientific data and also any operating experience and good regulatory practices among national nuclear authorities. The resulting principles and standards will, therefore, be reflected in national legislation due to their broad acceptance at international level. Additionally, the Statute provides for the application of these developed standards in three distinct areas:

- For the Agency's own operations, where the standards are automatic and obligatory
- For national operations carried out with the Agency's assistance or association, when the standards are applied through an agreement concluded between the state requesting assistance and the Agency
- The standards may also serve as guidelines for any national regulatory authority

Article XI of the Statute provides for the arrangements and procedures for the application of the standards.

The amount of control exercised by the Agency is dependent on which of the above types of operations is being considered. For example, for its own operations and also for Agency-assisted operations, control is direct and the Agency enjoys specific rights and responsibilities to ensure the effectiveness of the application of the standards.[15] These include the rights of examination and approval of both equipment and facilities, and the right of entry into property by inspectors to verify compliance. The Agency enjoys the same rights with respect to non-assisted operations, but the practice has been not to make extensive use of these rights when the Agency has been requested to apply safeguards.[16] Although the standards are advisory only, except in those cases in which the Agency has ownership or provides assistance, some would say that they enjoy considerable importance.[17] However, others would claim that the lack of enforceability creates a major problem.[18]

6.5 OECD NUCLEAR ENERGY AGENCY

The OECD has been involved with nuclear energy matters through the Nuclear Energy Agency, whose aims, although similar to those of the IAEA, include encouraging the adoption of common standards. The Agency is also involved in collaboration with the IAEA on the development of standards regarding radiation protection and waste management. The OECD also initiated a convention on third-party liability[19] and a multilateral consultation procedure for the dumping of radioactive waste at sea.

The Statute of the Nuclear Energy Agency takes the form of a Decision originally adopted by the Council of the Organisation for European Economic Co-operation on December 20, 1957. It was subsequently approved by the OECD Council on September 30, 1961.

Article 8 provides that the Agency shall:

1. Contribute to the promotion, by the responsible national authorities, of the protection of workers and the public against the hazards of ionising radiations and of the preservation of the environment;
2. Contribute to the promotion of the safety of nuclear installations and materials by the responsible national authorities;
3. Contribute to the promotion of a system for third party liability and insurance with respect to nuclear damage.

These competences as to safety, radiation protection, and liability, are exercised in a variety of ways:

- It provides a forum for confidential meetings among experts from national authorities, in which they can exchange experiences, views, and opinions regarding the technical and socio-economic aspects of nuclear energy.
- The OECD Council has mandated the Steering Committee for Nuclear Energy[20] to make recommendations or give advice to Member States on all questions within their competence. Although the recommendations of the Agency are not binding on member governments, this does not affect their efficiency.

The process of international consultation and cooperation that precedes the recommendations ensures that the norms agreed to by the national experts are accurately transposed into national legislation.

6.6 CONTROL OF NUCLEAR RISK

6.6.1 INTERNATIONAL OBLIGATIONS

In the absence of any treaty commitment to control the environmental risks of nuclear power, there is, nevertheless, evidence of an obligation to minimise nuclear risks and to prevent injury to other states, also to prevent radioactive pollution of the global environment.[21] There are agreed standards of safety and radiation protection as regards to nuclear-powered merchant ships and satellites.[22] Additionally, the same principle is encompassed in the 1989 Basel Convention on the Control of Transboundary Movement of Hazardous Wastes.[23] Hitherto, the military use of nuclear power falls outside these rules, but even that is now being challenged in the UK courts.[24]

Regional Treaties concerning radioactive waste disposal exist in the Antarctic,[25] in the South and South-East Pacific,[26] also in Africa.[27] All of these precedents show that the emission of radioactive substances into the environment of common spaces is presumed to be prohibited,[28] with the exception of acceptable low-level radioactivity when proof of harm caused would be needed.

The 1963 Nuclear Test Ban Treaty prohibits weapons test explosions in the atmosphere, outer space, at sea, in Antarctica, or in any circumstances where any radioactivity debris may spread beyond the territory of the state carrying out the tests.[29] The only way that all these requirements can be complied with is to conduct the tests underground and ensure that no polluting material escapes. However, not all nuclear powers are party to the Treaty and in the *Nuclear Test* case,[30] the International Court of Justice (ICJ) declined to decide whether the Treaty was part of customary international law. Nevertheless, in practice, subsequent tests have complied with the 1963 Treaty. Given such state practice and the weight of international opinion against the deliberate release of radioactive material into the atmosphere, it could be argued that a prohibition of nuclear testing, other than underground, could be founded in international customary law.

The *Nuclear Tests* cases also raise the question of what constitutes 'serious harm' from the deposit of radioactive particles on the territory of another state. The problem is that injury is not always immediate or apparent. In the *Nuclear Tests* cases the claim was based on a violation of territorial sovereignty rather than actual harm caused; however, Handl concludes that material injury is necessary.[31] It then becomes necessary to develop international standards of radiation exposure, based on evidence of long-term effects, in order to establish an agreed threshold of harm without regard to immediate injury.[32]

The cumulative effect of such evidence is that states do have an international responsibility based on customary law to ensure that the nuclear activities based within their territory are conducted safely.

6.7 RADIATION PROTECTION STANDARDS

The IAEA's revised Basic Standards for Radiation Protection[33] are derived from recommendations provided by the International Commission on Radiological Protection (ICRP).[34] The latest standards were considered and agreed to by the NEA, the World Health Organisation (WHO), and the International Labour Organisation (ILO). These provided a very broad acceptance at international level, and have subsequently been adopted and applied almost universally to nuclear operations, notwithstanding some differences due to national implementation regulations and national policies. The IAEA has set up Radiation Protection Advisory Teams (RAPTS) to assist states in the implementation of the standards. These teams assess, on request, the quality of radiation protection in the requesting state, in addition to determining the immediate and future radiation protection needed. The teams also define any long-term strategy for technical assistance.

Since the establishment of the basic standards, the activities of the NEA have focused on the scientific bases for radiation protection regulations in addition to the interpretation and application of principles and concepts as developed by the International Commission on Radiation Protection (IRCP). These cover the protection of workers in the nuclear industry and the general public from accidental releases of radioactive material. This work is the responsibility of the Committee on Radiation Protection and Public Health (CRPPH), which examines national radiation policies on a regular basis. Advice is then given on the risks of any new applications of nuclear energy. In support of its technical advice, the NEA supports a continuing programme of research into not only health and safety from the technical basis but also fundamental assumptions and philosophy.

6.8 STATE RESPONSIBILITY FOR NUCLEAR DAMAGE AND ENVIRONMENTAL HARM

Sovereign states enjoy supreme authority and absolute rights within their jurisdiction. However, all rights carry responsibilities with them, which in the international law means liability arising from the breach of an obligation. In the field of the protection of the environment, the *Trail Smelter Arbitration* held that: '. . . no State has the right to use or permit the use of its territory in such a manner as to cause injury by fumes in or to the territory of another or the properties or persons therein, when the case is of serious consequence and the injury is established by clear and convincing evidence'.

A state is also responsible in cases of breach of its obligations under treaty or its wrongful acts against another state or other entity having international personality. However, any harm suffered by individuals cannot be considered as that by another state as individuals have no *locus standi* before an international tribunal.[35] Nevertheless, the claim may be presented on their behalf by their own state.

In the development of international environmental law, the question of state responsibility was raised, especially when an accident with trans-boundary harm occurs. A number of the international agreements and declarations addressed the scope of state liability.[36] By ratifying an international convention, the State undertakes to conform with its requirements, and failure to do so will give rise to international responsibilities. However, the scope of those responsibilities largely depends on three questions:

1. What is the nature of the trans-boundary harm?
2. What is the extent of state responsibility for trans-boundary harm arising from private activities?
3. What is the type of activity causing the harm?

6.9 THE TRANSPORTATION OF RADIOACTIVE MATERIALS

The regulation of the carriage of radioactive materials is a complex amalgamation of international conventions, bilateral agreements, domestic law, licensing agreements, and codes of practice. However, the responsibility lies predominantly with the state from which the materials are being shipped to ensure that not only are all the general treaty obligations observed but also that there is effective insurance cover.

6.10 THE IAEA REGULATIONS

The first set of IAEA Regulations was published in 1961, and these have been effectively updated and amended since then in the light of extensive operating experience. They provide for basic standards for the transportation of radioactive materials with the main objective being the protection of the public, transport workers, and property. The Regulations only apply to materials and facilities where the IAEA may have an involvement on some level. However, they are also considered to be a model of good practice, and as such are recommended to all states and international organisations.

The Regulations cover such areas as the exemption of small quantities of radioactive materials from the packaging and labelling requirements; the simplification of the packing requirements for low-specific activity material; the design criteria and performance tests for packaging for normal and special forms of radioactive and fissile materials; and the administrative controls, including the competent authority approval of package design and shipment, transport documents, information for carriers, and the notification procedures.

6.11 THE CIVIL LIABILITY CONVENTIONS

There are a growing number of conventions on liability in this field of law and yet there are liability problems that are still insufficiently solved. This is especially true when considering the harmonisation of nuclear liability and the levels of compensation.

6.11.1 The Paris Convention on Third-Party Liability in the Field of Nuclear Energy, 1960[37]

This Convention applies to nuclear incidents within the OECD states. It provides for absolute liability that is channelled exclusively to the operator of the nuclear installation, with all other potential defendants being protected. Liability is limited and if the national legislation provides for compensation by the operator for the means of transportation, it must be on the condition that the right of other victims is not reduced as a consequence of the extension of liability. The Convention has been updated by any number of Protocols since its original signing, mainly to harmonise it with other conventions in the field.

6.11.2 The Vienna IAEA Convention on Civil Liability for Nuclear Damage, 1963

This Convention provides a scheme comparable to the Paris Convention but it is intended to have a global application. Both treaties harmonise important aspects of liability for nuclear accidents and simplify the procedures for obtaining redress. Limits are placed on the amount and duration of liability but the operator is required to maintain an insurance policy or a financial security for the full amount of his liability. To simplify this the Standing Committee on Atomic Risk of the European Insurance Committee has formulated a model agreement between the insurers concerned and a model certificate of financial security.

6.11.3 The Brussels Convention Relating to Civil Liability in the Field of Maritime Carriage of Nuclear Material, 1971

Under both the Paris Convention Article 6(b) and the Vienna Convention Article 11(5), shipowners were liable, in addition to the operator of the nuclear installation, if their vessels were involved in a nuclear accident. This position made it almost impossible for shipowners to obtain insurance for the carriage of nuclear material. It was this position that was remedied by the Brussels Convention.

Under Article 1 of the Convention: '. . . any person who by virtue of an international convention or national law in the field of maritime transport might be held liable for damage caused by a nuclear incident shall be exonerated from such liability'.

However, where damage is caused by an act of omission done with intent to cause damage, then the liability is not affected. In addition, this Convention only affects those states that are party to the Convention. The obligations of third-party states that are not party to the Convention remain unaffected.

All of these treaties attempt to harmonise certain aspects of liability by creating a common scheme for loss distribution, based on absolute liability, which is channelled to the operator.

6.11.4 The General Scheme

The overall scheme of all the conventions is based on similar principles. Firstly, liability is absolute with no proof of negligence required as a condition of liability.[38] Secondly, the liability is channelled exclusively to the operator of the nuclear installation or ship and all other potential defendants are protected.[39] Also, in certain cases a carrier or handler of nuclear material may be treated as an operator. Thirdly, in return for accepting absolute liability, operators have a limitation placed on the amount and duration of that liability.[40] Fourthly, compulsory insurance or security must be held by the operator, and guaranteed by the state of installation or registry.[41] Finally, the rules determine which state has jurisdiction over claims and preclude all other civil proceedings.[42]

Under the conventions, states are given discretion in the areas of periods of limitation; insurance and liability; the definition of nuclear damage; and the discretion not to relieve operators of liability in cases of grave natural disaster.[43] In fact, both Germany and Austria reserved the right to exclude Article 9 and, therefore, make liability absolute.[44]

The principal objective of the conventions is, therefore, to set minimum international standards that are flexible enough to be adaptable to a variety of jurisdictions with differing legal, social, and economic systems.[45]

The conventions cover most potential sources of nuclear damage. Both the Paris and Vienna Conventions apply to broadly defined 'nuclear installations'. These include reactors, reprocessing plants, manufacturing and storage facilities, where nuclear and radioactive materials are used or produced.[46] They also apply to the transportation or handling of nuclear waste.[47] The Brussels Convention, alternatively, covers nuclear-powered ships, their fuel, and incidental waste, but not the carriage of nuclear material by sea.[48] This is covered by other conventions.

6.11.5 Claims under the Convention

The conventions simplify the jurisdictional issues involved in bringing trans-boundary civil actions. Determining which state has jurisdiction over claims against operators or their insurers, the location of the nuclear incident causing the damage, or exceptionally, the location of the installation itself are the deciding factors while deciding the jurisdictional issue.[49] This extended definition was included to cover incidents caused by material in transit. Should there be cases concerning multiple jurisdictions, these are to be dealt with by agreement of the parties under the Vienna Convention[50] or by tribunal under the Paris Convention.[51] The tribunal would then decide on the court that was the 'most closely related to the case', particularly having regard to cases involving ships, when both the licensing state and the state or states where the damage occurred have jurisdiction.[52]

Additionally, all judgements decided by courts that are competent in accordance with the conventions must be recognised and enforced by all other Member States.[53] For all practical purposes in Western Europe, most judgements are normally recognised under EC treaties;[54] nevertheless, for states who are not party to the EC treaties, it is an important guarantee of access to compensation funds for trans-boundary damage.

An action must be brought within 10 years of the incident. However, as the effects of nuclear radiation on human health may not become apparent for many years after the event, the 10-year limitation may leave victims without redress. To mitigate this situation, a number of states have adopted a limitation period of up to 30 years.[55]

6.12 ASSESSMENT OF THE NUCLEAR CONVENTIONS

While considering the extent to which the nuclear conventions protect the environment, one problem that is a common feature is the narrow anthropocentric definition of damage.[56] Both the Brussels and the Vienna Conventions provide for the extension of the definition, but few states have done so without broader consensual agreement.

The problem of this narrow definition was made apparent after the nuclear accident at Chernobyl when extensive ecological damage to wildlife was recorded. The outcomes of any nuclear accident at sea would not easily fall into the terms of the convention. The nuclear conventions should be more realistic in their approach to environmental damage if the true costs of damage caused by nuclear incidents are to be borne by the nuclear industry and so reflect 'the polluter pays' principle.[57]

A further problem concerning the common scheme of the conventions is the lack of widespread international support. Currently, it is only the Western European countries, as parties to the Paris Convention, who have embraced an international agreement on civil liability. The spread of nuclear energy sources and the effects of major incidents such as Chernobyl, highlight this lacuna in the international regulation of nuclear power.

6.13 DEVELOPMENTS SINCE CHERNOBYL

Prior to the accident at Chernobyl, international nuclear law had addressed a range of issues. The early legislation had addressed liability for damage caused by nuclear incidents and also the protection of workers in the industry.[58] In addition, although this book does not include the law in this area, a number of treaties and agreements addressed atmospheric and other nuclear testing and the placing of nuclear weapons. The non-proliferation of nuclear weapons was also provided for under The Treaty on Non-Proliferation of Nuclear Weapons, July 1968, which came into force in March 1970 and was extended indefinitely in 1995.[59] Further, the issues of physical protection of nuclear material and border area cooperation[60] have been the subject of international agreement. The disposal of radioactive waste at sea has also been the subject of regulation or prohibition under treaties and other international agreements.[61]

As far as liability is concerned, following the incident at Chernobyl, states and the IAEA made a commitment to improve the existing rules on liability for nuclear damage. However, nearly 15 years after Chernobyl, the international community has been unable to adopt amendments to the 1960 Paris Convention or the IAEA Convention, other than a Joint Protocol to link them. The previous problems, as identified, remain. The two Conventions fail to provide adequately to protect the environment. As expressed by Sands,[62] they allow parties to apply absurdly low ceilings on the levels of financial liability, which amounts to a subsidy, and so protects the industry against claims. In addition, plaintiffs are required to bring an action in the state where the incident occurred rather than the state where the damage was suffered.

Having outlined those areas in which the lacuna remains, it must be noted that efforts have been made particularly in the areas of notification and assistance in the event of an accident. Prior to the incident at Chernobyl, there was only limited appreciation of the potential trans-boundary impacts of major nuclear accidents. Bilateral agreements between adjacent states on whose border the nuclear installation was situated were the predominant instruments for notification and protective actions.[63] Essentially, these agreements provided for the notification of an emergency, with some requirement for cooperation in the event of an accident.[64]

The IAEA Board of Governors immediately convened a meeting of the Group of Experts to consider a draft international agreement on early notification and assistance in the event of nuclear accidents and radiological emergencies. Within a month, the Group of Experts had agreed on two texts for submission to the Board, and from thence to the General Conference to be held the following month. The General Conference adopted the Convention on Early Notification of a Nuclear Accident[65] and the Convention on Assistance in Case of a Nuclear Accident or Radiological Emergency.[66] The Conventions came into force in October 1986 and February 1987, respectively.

The Early Notification Convention applies should there be an accident at a designated facility where a release of radioactive material either has occurred or is likely to occur and which has resulted or may result in an international trans-boundary release of potential radiological significance (Article 1(1)). This Article is pre-emptive and primarily covers incidents at nuclear reactors, fuel cycle facilities, and waste management facilities. It also covers the transport and storage of fuel or waste; the manufacture, use, storage, disposal, and transport of radioisotopes for agricultural research and industrial use, in addition to the use of radioisotopes for power purposes in space objects (Article 1(2)). The state in which the installation is situated is under a duty to notify those states that are or may be affected, and provide details of the accident in addition to any information that may mitigate the effect of the radiological consequences (Articles 2 and 5).

Problems already identified with this Convention[67] include the lack of a duty of parties to establish monitoring networks.[68] Additionally, the non-inclusion of military facilities leaves numerous installations without reporting obligations. Finally, the degree of discretion awarded to the Contracting Parties on when a notification must be made, and what releases are of radiological significance to other states.

On the more positive side, the speed with which the IAEA reacted to the Chernobyl accident, and the inclusion of the requirement to report to all states who may potentially be affected by trans-boundary damage, and not just other Contracting Parties, is to be welcomed.

The Assistance Convention requires the cooperation between States Parties and the IAEA to facilitate prompt assistance in the event of a nuclear accident.[69] A duty is imposed upon the State Party to whom the request is made to reply promptly and identify what, if any, assistance may be afforded by the said State Party.[70] However, there is no duty under the Convention to request assistance, accept any or all of the assistance offered, or supply assistance upon demand. The main obligations on a State Party are to both notify to the IAEA and other States Parties, the competent authority, and contact points within its jurisdiction, and also to respond promptly to requests for assistance.

6.14 NUCLEAR WASTE

After a certain number of years of use, radioactive materials become decontaminated and therefore need to be disposed of as waste. The dumping of waste has been covered by a number of international agreements and also domestic regulation; however, this activity presents many obvious hazards including explosion and radiation. The need to find a safe medium for the disposal of radioactive waste material continues to exercise the minds of the international community. A number of nuclear states including the United Kingdom, United States, and Japan dump their radioactive waste at sea.

Although Article 25 of the High Seas Convention 1958 requires that states take measures to prevent pollution of the high seas from the dumping of radioactive waste material, it was never anticipated that this should be a complete ban.[71] It was left to the IAEA to set the standards. Even the London Dumping Convention only prohibited the dumping of 'high level' radioactive material, defined by the IAEA as unsuitable for this kind of disposal.[72] These guidelines have now been accepted as a minimum standard for national regulation.[73] However, some states apply their own more stringent standards as a growing number of least developed country (LDC) parties and pressure from non-governmental organizations (NGOs) led to the adoption of a moratorium on all radioactive dumping at sea in 1983.[74] However, the moratorium is voluntary and some parties, including the United Kingdom, continue to reserve the right to resume dumping of low-level radioactive waste.[75]

6.15 EURATOM AND THE EUROPEAN DIMENSION

Similar to the European Coal and Steel Community (ECSC) and the European Economic Community (EEC), the EURATOM Treaty was a child of the 1950s. It was the ideal of European integration and the belief that nuclear energy 'represents an essential resource for the development and invigoration of industry and will permit the advancement of the cause of peace'.[76] The potential problems were thought to be containable and in the post-war years, the tone of the EURATOM Treaty was progressive and optimistic.[77] The objective of the Treaty was to build up a 'powerful nuclear industry' in Europe and a framework was established under the Treaty in which the nuclear industry could develop.

Under Article 2 of the Treaty the Community must:

1. Promote research and ensure the dissemination of technical information;
2. Establish uniform safety standards to protect the health of workers and of the general public and ensure that they are applied;
3. Facilitate investment and ensure, particularly by encouraging ventures on the part of undertakings, the establishment of the basic installations necessary for the development of nuclear energy in the Community;
4. Ensure that all users in the Community receive a regular and equitable supply of ores and nuclear fuels;
5. Make certain, by appropriate supervision, that nuclear materials are not diverted to purposes other than those for which they are intended;
6. Exercise the right of ownership conferred upon it with respect to special fissile materials;

7. Ensure wide commercial outlets and access to the best technical facilities by the creation of a common market in specialised materials and equipment, by the free movement of capital for investment in the field of nuclear energy and by freedom of employment for specialists within the Community;
8. Establish with other countries and international organisations such relations as will foster progress in the peaceful use of nuclear energy.

Although these tasks are described variously in the 10 chapters of Title II of the Treaty, it is Chapter III that is the most relevant to this book.

The basic principle is set out in Article 30:

> Basic standards shall be laid down within the Community for the protection of the health of workers and the general public against the dangers arising from ionising radiations.
>
> The expression 'basic standards' means:
>
> a. Maximum permissible doses compatible with adequate safety;
> b. Maximum permissible levels of exposure and contamination;
> c. The fundamental principles governing the health surveillance of workers.

These were the basic standards as established in 1959 and have subsequently been updated by Directive, but it is for the Member States to ensure that they are applied under normal Community rules.

However, uniquely to the EURATOM Treaty and without parallel in the whole of Community law, Article 38(2) provides:

> In cases of urgency, the Commission shall issue a directive requiring the Member States concerned to take, within a period laid down by the Commission, all necessary measures to prevent infringement of the basic standards and to ensure compliance with regulations.

The Commission has never applied this provision, even at the time of the Chernobyl accident.

Furthermore, under Chapter III, both Articles 34 and 37 require that Member States obtain from the Commission their opinion as regards certain nuclear projects that may have a detrimental effect on health and safety. Article 34 concerns 'dangerous experiments' and may only be carried out after the Member State has taken 'additional health and safety measures, on which it shall first obtain the opinion of the Commission'. It is questionable that if such requirements had covered the experiments that led to the accident at Chernobyl, the accident may well have been prevented.

Article 37 concerns the disposal of radioactive waste and provides that:

> Each Member State shall provide the Commission with such general data relating to any plan for the disposal of radioactive waste in whatever form as will make it possible to determine whether the implementation of such plan is liable to result in the radioactive contamination of the water, soil, or airspace of another Member State.
>
> The Commission shall deliver its opinion within 6 months, after consulting the group of experts referred to in Article 31.

The European Court of Justice (ECJ) considered this provision pursuant to a reference for a preliminary ruling from an administrative tribunal in Strasbourg. In 1986, French inter-ministerial orders were adopted authorising the disposal of radioactive waste from the Cattenon nuclear power station in the Moselle, prior to the provision of data concerning the disposal operation to the Commission. This was provided some 2 months later after the authorisation had been granted, although prior to any actual disposal.

The ECJ adopted a purposive approach in construing Article 37 and stressed the importance of the Commission's opinion. Given the expert assistance upon which that opinion was based and the wider implications for the development of the nuclear industry throughout the Community, the Court held that it must be brought to the notice of the Member State that the opinion of the Commission should properly be taken into account by the Member State and further that it would be more difficult to take suggestions into account if authorisations had already been granted. Finally, the Court emphasised the significance of the Commission's opinion for those assessing the merits of legal action against a decision to authorise disposal. This is an important case, particularly in respect of the continued nuclear testing carried out by France.[78]

EURATOM has the advantage over the IAEA inasmuch as it can give legal force to its safety measures and it benefits from a wider consultation requirement in the event of trans-boundary risk. However, faced with the reluctance of some Member States to allow the Community to regulate nuclear power more comprehensively, the main protection against nuclear risks is that of consultation and notification. Consequently, as far as the regulation of environmental nuclear risks are concerned, the EURATOM Treaty has proved to be little more effective than the IAEA Statute, despite its apparent advantages.

6.16 NUCLEAR INSTALLATIONS IN THE UNITED KINGDOM

A number of regulatory bodies in the United Kingdom are concerned with operations involving radioactive substances. Consents for discharges from nuclear installations are obtained from the Ministry for Agriculture Fisheries and Food in England, the Environment and Heritage Service in Northern Ireland, the Secretary of State for Wales, and the Secretary of State for Scotland. The Health and Safety Executive (HSE) is responsible for the exposure of workers to ionising radiation and the Environment Agency is responsible for the discharges of radioactive waste.

The Nuclear Installations Act 1965[79] sets out the regime for the authorisation and control of the operation of nuclear facilities, with details set out in the Nuclear Installations Regulations 1971.[80] The facilities covered include installations for the manufacture of fuel elements from enriched uranium, plutonium, or any material containing either of the above:

1. Any installation for the production of enriched uranium or of any alloy, chemical compound mixture, or combination containing enriched uranium or any material containing enriched uranium.

2. Any installation for the production of plutonium or of any alloy, chemical compound mixture, or combination containing enriched plutonium or any material containing plutonium.
3. Any installation for the incorporation of enriched uranium, plutonium, or any material containing either of the above, into any device that forms part of a nuclear assembly or that is destined for use in a reactor.
4. Any nuclear assembly in which a controlled chain reaction can be maintained with an additional source of neutrons, which contains enriched uranium, plutonium, or any material containing either of the above.
5. Any installation for the reprocessing of irradiated nuclear fuels.
6. Any installation designed for the storage of fuel elements; irradiated nuclear fuel; bulk quantities of any other radioactive matter that has been produced for use in nuclear fuel.
7. Any installation for the treatment of irradiated matter by extraction of plutonium or uranium and any treatment of uranium that increases the proportion of isotope 235 in that matter.

The statute gives the HSE the power to not only grant a licence and attach any conditions it deems necessary to ensure safety but also the power of revocation of the licence at any time. If the licence has been revoked, surrendered, or has expired, then a duty remains with the former licensee and only expires when the HSE deems that there is no longer any danger from ionising radiation from anything on the site, or a new licence has been granted for the site.[81] The HSE is also responsible for consulting public bodies prior to the grant of a licence.

The Statute also makes extensive provision for the inspection of sites[82] and for compensation for any damage or injury caused. The details of a licence must remain on the public register for 30 years after the end of the period of duty.[83]

6.16.1 Radioactive Waste

In the 1950s, when the EURATOM Treaty was drawn up, nobody was very concerned about the wastes produced by the operation of a nuclear power plant. Certainly no one foresaw that the management of radioactive waste would play such a crucial role in the future of the nuclear sector or in the energy/environment debate. Consequently, there is limited reference to the subject in the Treaty. Nevertheless, current Community activities grew out of the application of Chapter I of the Treaty,[84] and, to a lesser extent, Chapter II.[85] These activities were carried out by DG XII (research) and the Joint Research Centres. Given the growing importance of this area, DGXII became increasingly involved in aspects of policy and strategy as well as pure research. Eventually, by the 1990s, the non-research activities were transferred to the more appropriate DG XI,[86] therefore, highlighting the overriding principles of protection of man and the environment. A Council Resolution on December 19, 1994 helped define the Community strategy in this field and a new Plan of Action from the year 2000. This was updated by the Council Directive 2011/70/Euratom establishing a Community framework for the responsible and safe management of spent fuel and radioactive waste.

The Radioactive Substances Act 1993 consolidated the law relating to the use and handling of radioactive substances in the United Kingdom. It provides for the regulation of the disposal of radioactive waste, from whatever source, and provides for the registration, with the appropriate Agency,[87] of any use or disposal of radioactive materials.

Sections 13 and 14 provide for the prohibition of the disposal and accumulation of radioactive waste unless an authorisation has been granted. The exemptions include where the accumulation of the radioactive waste is on a nuclear site that is licensed under the Nuclear Installations Act 1965. Any disposal of waste from a nuclear power station must be authorised under the Act and includes operational emissions of wastes to air, water, and land. This duty is extended to persons who receive radioactive waste for disposal.[88] However, the accumulation of radioactive waste does not require authorisation, if consent to dispose of the waste has been given and the accumulation was prior to the disposal.[89]

6.16.2 Transport of Radioactive Substances

The Radioactive Material (Road Transport) Act 1991 provides for the Secretary of State to make regulations to ensure the safe transport of such materials.[90] The 1996 Regulations require that all such material transported by road comply with the very detailed technical regulations for such transport. The statute also provides for wide powers of inspection and may issue a prohibition notice on all those who are in breach or are likely to be in breach of the Regulations.[91]

Under the EURATOM Directive,[92] there is a requirement that Member States set up a system of monitoring and control over shipments of wastes across national frontiers. This is transposed into UK law under the Transfrontier Shipment of Radioactive Waste Regulations 1993.[93] The Regulations apply to all shipments into and out of the Community in addition to movements between Member States. Essentially, the Regulations require the producer of the waste to give prior notification to the appropriate agency in each state through which the radioactive waste will travel as well as the appropriate agency in the state of disposal. The receiving state must also give assurance that they have the proper means of disposal prior to the waste starting on its journey. The originating state must be notified of the arrival of the waste at its final destination. This is to prevent the recurrence of situations where states who were unaware that such waste was to cross their borders refused to allow the consignment through upon its arrival at the border.

6.17 DECOMMISSIONING OF NUCLEAR INSTALLATIONS IN THE EUROPEAN UNION

The guiding principles for 'the protection of the health of workers and the general public against the dangers arising from ionising radiation have been established in Article 2b of the EURATOM Treaty, thus leading to Chapter 3 calling for the editing of Basic Safety Standards.

These were worked out by the Commission and their last version was published in May 1996 and are now updated under the 2011/70/Euratom Directive of July 19,

2011.[94] Further important documents on radiation protection from the Commission that are also applicable during decommissioning activities in the broadest sense are the Council Directive on the operational protection of outside workers[95] and the recommended radiological protection criteria for the recycling of metals from the dismantling of nuclear installations.[96]

Regarding the protection of the workers against the conventional risks incurred during decommissioning works, it is for each Member State to have in place domestic legislation and rules for general industrial application.

As far as the definition of decommissioning works are concerned, a list of specialities is identified.[97] These activities were evaluated by the Group of Experts[98] who considered that modern radiation protection must involve the use of the as low as reasonably achievable (ALARA) principle. This means that even if the incurred individual doses are below national, legally binding dose limits, they should also follow ALARA principle. However, economic and social factors must also be taken into account. Under Article 6 of the Council Directive,[99] each Member State shall ensure that ALARA is implemented. Merely issuing a regulation or changing a legal text does not implement ALARA—it is the teamwork done at the plant under the responsibility and support of the management and the supervision of the national authorities.

6.18 DECOMMISSIONING IN THE UNITED KINGDOM

The Government departments involved in the development of policy on decommissioning in the United Kingdom are the Department for Environment, Transport and the Regions (DETR), the Welsh Office, the Scottish Office, the Department of Trade and Industry (DTI), the Ministry of Defence, and the Ministry of Agriculture, Fisheries and Food. The Regulations covering the decommissioning of nuclear facilities are laid down in the site licence, which is issued by the HSE, who are also responsible for their enforcement. The licensee is responsible for the development of decommissioning strategies, plans, and the technology, and also for the implementation of decommissioning.

The site licence is all embracing, and covers all aspects of activities on the site, including both the operation of the facilities and their decommissioning. Therefore, the operating licence covers decommissioning operations, and is not transferred at the start of decommissioning as no separate decommissioning authority exists. Under the general licence conditions, the plant design now takes decommissioning into account.[100]

The published policy on decommissioning is that a case-by-case system is used. The period for decay is decided on a case-by-case basis between the licensee and the regulators. The IAEA Stages are not defined nationally as each operator defines how closely his strategy corresponds to the IAEA Stages. It is not even obligatory to achieve Stage 3 and the end result of a specific project is negotiable.

The release of material is regulated by statutory instruments.[101] The radiometric checks are subject to an inspection regime that includes the management system associated with its release and radiometric analysis of the materials by a third party. It is also standard practice for appropriate financial arrangements to be put in place to cover the costs of decommissioning civil nuclear plants and as such, nuclear companies have a duty to make full provision for this in their accounts.

There is presently no legal requirement for public involvement, although this will change when Directive 97/11/EC comes into force. However, nuclear companies do maintain contacts with the local communities and try to keep them informed.[102]

ENDNOTES

1. President Eisenhower (1945), *Atoms for Peace Address*, GAOR 8th Session, 470th meeting, paras 79–126; Agreed Declaration on Atomic Energy, Washington (1945) 1 *UNTS* 123 (US, Canada, UK); UNGA Resolution 1(1), 1945.
2. Statute, Articles 111(1)–(4), amended (1961) 471 *UNTS* 334.
3. UNGA Resolution 32/50, 1977.
4. *A Programme of Nuclear Power*, HMSO: London, Cmd. 9389, 1955.
5. Ibid, p. 12.
6. A/Conf. 48/14/Rev. 1 (1972), Rec. 75, *Action Plan for the Human Environment.*
7. LDC Resolution 14(7), 1983 and Resolution LDC 21(9), 1985. Twenty-five states voted for Resolution 21(9), six voted against, and six abstained. At their 13th Consultative meeting in 1990, the parties to the LDC called for a disposal of radioactive wastes into seabed repositories to be included in the moratorium.
8. See Chapter 9.
9. *IAEA General Conference*, Special session, 1986, IAEL/GC (SPL,1)4.
10. European Parliamentary Assembly Rec. 1068 (1988).
11. Birnie & Boyle (2002), *International Law and the Environment*, Chapter 9, pp. 452–500.
12. MARPOL Convention, 1973 and 1978; UNCLOS 1982 Articles 211, 217, 218, and 220.
13. Jack Barkenbus (1987), 'Nuclear power safety and the role of international organization' at p. 482 in *International Organization* 41(3), June 1987, pp. 475–490.
14. See Chapter 5.
15. Article XII of the MARPOL Convention 1978.
16. Safeguards are measures of verification to ensure that assistance provided by the Agency or under its supervisory control is not used in such a way as to further any military purpose.
17. Reyners & Lellouche (1988), 'Regulation and control by international organisations in the context of nuclear accident', Cameron et al., Editors, *Nuclear Energy Law After Chernobyl.* Graham & Trotman publishers and International Bar Association.
18. Birnie & Boyle, ibid.
19. See Section 6.11.1.
20. The Steering Committee is the governing body of the Agency.
21. In the dispute over nuclear testing in the Pacific, France accepted 'its duty to ensure that every condition was met and every precaution taken to prevent injury to the population and the fauna and flora of the world' (note to New Zealand of 19 February 1973, *French Nuclear Testing in the Pacific*); Note 88, Birnie & Boyle, ibid, p. 358.
22. Committee on the Peaceful Uses of Outer Space, *Report of the Legal Subcommittee*, UN Doc. A/AC. 105/430, Annex III; Safety of Life at Sea Convention 1974, Annex, Chapter 8 and attachment, 19 April 2002.
23. See Section 6.16.2.
24. 'A local peace campaigner and the reading-based national nuclear awareness-raising group (NAG) have been granted permission to bring a case for judicial review to challenge the continued manufacture of trident warheads. The decision being challenged through judicial review is that of the Environment Agency to grant authorisations for radioactive discharges to the Atomic Weapons Establishments sites at Aldermaston and Burghfield', Press Release, 10 May 2000.
25. 1959 Antarctic Treaty, Article 5.

26. 1986 Noumea Convention for the Protection of the Natural Resources and Environment of the South Pacific, Article II; 1989 Protocol on the Protection of the South East Pacific Against Radioactive Pollution.
27. The Bamako Convention on Transboundary Movement of Hazardous Waste in Africa.
28. Kirgis (1972) *AJIL* 66.
29. *Treaty Banning Nuclear Weapons Tests in the Atmosphere, in Outer Space and Under Water*, 1963.
30. *Nuclear Test Cases (Australia v France)*, ICJ Rep. (1973), 99 (1973) (Interim Measures); ICJ Rep. (1974) (Jurisdiction); *New Zealand v France*, ICJ Rep. 135 (1973) (Interim Measures).
31. Handl (1975), *AJIL* 69, 50.
32. Handl (1988), *RGDIP* 92, 55.
33. *Basic Standards for Radiation Protection*, Safety Series no. 72, 1082nd edition. IAEA: Vienna.
34. This is a private association of experts set up to provide guidance in the field of radiation protection.
35. Statute of the International Court of Justice, Article 34.
36. Principle 21 of the Stockholm Declaration on the Human Environment provides 'States have, in accordance with the Charter of the United Nations and the principles of international law, the sovereign right to exploit their own resources pursuant to their own environmental policies, and the responsibility to ensure that activities within their jurisdiction or control do not cause damage to the environment of other States or of areas beyond the limits of national jurisdiction'.
37. This Convention was agreed under the auspices of the OECD and came into force on 1 April 1968. See Cmnd. 1211.
38. Certain exceptions such as war or natural disasters may be allowed under the Vienna Convention, Article IV; Paris Convention, Articles 3 and 9; and Brussels Convention on Nuclear Ships, Articles II and VIII. The negligence of the victim may also mitigate the degree of liability.
39. Vienna Convention, Article II; Paris Convention, Article 3; Brussels Convention on Nuclear Ships, Article II.
40. Vienna Convention, Articles V and VI; Paris Convention, Articles 7 and 8; Brussels Convention on Nuclear Ships, Articles III and V.
41. 1963 Convention Supplementary to the Paris Convention, with additional Protocols.
42. Vienna Convention, Article VIII; Paris Convention, Article 13; Brussels Convention on Nuclear Ships, Article X.
43. Vienna Convention, Article IV(3)(b); Paris Convention, Article 9.
44. This is effected in state law under the FRG Atomic Energy Act 1985.
45. International Expert Group on Nuclear Liability. *IAEA Conference on Civil Liability*. Vienna, 7th–11th July 2003, p. 67.
46. Vienna Convention, Article I; Paris Convention, Article I.
47. Vienna Convention, Article II; Paris Convention, Article 4.
48. Article XIII.
49. Vienna Convention, Article XI; Paris Convention Article 13.
50. Article XI(3).
51. Article 13(c).
52. Article X.
53. There are limited exceptions under Vienna Convention, Article XII; Paris Convention, Article 13(d); and Brussels Convention on Nuclear Ships, Article XI(4). However, these do not permit reconsiderations of the merits of the case.
54. 1968 and 1978 Conventions on Civil Jurisdiction and the Enforcement of Judgements.

55. FRG Atomic Energy Act 1985; UK, Nuclear Installations Act, 1965; Switzerland, Act on Third Party Liability, 1983.
56. 'Damage' is restricted to loss of life, personal injury, or loss or damage to property.
57. See Birnie & Boyle, p. 410.
58. ILO Convention (no. 115) concerning the protection of workers against ionising radiations, June 1960. In force from June 1962, 431 *UNTS* 41.
59. See Ambassador Tuiloma Neroni Slade (1996), 'The 1995 NPT review and extension conference', *Review of European Community and International Environmental Law* 5(3), 246–252.
60. A good example is the agreement between France and Belgium on Radiological Protection concerning the Installations of the Nuclear Power Station of the Ardennes, March 1967, 588 *UNTS* 227.
61. The High Seas Convention, 1958, Article 25(1); the Roraonga Treaty, 1985, Article 7; the Noumea Convention, 1986, Article 10(1).
62. Philippe Sands (1996), 'Observations on international nuclear law ten years after Chernobyl', *Review of European Community and International Environmental Law* 5(3), 199–204.
63. *Response to a Radioactive Materials Release Having a Transboundary Impact*, IAEA Safety Guides, Safety Series no. 94. STUI/PUB/814, 1989. IAEA: Vienna.
64. Birnie & Boyle, p. 459.
65. *Convention on Early Notification of a Nuclear Accident*, Legal Series no. 14, 1987. IAEA: Vienna.
66. *Convention on Assistance in the Case of a Nuclear Accident or Radiological Emergency*, Legal Series no. 14, 1987. IAEA: Vienna, 61.
67. Simon Carroll (1996), 'Transboundary impacts of nuclear accidents: are the interests of non-nuclear states adequately addressed by international nuclear safety instruments?' *Review of European Community and International Environmental Law* 5(3), 205–210.
68. Birnie & Boyle, p. 471.
69. Article 1(1).
70. Article 2(3).
71. International Law Commission (1956), *II Yearbook ILC*, United Nations, 286.
72. *Definitions and recommendations for the convention on the prevention of marine pollution*, Safety Series no. 78, Adopted at the 10th Consultative Meeting of the LDC, 1986. Vienna.
73. Birnie & Boyle, pp. 463–4.
74. LDC Resolution 14(7), 1983; LDC Resolution 21(9), 1985.
75. 'The Government have decided not to resume sea-disposal of drummed radio-active waste, including waste of military origin. Nonetheless, the Government intend to keep open this option for large items arising from decommissioning operations, although they have taken no decisions about how redundant nuclear submarines will be disposed of'. Hansard (1989), HC Debs, vol. 153, col. 464.
76. Preamble to the EURATOM Treaty.
77. Jurgen Grunwald, 'The role of EURATOM', Cameron et al., Editors, *Nuclear Energy Law After Chernobyl*. Graham and Trotman: London.
78. The geographic scope of the EURATOM Treaty is determined by Article 198, which states, 'Save as otherwise provided, this Treaty shall apply to the European Territories of Member States and to non-European territories under their jurisdiction'. It, therefore, applies to French Polynesia. Additionally, the Commission has accepted that Article 34 applies to military as well as civil activities. See Diemann & Betlem (1995), 'Nuclear testing and Europe', *National Law Journal* 145, 1236.
79. As amended by the 1969 Act.
80. S.I. 1971 no. 381.

81. Section 5(3).
82. The inspections are carried out by the HSE's Nuclear Installations Inspectorate.
83. Section 6.
84. On promotion of research.
85. On dissemination of information.
86. The DG for the Environment.
87. Since 1995, this has been the Environment Agency in England and Wales, the Scottish Environmental Protection Agency in Scotland, and the Alkali and Radiochemical Inspectorate in Northern Ireland.
88. Section 13(3).
89. Section 14(2).
90. Section 2.
91. Section 4.
92. EEC Directive 92/3.
93. S.I. 1993 no. 3031.
94. Euratom 96/29 (1996), Council Directive of 13 May 1996 laying down basic safety standards for the health protection of the general public and workers against the dangers of ionising radiation (O.J. L159 of 29 June 1996, p. 1).
95. Euratom 90/641 (1990), Council Directive of 4 December 1990 on the operational protection of outside workers exposed to the risk of ionising radiation during their activities in controlled areas (O.J. L349 of 13 December 1990, p. 21).
96. European Commission (1998), Radiation Protection 89—Recommended radiological protection criteria for the recycling of metals from the dismantling of nuclear installations, Rep.113, Brussels.
97. Cregut & Roger, *Inventory of Information for the Identification of Guiding Principles in the Decommissioning of Nuclear Installations*, Rep. EUR 13642 EN, Commission of the European Communities: Luxembourg, 1991.
98. European Commission (2007), 'Nuclear safety and the environment', Supporting documentation for the preparation of an EC Communication on the subject of decommissioning nuclear installations in the EU, P. Vankerckhoven, DG XI.C2.
99. Council Directive 96/29/EURATOM, Article 6.
100. For example, the design work and the overall safety case for Sizewell B power station take account of decommissioning.
101. 1986, no. 1002; 1992, no. 647.
102. Each of the Magnox Electric sites has Local Community Liaison Councils. At Trawsfynydd, a specific consultation exercise on decommissioning with the local community was conducted following the formal shutdown of the station.

Section III

Section III considers some of the main energy producer/user jurisdictions within which any energy company may operate. It looks at the more developed systems around the world and identifies some areas of good practice. It also looks at the new and emerging economies where the regulation of the energy sector is evolving but is still not yet complete. A number of case studies are used to identify problems and solutions that any enterprise operating in the energy sector needs to be aware of.

7 Energy Law in the United States

The first jurisdiction to be covered in this section of the book is the United States.[1]

Energy policy in the United States can be likened to 'catching smoke' as at one level the United States has no full scale, comprehensive, or coordinated national energy policy other than reports and several pieces of legislation called *energy policy* that address specific industries or specific problems. However, the United States has, in fact, a traditional identifiable energy policy built up over a number of years.

The main characteristic of this traditional policy is fossil fuel–based and supports large-scale, centralised energy industries. Each Presidential administration has expressed its own preferred energy policy reflecting the economic and political circumstances of the time. The energy policies of Presidents Nixon and Ford did not last much beyond their immediate causes, but the Carter and Reagan policies were similar in that they were both inconsistent with what Tomain and Cudahy describe as the dominant model.[2]

The dominant energy policy model requires support of the conventional resources and recognition that parts of the energy sector require regulation due to their dominance in the energy market. A mixed market economy requires a stable energy source and distribution to meet a constant energy demand. The US domestic energy policy is based on the assumption that a link exists between the level of energy production and the gross national product. As a consequence, domestic energy policy supports large-scale and centralised energy producers who rely on fossil fuels.

According to Tomain and Cudahay, this dominant model has the following general goals:

1. To assure plentiful energy supplies
2. To maintain reasonable prices
3. To limit the dominant market players
4. To promote competition between different types of fuel
5. To support conventional fossil fuels
6. To allow decision and policy making in the energy sector to develop through a federal state regulatory system

7.1 FEDERAL REGULATORY BODIES

The energy sector in the United States is regulated by the Department of Energy (DOE). The Department also contains within it a quasi-independent regulatory agency called the *Federal Energy Regulatory Commission* (*FERC*).

The DOE is a cabinet-level department that oversees the US policies on energy and safety in handling nuclear material. The remit of the DOE is broad and includes the US nuclear weapons programme, nuclear reactor production for the US Navy, energy conservation, all energy-related research, radioactive waste disposal, and domestic energy production.

The FERC is charged with implementing and enforcing the major statutes regarding the energy sector including the Federal Power Act, the Natural Gas Act, the Natural Gas Policy Act, and other federal energy legislations. However, the main federal statute governing procedures that must be followed by all federal agencies is the 1946 Administrative Procedure Act (APA) 5 USC §§551 et seq. This provides a procedural framework for the numerous agencies and groups provided for under the 'New Deal'.

Under the APA, agencies have two main areas of authority, one is adjudication of dispute resolution, whereby the APA provides for hearings for adjudication rather than legislative proceedings. The second is rulemaking, which includes ratemaking for electric and gas utilities. This is forward looking and is classified as rulemaking. The APA is applicable to all of the activities of the DOE but there are certain modifications of the APA-approved procedures under the DOE Organization Act. Under these modifications, if a proposed DOE rule is likely to have a substantial impact on the national economy, then there must be an opportunity for oral presentation of comments, data, and arguments. Likewise, if the Secretary determines that a substantial issue of fact or law exists, then again the opportunity for oral presentation must be available.[3] However, due process would be satisfied by the opportunity for submission of written comments. The Secretary of State is permitted to waive these requirements should strict compliance be likely to cause harm or injury to public health, safety, or welfare. However, these safeguards do not apply to the FERC, whose procedures are covered by the APA.

The FERC has assumed the powers and duties of the previous regulatory body; the Federal Power Commission. There are five commissioners appointed by the President and confirmed by the Senate, with no more than three from one particular political party. The FERC staff[4] are organised into a number of different offices, including an Office of Energy Projects.[5] The Office of Tariffs and Rates relates to electric, natural gas, and oil pipeline facilities and services.[6] However, the FERC has traditionally been involved in exercising its powers under the Federal Water Power Act, the Federal Power Act, the Natural Gas Act, the Natural Gas Policy Act, and other energy-related statutes.

In the area of ratemaking, sections 205 and 206 of the Federal Power Act provides for wholesale rates for the regulated electricity utilities to be regulated by the FERC. Similarly, the rates of regulated natural gas companies are subject to FERC regulation under sections 4 and 5 of the Natural Gas Act.

It was the Supreme Court decisions in *United Gas Pipe Line Co. v Mobile Gas Service Corp.*,[7] and *FPC v Sierra Pacific Power Corp.*[8] that produced the Mobile Sierra doctrine, which declares that rates negotiated in a freely negotiated contract are presumed to be 'just and reasonable' as prescribed by the Federal Power Act and the Natural Gas Act, and therefore, the FERC cannot impose changes unless the contract price 'seriously harms the public interest'. The principle was recently reaffirmed in *Morgan Stanley Capital Group, Inc v Pub. Util. Dist. No. 1*.[9] This case arose from the inflated prices emerging from the California energy crisis due to illegal activities causing a market dysfunction. The Court concluded that 'the mere fact of a party's engaging in unlawful activity . . . does not deprive its forward contracts of the benefit of the Mobile Sierra presumption'.

Although there was an effort to consolidate the principal energy programmes of the government under the DOE Organization Act, a number continued to remain outside the new department, the most prominent being the Nuclear Regulatory Commission (NRC). The NRC had previously come out of the former Atomic Energy Commission (AEC) because of concerns that it was required to both regulate and yet promote the nuclear industry. This created a conflict of interest, and so its functions were divided between the NRC and the Energy Research and Development Administration. It was also recognised that there were concerns about the safety of nuclear power plants and that this function should not be absorbed into a more broadly based regulatory body.

The US Environmental Protection Agency (EPA) also has a major impact upon the energy sector. The EPA has the duty to develop and implement environmental laws enacted by Congress. It is responsible for setting standards for a variety of the environmental programmes, but delegates to the states and tribes the responsibility for issuing permits and for monitoring and enforcing compliance. Given that the development and use of energy may have major environmental impacts, the work of the EPA has a great influence on the production, transportation, and distribution of energy. The mission of the EPA is to protect human health and the environment, and its principal duty is to assure an adequate and reliable source of energy without any adverse effects upon the environment. In 2007, in the case of *Massachusetts v EPA*, 549 US 497, the Supreme Court held that the EPA not only could but should, under its existing Clean Air Act (CAA) authority, regulate greenhouse gases to avoid undesirable climate change. Thereafter, the EPA issued regulations for electric utilities and other energy producers who were major emitters of greenhouse gases.

The Department of the Interior (DOI) is another major agency whose mission is the provision of wise stewardship of energy and mineral resources. The DOI manages about one-fifth of the land in the United States and about 28% of the nation's energy production through federally managed lands and offshore areas. The DOI encompasses a number of bureaus, including the Bureau of Reclamation (BOR), the Bureau of Land Management (BLM), and the Office of Surface Mining (OSM).

The Department's Minerals Management Service was roundly criticised as a result of the Deepwater Horizon oil spill off the coast of Louisiana. The Service had long been accused of being captured by the oil and gas companies to whom it granted permits for offshore production. It was reorganised and renamed the Bureau of Ocean Energy Management, Regulation and Enforcement.

7.2 ENERGY REGULATION BY THE STATES

The historical roots of state regulation came out of the common law and the occupational guilds in England to provide service to all comers at reasonable prices. When the common law courts displaced the authority of the Guilds, manors, and towns, certain occupations, which became known as *common callings*, were distinguished due to the fact that they were 'affected by a public interest' and were thus regulated by the English Parliament.

This principle was developed in colonial America and prices for products such as beer, bread, and corn were regulated. After the 1812 war, it was decided that competition was the best form of control for the common welfare, but after the Civil War, the principle of the public interest was reassessed under the general view that competition did not seem to be working. In the 1870s, a commission was set up to regulate the monopolies of the railroads and in 1877, the Supreme Court decided in the case of *Munn v Illinois*[10] that the state of Illinois could regulate the rates of grain elevators in Chicago under Lord Hale's theory that these were businesses 'affected with a public interest'. After 1890, however, the tide of public and judicial opinion turned yet again and *laissez faire* ruled once again. Nevertheless, certain enterprises called *public utilities* were still subject to public regulation.[11]

This power remains with the states through public service commissions and is of major importance in the energy sector. There was, in fact, no legal authority for federal regulation of electric power until 1935 and the Federal Power Act.

It is the state commissions who have the authority to issue licences, franchises, or permits for the initiation of service, and for construction or decommissioning of facilities. The commissions may require prior authorisation for rate changes, and may suspend proposed rate changes. They may prescribe interim rates and initiate rate investigations. In addition, most commissions have the authority to control both the quality and quantity of any services provided. They may prescribe uniform systems of accounts and require annual reports.

State control over efforts to reduce the emissions of greenhouse gases to mitigate climate change is an important area for the energy sector. Some states have set up cap-and-trade systems in addition to building codes for energy efficiency. California is in the forefront, having established its own environmental standards, including a renewable portfolio standard (RPS) and building and appliance efficiency codes. In 2006, the California legislature adopted robust greenhouse gas restrictions designed to bring emissions down to 1990 levels by 2020. In the 2010 election, the voters of California rejected a proposition to repeal the legislation.

7.2.1 Interrelationship between Federal and State Policy and Regulations

The boundary lines between federal and state regulation of public utilities is now well established. The Federal Power Act preserves state jurisdiction through the regulatory commissions. In the case of natural gas, however, federal regulation is inclusive as it covers gas transmission and only local distribution is reserved to the states. Electric power is primarily under state control but due to further deregulation, electricity is increasingly under federal oversight.

7.3 CONSTITUTIONAL PRINCIPLES AND REGULATORY JURISDICTION

When considering the allocation of regulatory powers between the Federal government and the states, according to Tomain and Cudahy, certain constitutional principles must be recognised. The main principles to be considered are the Commerce Clause, the Supremacy Clause, and the Takings Clause of the Fifth Amendment to the states.

To consider the Commerce Clause of the US Constitution first, this gives Congress the power to regulate commerce with foreign nations, between the states and with the Indian tribes. It is an affirmative grant of Congressional authority, which results in a limitation on the state regulatory powers. It restricts the states from enacting legislation that may be anticompetitive as regards out-of- state actors. This is of particular interest to the energy sector, which primarily produces energy resources and sells them locally; these are then transported and sold out of state. Therefore, these transactions are regulated by two or more regulatory bodies, which may well result in market disruption.[12]

Congress, however, does not have authority under the Commerce Clause to regulate anything that is not related to commerce and no authority to do anything about commerce, except to regulate it.[13]

The Supremacy Clause of the Federal Constitution provides that the laws of the United States shall be the supreme law of the land. As an act duly and legally passed by Congress, becomes part of the law of the United States, any state legislation that is in conflict with that law is null and void.[14] However, this is only the case when the scope of the federal statute actually states that Congress intended the federal law to cover the whole field in question.[15] In matters concerning the electricity sector, Congress has taken great care to avoid infringing the prerogatives of the state commissions and their jurisdiction, notwithstanding the exclusive jurisdiction over transmission in interstate commerce of FERC under the Federal Power Act.[16]

The Fifth Amendment to the Constitution is also applicable to the states under the Fourteenth Amendment, which provides that private property shall not be taken for public use without compensation. Regulatory actions, within the energy sector, may permanently deprive private property owners of its use.[17] However, should the new regulation or permit merely deny the owner part of the use of his land and it could be shown that the owner did not enjoy the full rights, including those covered by the regulation or permit, then the state may resist any payment of compensation, but this would be in very limited circumstances.[18]

In *Duquesne Light Company v Barash*, 488 US 299 (1989), the light company had made expenditure for electricity- generating facilities that were planned but not completed. The state fixed the rates for electricity without including the expenditure.

The Supreme Court held that:

> The Supreme Court of Pennsylvania held that such a law did not take the utility's property in violation of the Fifth Amendment to the United States Constitution. We agree with that conclusion and hold that a state scheme of utility regulation does not "take" property simply because it disallows recovery of capital investments that are not "used and useful in service to the public.

Notwithstanding this decision, 'takings claims' are a major arrow in the quiver of the energy lawyer acting for a regulated company.

7.4 REGULATION FOR THE DIFFERENT ENERGY SECTORS

The US energy policy with regard to oil has been one of 'light touch regulation'. This is mainly because the petroleum sector purports to be fully competitive and no one company enjoys a monopoly.

In 1970, domestic oil production peaked and thereafter the United States was reliant on imports, which were relatively cheaper to produce. However, since Presidents Nixon's Project Independence, the policy has been to be free of oil imports. This has been difficult to achieve as the exploration for and production of petroleum presents a number of environmental problems, particularly after the BP disaster in the Gulf of Mexico. Additionally, partisan political disagreements over opening up public lands, such as the Arctic National Wildlife Refuge (ANWR) to oil and gas interests has slowed domestic development.

7.5 STATE REGULATION OF OIL AND GAS

A characteristic of oil *in situ* is that it does not recognise the boundaries of the surface owner. Wells must be drilled to explore for the location of the oilfields and the extent of those fields may not be wholly known, but merely estimated. Therefore, special property law regimes have been developed in the United States. No one owner will be able to finance an oil exploration project and so will grant a license for the mineral rights, at a fee, to an exploration company, who will acquire a number of licences from adjoining property owners. Originally, as in most common law countries, the United States operated the *ad coelum* doctrine, which meant that the surface owner owned all that from the surface of the land to the centre of the earth and to the sky. This did not work well for oilfields and migratory minerals because of the uncertainty of the parameters of the property right, and so the United States adopted the rule of capture, which operates in the global commons of the high seas. Ownership is not complete until the operator 'takes the property into possession'. This rule inevitably created its own problem inasmuch as an operator may drain, by taking into possession the neighbour's oil as well as that from the property where the drilling rig was placed without liability. In the case of *Barnard v Monongahela Natural Gas Co.*,[19] it was held that once extracted, the title to the oil was vested in the person who had captured those resources, as long as the operator did not trespass on the surface of the neighbour's land.

This had the effect of encouraging overproduction due to the race to capture the oil. The rule of capture has now been modified and applied differently in different states. For example, the state of Texas adopted the principle of correlative rights, which permits the adjacent surface owner the right to a fair and equitable share of the oil and gas resources in conjunction with the right of protection from negligent damage to his property.

During the early part of the twentieth century, the overproduction of oil exceeded the demand of consumers and also the capacity of the pipelines to transport it. The oil was, therefore stored in 'lakes' on the surface of the land, which caused contamination and was subject to fire and seepage.

The states tried to mitigate this situation through the state oil and gas conservation legislation, which introduced provisions for pooling, unitisation, well spacing, and prorationing. Texas, Louisiana, and Oklahoma attempted to keep the oil industry viable by passing legislations to insure that the market was not flooded.

However, since the 1970s, with domestic production having peaked, all states have introduced a maximum permitted oil production level in various forms. In addition, the states have introduced a maximum efficient rate regulation (MER).

7.5.1 Federal Regulation

By the 1970s, imports of cheap oil also exceeded domestic production. The states of the Middle East came to the conclusion that they were getting very little in return for their finite mineral resources and formed the Organisation of Petroleum Exporting Countries (OPEC) and took more control of their natural resources (Chapter 3). By the 1950s, the major international oil companies were importing large amounts of still relatively cheap oil, thus enjoying a competitive edge over the domestic independents. This caused concern for the Federal government as the situation was perceived to be damaging the domestic oil producers. In 1959, the Federal government set up the Mandatory Oil Import Programme (MOIP) by Presidential Proclamation under the original Trade Agreements Extension Act 1958.[20] This had the effect of raising domestic oil prices and depleting the national oil reserves.

7.5.2 Federal Public Land

It is the Department of the Interior (DOI), through the Bureau of Land Management (BLM) that has jurisdiction over oil and gas deposits in public land, which cover over 600 million acres and consists of half of the land west of the Mississippi river.

Originally, it was the 1920 Minerals Leasing Act[21] as amended by a number of different sections, which classified oil as 'a leasable mineral'. The statute also reserves a royalty to the government on the production of oil and gas located on public lands. This statute was reformed in 1987 by the Federal Onshore Oil and Gas Leasing Reform Act.[22] The reforms took the shape of changing the leasing requirements, in particular in respect of the requirement that all federal onshore lands must be offered for lease under competitive conditions. Both of these statutes currently operate in unison, and federal-owned lands that may not be leased must be identified.

It is the Secretary of the Interior who administers the Mineral Leasing Act through the BLM, which issues leases, both competitive and non competitive, for oil, gas, and any other mineral that comes under the jurisdiction of the DOI. In some circumstances, the BLM may issue leases for lands under the jurisdiction of the Forest Service, an Agency of the Department of Agriculture.

Those lands that are available are open for bidding, with the lease going to the highest bidder. Once the lease has been granted, for a price, there is an annual rental that increases after the first 2 years.[23] Originally established by Congress, these rentals are raised by Regulation. Should oil or gas be discovered within the leased area, the lessee will pay a minimum royalty of not less than what the rental would be, which normally works out at 12.5% royalty to the Federal government on all discovered resources, with the state receiving half of the income. There is one exception to this rule, which is Alaska, where the state receives 90% of the income.

Lease terms are 5 years for competitive leases, and if the lessee has not produced any oil or gas within that period, the lease automatically expires. If, however, production has not started but drilling has, then the lease may be extended for a further 2 years.

7.6 OFFSHORE OIL AND GAS

In the United States, offshore normally refers to the outer continental shelf (OCS). As we have considered earlier in this book, the continental shelf is part of the seabed that lies adjacent to the coastal state. This area is estimated to be approximately 100 billion acres under the jurisdiction of the United States. However, unlike offshore oil and gas development in Northern Europe, most of the realised offshore oil and gas come from off the coast of California and within the Gulf of Mexico.[24] In terms of total production, offshore only contributes 25% of the domestic oil produced. As offshore technology is developed in the North Sea and elsewhere in the world, the United States is looking further out into deeper waters to exploit known oil and gas reserves.

As was mentioned in Section I of this book, it was President Truman in 1945 who first signed a Proclamation on the 'Policy of the United States with Respect to the Natural Resources of the Subsoil and Sea Bed of the Continental Shelf', which was enacted into domestic law through the Submerged Lands Act (SLA)[25] and the Outer Continental Shelf Lands Act (OCSLA).[26]

Offshore jurisdiction is divided among the states under the SLA over the seabed up to 3 mi beyond the low watermark along the coast,[27] and federal jurisdiction under the OCSLA, which defined the OCS as 'all submerged lands lying seaward and outside of the area of lands'.[28] This would, under international law, extend to 200 nautical miles beyond the coastal baseline. See also cases *United States v Louisiana*[29]; *United States v Texas*[30]; *and United States v California*.[31] Congress itself was more specific inasmuch as under the Statute, the Secretary of the Interior is responsible for outer continental (OC) leases, which are granted by the Minerals Management Services in the Department.

It was in 1978 under the Amendment to the OCSLA that the states were given greater authority over leasing as the amendment provided more coordination with the 1972 Coastal Management Act.[32] Under the Amendments, the Secretary of the Interior is required to conduct environmental studies for the areas as well as the management of the leased areas. The Secretary is also required to collect, maintain, and publish data resulting from exploration, development, and production on the OCS, for public dissemination.

The core of the OCSLA amendments is the leasing system. This is similar to the licensing systems operating in the North Sea but with some differences. Both regimes, however, try to balance a number of competing interests including commercial, public, and environmental interests.

Firstly, an environmental impact assessment is called for under the 1969 National Environmental Policy Act.[33] Once the final environmental impact statement (EIS) has been published, a Notice of Proposed Sale is then followed by a notice under the bidding system. There are a number of systems that the Secretary may use under the 1978 Amendments, which involve both cash payments and percentage royalties. The leases provide the authority for exploration and development and the 'bids are solicited on the basis of a cash bonus with a fixed royalty, a royalty bid with a fixed cash bonus, or various combinations of the two'.[34] This system differs from the UK and Norwegian systems inasmuch as the governments set the royalties and the corporation tax, so that the bid by the oil company is just that; a bid for the licence, as in an auction.

The US system is designed to balance a number of competing interests including, economic, social, and environmental interests.[35]

Although the Federal government has wide powers regarding leases, it also has obligations. In the case of *Mobil Oil Exploration and Producing Southeast, Inc v United States,* 530 US 604 (2000), two companies paid $156 million to the government for lease contracts that gave them the right to explore for and develop oil off the coast of North Carolina. However, the government denied them the opportunity to acquire additional permits that were necessary. The oil companies sued for restitution of the monies paid for the leases they were *de facto* unable to operate. The Supreme Court held that the government broke its promise and thus repudiated the contract and the fees were returned to the oil companies.

7.7 ENVIRONMENTAL CONCERNS

It was Rachel Carson's book, *Silent Spring*,[36] which heightened the awareness of the American people to environmental concerns, in addition to the Santa Barbara oil spill on January 28, 1969.[37] A plethora of legislation with regard to the protection of both onshore and offshore flora and fauna followed thereafter.

7.7.1 Areas for Concern

It is the pristine areas of the National Parks that are of most concern. The Arctic National Wildlife Refuge (ANWR) is one such area of concern. The ANWR, with 19 million acres that sit hundreds of miles north of the Arctic Circle next to the Arctic Ocean, with rolling tundra, migratory birds, and a very diverse wildlife, was designated a federal wilderness area by President Eisenhower. There are lagoons, barrier islands, large bays, estuaries, and thaw lake wetlands. The ANWR is considered by the World Wildlife Federation to be one of the few places on the Earth that protect the complete spectrum of sub-Arctic and Arctic habitats. Jeep tracks from decades ago are still visible on the ANWR tundra. On the other hand, there would be 2000 acres of oil pipelines and facilities projected to explore and exploit the 1.5 million acres of ANWR, which are recognised as containing potential oil and natural gas reserves. The DOI carried out an environmental impact assessment and in the statement produced concluded that oil development in the coastal plain would have serious impacts on the wildlife and the ecosystem. The previous George W. Bush administration was keen to go ahead with the development, reflecting the partisan energy debates in the United States. To put this into perspective, the total quantity of recoverable oil within the entire assessment area is estimated to be between 5.7 and 16 billion barrels. According to the Energy Information Agency, the ANWR would only produce 8,000,000 barrels a day once it reaches its peak of production 10 years after the start of production. The Environmental Impact Assessment estimated that this production would reduce the net share of oil imports in 2020 from 62% to 60%.

A further area of concern is offshore continental shelf exploration. The Department of the Environment has authority to develop the OCS in order to increase domestic production. However, the ecological dangers of offshore development are apparent and concerns have been expressed as to whether the environmental impact statements for offshore leasing satisfy the requirements of the Natural Environmental Policy Act (NEPA).

7.8 CASE STUDIES

In the case of *Commonwealth of Massachusetts v Andrus*,[38] the issue before the US Court of Appeals for the First Circuit was a preliminary injunction, which had been issued by the district court, which enjoined the Secretary of State for the Interior from proceeding with the proposed sale of oil leaseholds in the OCS off the coast of New England. The basis of the injunction was that necessary safeguards against oil spills were lacking. In response to this decision, Congress amended the Outer Continental Shelf Lands Act to create an oil spill fund.

The problem of oil spills, which can occur as a result of a number of operational activities, presents significant environmental risks. The main risks are blowout, pipeline leak, or tanker oil spill.

The tanker oil spill of the *Exxon Valdez* in Prince William Sound in March 1989 is probably the most notorious. The tanker was loaded with 53 million gallons of oil and was bound for Long Beach in California. The Captain had been drinking and the tanker ran aground at Bligh Reef, having ruptured eight of its 11 cargo tanks. 10.8 million gallons of crude oil were dumped into Prince William Sound. Fortunately, no lives were lost at the grounding but four deaths were attributed to the clean-up efforts.[39] However, the economic loss with regard to damage to fisheries, subsistence livelihoods, tourism, and the loss of wildlife was incalculable as the oil spill extended nearly 500 mi for 2 months after the grounding. Exxon were originally awarded $4.5 billion against them in the form of punitive damages.[40] However, this was held to be excessive by the US Supreme Court (USC) who remanded the case to the ninth Circuit. The lower court revised the damages to just over $500 million as the company had paid out large sums in clean-up costs.[41]

As far as oil spills are concerned, a number of cases have been decided with regard to clean-up costs by individuals, legal persons, and government.[42] Environmentalists obtained an injunction against the DOI from issuing leases for exploration of Alaska's North Slope prior to the EIS being supplemented in the case of *North Slope Borough v Andrus.*[43]

In response to the *Exxon Valdez* disaster, Congress addressed the problem of oil spills further in the Oil Pollution Liability and Compensation Act 1990.[44] This is a major piece of legislation, which requires reporting and clean-up, assigns liability, and addresses preventative action. Liability is initially placed with the owner or operator of the vessel or installation. This has the effect of reversing the burden of proof, and it is for the owner/operator to establish a statutory defence if the third party intervenes. In return for accepting this strict liability, the statute also contains limits on the liability. Owners and operators must have insurance to offset this liability or a surety bond, letter, or letter of credit. Under the statute, there are certain requirements with regard to preventative actions.

7.9 CASE STUDY OF THE BP OIL SPILL

The *Deepwater Horizon* oil spill is the most recent offshore disaster concerned with blowout. On April 20, 2010, a blowout of BP's Macondo well in the Gulf of Mexico led to the deaths of 11 workers on Transocean's *Deepwater Horizon* drilling rig with a further 17 injured. The fire, which was fed by hydrocarbons from the well, continued for 36 hours until the rig sank. Hydrocarbons continued to flow from the reservoir through the wellbore with the release of an estimated 4.9 million barrels of oil. The well head itself was located in a depth of water over 1500 m.

The number of actors involved in the disaster demonstrates the complexity of ownership of legal rights under the activity of the exploitation of oil. BP Exploration & Production Inc. was the lease operator of Mississippi Canyon Block 252, which contains the Macondo well. BP is owned 55% by British Petroleum and 45% by Amoco. The operating installation was owned by Transocean, which is a company registered in Switzerland.

The specific operation was being carried out by Halliburton, an American company who owned the seabed drilling bit and were responsible for the sealing activity. Unfortunately, this was carried out with a substandard concrete mix. A further provider was Cameron International who were the manufacturers of the blowout preventer.

So who does the 'damaged' party sue? Under the principle of 'Channelling', it is the primary owners of the operating licence who are responsible for all damages, but they may recover through contract law or tort if it is considered that those further down the 'chain' are at fault. In this respect, BP have registered three law suits against their contractors. BP is claiming $40 billion from Transocean, over claims that they acted negligently. BP have also registered claims against Halliburton and Cameron International. The respondent companies are claiming indemnity under their contracts.

Under the US legal system, it is likely that responsibility will be attributed mainly to those companies that had final decision making capacity, and if this is the eventual outcome it suggests that service companies and operators may have serious discussions in the future over the ultimate decision maker and where liability in this area lies.

The Oil Pollution Act 1990[45] is the main statute under which the primary case was brought and the first tasks were to establish procedures for identifying the party legally responsible for an oil catastrophe, provide a method by which the victims may file claims for damages against this party, and require industry and government to develop contingency plans for responding to such an event. The Act also caps liability of companies operating onshore and offshore facilities and oil tankers.

The trial in New Orleans is officially called *Multilitigation-2179* (MDL), and it consolidates 535 lawsuits originally filed all over the country. More than 111,000 individuals and businesses have filed notice to take part in MDL.

The plaintiffs include:

- Fishermen
- Seafood processors
- Restaurants
- Coastal landowners
- Individuals who were harmed by dispersants or oil
- The Federal government
- Gulf Coast states
- Municipalities including some from Mexico

The Gulf Coast Claims Facility (GCCF) was set up by BP as a $20 billion compensation fund in response to President Obama declaring that BP should pay.

Any plaintiffs who reach a settlement with GCCF must withdraw from MDL.

This trial does not address shareholder suits, which will be handled in Houston, nor any criminal charges.

The defendants are:

- BP, which held the lease on the Macondo well
- Transocean, which owned *Deepwater Horizon*
- Halliburton, which poured the cement lining into the well
- Cameron, which manufactured the blowout preventer that malfunctioned during the crisis
- Nalco, which manufactured the dispersants that are alleged to have made people sick and to have harmed the environment.

The President's Commission, which investigated the *Deepwater Horizon* oil spill, developed a series of recommendations for future oil and gas operations in deepwater offshore.

Although oil pollution will continue to be a cause for environmental concern the production and use of oil will necessarily play a major part in the energy mix both in the United States and worldwide.

7.10 COAL

Tomain and Cudahy describe coal in three words: abundant, dirty, and dangerous. This description would fit most coal operations throughout the world and the law and state policies reflect this evaluation. In the United States, coal accounts for 33% of the energy produced and accounts for over 50% of the electricity produced. One of the main problems with energy produced from coal is the negative external effect on the environment. Last, but not least, is the fact that coal mining is detrimental to human health as well as the environment. This was evidenced by the explosion of Massey Energy's Upper Big Branch mine in West Virginia, which killed 29 workers and became the United States' worst mining accident in the last 40 years.[46] Such health and safety issues have had notable effects on the regulation of the coal industry in addition to the industry investment in the development of 'clean coal' technology.

Coal-fired power stations are the largest contributors of greenhouse gas emissions in the United States,[47] and as regulators evaluate the potential costs of climate change regulation, the energy system in the United States could change profoundly.[48]

7.10.1 A Regulatory Overview

The Federal government owns approximately one-third of the coal reserves on federal lands and manages the activity under the DOI, through different agencies. It is the BLM who issue the licences for coal mining under an extensive statutory and regulatory process.

The key federal statutes are the Mineral Lands Leasing Act 1920[49]; the Federal Coal Leasing Amendments Act 1976[50]; the Mineral Leasing Act for Acquired Lands[51]; and the Federal Land Policy and Management Act 1976.[52]

7.10.2 Health and Safety in Mines

Coal mining is a very risky business and with over 2000 coal mines in the United States employing over 133,000 miners,[53] some have described underground mining as the most hazardous occupation in the United States. Fatalities can be caused by mine cave-ins, or explosions; and long-term diseases such as 'black lung' are threats to miners.

Federal legislation addressing the health and safety aspects of coal mining include the Federal Coal Mining Safety Act[54] and the 1969 Federal Coal Mine Health and Safety Act, as amended under the Federal Mine Safety and Health Amendments Act 1977,[55] and the Black Lung Benefits Act 1972,[56] along with the Coal Industry Retiree Health Benefits Act 1992.[57] The US government has, therefore, attempted to provide benefits for coal miners who are subject to debilitating diseases.

In the case of *Usery v Turner Elkhorn Mining Company*,[58] the Supreme Court held that the constitutional and retroactive effects of the legislation to establish a compensation system for coal miners affected by black lung disease as caused by long-term inhalation of coal dust was lawful.

Although health benefits for coal miners remained underfunded, when Congress passed the Coal Industry Retiree Health Benefits Act in 1992, the coal companies challenged the new statute under the takings legislation.[59] In *Eastern Enterprises v Apfel*[60] the Supreme Court found portions of the Act apportioning liability to a company that had ceased mining in 1965 to be invalid. This produced a split decision of 4–4 on the takings issue. Some justices found that such an imposition constituted a taking whereas others noted that takings cases generally dealt with specifically identifiable property that had been physically invaded; the Act in question was an imposition of a financial obligation on a company. As this was a split decision, numerous subsequent cases have been brought on such issues as assignment of liability and successor liability.[61]

Issues of funding for black lung disease remain both open and controversial.[62] The Department of Labor completed a 4-year revision of the rules in 2000.[63] The new rules both expanded the eligibility criteria for new claimants as well as for pending benefits claims. The new rules also allowed the reopening of previously denied claims in addition to streamlining and simplifying the claims procedures. The rules not only eased the burden of proof on the claimant but also limited the amount of evidence that the defendant company could submit in denying the claim. Inevitably, these new rules were challenged by mine operators and insurance companies and were upheld in part, but were reversed in part in *National Mining Ass'n v Department of Labor.*[64]

Federal laws regarding coal miner's health and safety are very broad based, with state legislation only able to make the standards more stringent. It is the federal inspectors who have the authority to conduct searches without warrants in addition to the power to close down mines that are not in compliance with federal standards.[65]

Under the federal health and safety laws, violators are also subject to both civil and criminal penalties, as the state must set health and safety standards at a level necessary 'for the protection of life and the prevention of injuries'. Standards for toxic materials that are part of the mining process must be set at levels 'which most adequately assure on the basis of the best available evidence that no miner will suffer material impairment of health or functional capacity even if such miner has regular exposure to the hazards dealt with by such standard for the period of his working life'. This final test is intended to achieve 'the attainment of the highest degree of health and safety protection for the miner'.[66]

7.10.3 Land Reclamation

Environmental legislation with regard to coal mining involves the redistribution of the benefits and burdens of the commercial activity. The costs of reclaiming land after mining has ceased, in addition to the costs of cleaning coal both before and during burning must be borne by someone. Although environmental costs, in principle, should be borne by the polluter as 'internalising the externalities', inevitably these costs find their way down the supply chain to the consumer.

Sixty percent of the coal mined in the United States is surface or open-pit mined,[67] which is the most environmentally devastating form of mining. Although surface mining is cheaper and less hazardous to the mine workers than deep pit mining, it causes soil erosion, water contamination, destruction of the flora, and disruption of fauna. Visually, it creates aesthetic degradation.

The health and safety of miners used to be the province of the states as also environmental regulations for land reclamation. However, state land reclamation was not rigorously enforced[68] and so Congress enacted two new statutes: the Resource Conservation and Recovery Act 1976 (RCRA),[69] which provided for standards for disposal sites and hazardous waste treatment; and the Surface Mining Control and Reclamation Act (SMCRA) 1977,[70] which provided for mined land to be restored to its original condition.

Surface or 'strip mining' has a particularly detrimental effect on the land and environment and so the SMCRA is one of the most comprehensive statutes regarding land use. Congress was concerned to 'assure that the coal supply essential to the Nation's energy requirements, and to its economic and social well being is provided and strike a balance between protection of the environment and agriculture productivity and the Nation's need for coal as an essential source of energy'.

Some of the major concerns among Congressmen were the various agricultural and environmental interests, and so the Act contained four key regulations to protect these interests as follows:

1. Potential miners must submit a detailed application to commence surface coal mining.
2. Coal companies must post a bond in order to ensure that reclamation costs are covered.
3. Highly detailed standards for reclamation must be satisfied.
4. The Act delegated regulatory enforcement to the Secretary of the Interior and individual state regulatory agencies.

Generally, the Act required mining companies to restore approximate contour and use capacity of the mined land, and to do this by stabilising the soil, redistributing the top soil, and replanting the site.

Although the SMCRA is a federal act, it is the state courts that have the exclusive jurisdiction to enforce its provisions.[71] However, this has not proved to be an easy relationship with the states, and the DOI has had to develop and enforce reclamation plans. In the case of *Save Our Cumberland Mountains, Inc. v Watt*,[72] the district court held that the Secretary of the Interior had a mandatory duty to enforce the SMCRA and ordered that he should do so. However, this decision was overturned on appeal due to improper venue in *Save Our Cumberland Mountains Inc. v Clark*.[73]

A current practice causing concern is that of 'valley filling'. This happens when the coal miners simply remove the soil and rocks from the top of the mountain and deposit them in the valley, creating easier access to the coal. The problem is that the 'fill' often finds its way into streams, stopping up and polluting rivers as well as depositing rock and soil in valleys. This practice was challenged in the case of *Bragg v Roberson*,[74] when the District Court Judge granted a summary judgement in favour of the plaintiffs to prohibit valley fill. Shortly afterwards, the Governor of West Virginia announced a revenue shortfall. The Fourth Circuit, on Appeal, considered the injunction and remanded back to the District Court with instructions to dismiss the citizens' complaint on the grounds of sovereign immunity as incorporated in the Eleventh Amendment.[75]

Legislation concerning mountaintop mining involves the Clean water Act, NEPA, and the SMCRA. The US Corps of Engineers has the authority under the Clean Water Act to issue permits that may allow for the dumping of 'fill' into navigable streams in cases of mountain top removal. A series of permits were issued by the Corps in the case of *Ohio Valley Envt'l Coalition v Arcoma Coal Co.*[76] A group of environmentalists petitioned to have the permits withdrawn and the US District Court ruled in favour of the plaintiffs on the basis that the SMCRA did not allow the deposit of valley fill in US waters. The Fourth Circuit reversed this decision and held that the Corps of Engineers had permitting authority and that they had exercised that authority in accordance with applicable environmental laws.[77]

7.10.4 Issues of Clean Air

Coal is usually burned to extract the stored energy within the fossil fuel. Coal combustion generates the four main sources of pollution, all of which are detrimental to the environment and are the main source of greenhouse gases that create the 'greenhouse effect' and so, climate change. There are over 600 coal-fired power stations in the United States, which pose a risk to human health and the environment. However, the number of deaths due to fine particulate pollution from coal plants is reducing mainly due to the federal and state clean air regulations as stated in *Clean Air Task Force; the Toll From Coal: An Updated Assessment of Death and Disease from America's Dirtiest Energy Source* (September 2010).[78] The figures were confirmed in '*Emissions of Hazardous Air Pollutants from Coal-Fired Power Plants* (Environmental Health & Engineering, March 2011).[79]

The Clean Air Act, 42 USC §§7401 et seq. as amended in 1990 forms the basic legislation governing controls for pollution from coal-powered power stations. The legislation relies on state planning and enforcement and has given rise to much case law.[80]

Under the Clean Air Act, National Ambient Air Quality Standards (NAAQSs) are set for each pollutant listed. It is then for the individual states to develop state implementation plans (SIPs) in order to achieve those goals. If a state plan is rejected by the EPA, then the EPA Administration may impose penalties and impose its own federal implementation plan (FIP). Although the regulations were challenged by both environmentalists for being too lax, and by the utilities for being too stringent, the District of Columbia Circuit declared that the EPA were right in the case of *Sierra Club v Costle.*[81]

Power plants built before 1971 are exempt from the CAA rules unless they are modified or reconstructed, in which case the New Source Performance Standards (NSPS) are to apply. In these cases, along with new plants, the applicable regulations require power plants to adopt pollution control technologies.[82]

7.10.5 Clean Coal Initiatives

Carbon capture and storage (CCS) is widely considered to be clean coal technology although this is by no means the only clean technology available. As far as CCS is concerned, costs remain the crucial factor.[83]

However, clean coal is subject to two DOE programmes. Firstly, the DOE's Clean Coal Power Initiative (CCPI) recognises that although coal is cheap, abundant, and will continue to play a major role in the US power mix, it is also dangerous and a dirty energy source. Therefore, the DOE has a programme of funding clean coal technologies.[84]

Secondly, the DOE also has a programme 'FutureGen 2.0', which is the second iteration of the clean coal initiative. This is a public–private initiative and aims to capture 1.3 million tons of CO_2 a year, which is 90% of the power plants' emissions.[85]

7.11 REGULATION OF THE ELECTRICITY SECTOR

As in other developed countries, the development of regulation of the electricity sector in the United States has followed a changing pathway, with command and control giving way to a more *laissez faire* regime as the courts maintained a hands-off stance towards private industry.[86] Later, during the era of the New Deal, the government played a more active role by promoting the use of electricity. Energy consumption became central to the New Deal's effort to strengthen and broaden the economy. Further influence in this period was the acceptance of the principle that natural monopolies had to be regulated in order to keep prices competitive.[87]

7.11.1 The Current Situation

The Public Utility Regulatory Policies Act of 1978 (PURPA) was put in place to encourage more energy-efficient and environmentally friendly energy production. The statute defined a new class of energy producer called a *qualifying facility* (*QF*). These QFs are usually small-scale producers of commercial energy who normally generate energy for their own use but have an occasional surplus to sell. Alternatively, they may be incidental producers who happen to generate usable electricity as a by-product of their main activities.

Utility companies are obliged to purchase energy from such facilities at rates favourable to the producers. This is intended to encourage more of such production, which helps to reduce emissions and contributes to energy security.

The US federal agency responsible for overseeing the wholesale energy market in the United States is the FERC. This commission was set up as part of the reorganisation of the DOE in 1977 and is responsible for regulating prices, terms, and conditions for the sale of energy among states and regions. It is also actively involved with the industry sector that transports energy from the generating facilities to the markets. FERC also inspects and licenses hydroelectric facilities and enforces the Federal Power Act.

In 2005, Congress passed sweeping reforms of the energy industry under the Energy Policy Act.[88] This statute contained a number of major policy innovations.[89]

A further development of the electricity market in the United States is the reaffirmation of the Supreme Court's doctrine in *United Gas Pipeline Co. v Mobile Gas Service Corp.*[90] This doctrine provides that rates agreed to in a freely negotiated contract are presumed to be 'just and reasonable' within the Federal Power Act. Therefore FERC cannot interfere unless the contract price 'seriously harms the public interest'.[91] This doctrine has been revisited more recently in *Morgan Stanley Capital Group Inc. v Pub. Util. Dist. No. 1*,[92] when the Court confirmed the Mobile Sierra presumption even with respect to the extraordinary prices agreed to during the California energy crisis. Later in *NRG Power Marketing, LLC v FERC*,[93] the Court held that the Mobile Sierra presumption applies even against third parties.

A further important decision in the Supreme Court was in the case of *Massachusetts v EPA*,[94] when the Court empowered the EPA to regulate greenhouse gases as the major by-product of energy generation. This means that if the EPA regulates greenhouse gases under the Clean Air Act, the agency may well dictate what type of technology electricity producers should use to control their emissions.

The FERC is also promoting the 'smart grid' in addition to more constant monitoring of consumption through 'smart metering'. The 2007 Energy Independence and Security Act required the Commission to develop a National Action Plan on the 'smart grid', which was published in 2010. The US electricity market is in for further changes still.

7.12 NUCLEAR POWER IN THE UNITED STATES

The regulation of nuclear power in the United States has always been a federal issue. After World War II, the move from military to commercial use did not remove the government from the regulatory process, and the Atomic Energy Act of 1946 created two regulatory bodies. The primary administrative agency is the AEC whose main functions were to encourage research and promote development of the technology for peaceful purposes. There was also the Congressional Joint Committee on Atomic Energy (JCAE), which was a 'watchdog committee' and was in existence from 1946 until it was abolished in 1977. By 1954, atomic energy policy was to encourage private commercial development under the Atomic Energy Act 1954.[95] This statute ended the Federal government's monopoly over non-military uses of nuclear energy and provided for private ownership of reactors under an AEC licensing regime. The 1954 Act is still the main body of law for the regulation of nuclear energy provision. One of the main protections for nuclear operators is the Price Anderson Act of 1957,[96] which limits industry liability while assuring compensation in the event of a nuclear accident. Inevitably, this was challenged but was declared constitutional in the case of *Duke Power Co. v Carolina Environmental Study Group, Inc.*[97] the 1957 Act was renewed every 10 years until 2005, when the Act was extended until 2025 under the 2005 Energy Policy Act. This creates an insurance fund subscribed to by the nuclear plant owners, which is to be used in the event of a nuclear accident. Till now, such insurance payments of $71 million have been paid out as a result of the Three Mile Island claims.

7.12.1 Case Study of Three Mile Island Disaster

At the Three Mile Island nuclear power plant in Pennsylvania, a few seconds after 04.00 on March 28, 1979, the pumps supplying feed water to steam generators in the containment building of General Public Utility's Unit 2 closed down. This led to the emergency feed water pumps coming into operation; however, a closed valve in each line prevented water from reaching the generators. Unfortunately, the operators did not notice that this led to another critical valve, the pilot-operated relief valve (PORV), being stuck open. The plant operators thought that the PORV valve would have closed after 13 seconds. This open valve sent the critically needed cooling water to the containment building floor rather than the reactor core. The steam generators boiled dry causing the reactor coolant to heat and expand. Two large pumps automatically poured coolant into the reactor chamber while the pressure dropped.

Further operator errors meant that the pumping system was manually shut down. As water boiled into steam, the reactor failed to cool and eventually began to disintegrate. Thousands of gallons of radioactive water was negligently pumped into an adjoining building and so resulted in the accident that marked the end of the developmental period of nuclear power in the United States.[98]

Needless to say, there was much litigation following this accident.[99] There were also class actions under the Price Anderson Act, which resulted in payments for living expenses of families who had to evacuate a 5-mi area around the plant. In all, the estimated settlements came to approximately $25 million.[100] Since Three Mile

Island, there have been other minor nuclear reactor events, which are required to be reported to the NRC and can be viewed at the NRC web site.[101]

7.12.2 Reforms

The Three Mile Island incident changed the whole character of the method of assessing risk.[102] After the 1979 accident, the NRC increased safety inspections, stepped up enforcement, and developed emergency preparedness procedures.[103]

The Energy Policy Act 1992[104] provided for streamlining the licensing process for nuclear power plants, supporting research regarding new reactor technologies, and addressing permanent high-level waste storage. After the September 11 terrorist incident, the NRC turned its attention to the security of nuclear plant and adopted new regulations imposing additional security standards on licensees.[105] Nuclear power continues to attract the attention of both Congress and the Executive, and nuclear power is a key element of President Obama's energy planning. At the President's direction, the DOE Secretary Steven Chu appointed the Blue Ribbon Commission on America's Nuclear Future to study and report on methods of disposal of nuclear waste. The Commission submitted its report in January 2012 with a number of recommendations.

> The need for a new strategy is urgent, not just to address these damages and costs but because this generation has a fundamental ethical obligation to avoid burdening future generations with the entire task of finding a safe, permanent solution for managing hazardous nuclear materials they had no part in creating.

The strategy outlined in the Commission Report contains three crucial elements:

1. A consent-based approach to siting future nuclear waste storage and disposal facilities, noting that trying to force such facilities on unwilling states, tribes, and communities has not worked.
2. The responsibility for the nation's nuclear waste management programme be transferred to a new organisation that is independent of the DOE and dedicated solely to assuring the safe storage and ultimate disposal of spent nuclear waste fuel and high-level radioactive waste.
3. Changing the manner in which fees being paid into Nuclear Waste Fund (about $750 million a year) are treated in the federal budget to ensure they are being set aside and available for use as Congress initially intended.

The Report also recommends immediate efforts to commence development of at least one geological disposal facility, as well as efforts to prepare for the eventual large-scale transport of spent nuclear fuel and high-level waste from current storage facilities; and that the United States continues to provide support for nuclear energy innovation and workforce development, as well as strengthening its international leadership role in efforts to address safety, waste management, non-proliferation, and security concerns.

> These are all important questions that will engage policy makers and the public in the years ahead. However, none of them alters the urgent need to change and improve our strategy for managing the high-level wastes and spent fuel that already exist and will continue to accumulate so long as nuclear reactors operate in this country.

The Commission stresses that what is needed is a sound waste management approach that can lead to the resolution of the current impasse, and can be applied regardless of what sites are ultimately chosen to serve as the permanent disposal facility. The United States has currently more than 65,000 t of spent nuclear fuel stored at approximately 75 operating and shutdown reactor sites. Another 2000 t are being produced each year. In addition, the DOE stores 2500 t of spent fuel from past weapons programmes at a number of government sites.

The Commission's full Report can be found at www.brc.gov.

Currently and for some time in the future, nuclear power is not economically viable. Even with the costs associated with nuclear power, too much time, money, and effort has been committed to nuclear technology in the United States by both the public and the private sectors to make a full retreat from nuclear power politically possible. Nuclear-generated power is more environmentally sensitive than fossil fuel alternatives, particularly coal, and would reduce the dependence of the United States on oil. But how to make nuclear power safe, clean, and economical is the challenge.

In the Massachusetts Institute of Technology (MIT) study, *The Future of Nuclear Power*,[106] it was concluded that baseline costs of nuclear power are greater than those of electricity generated by either coal or natural gas. To make nuclear power economically competitive, costs would have to be assigned to the carbon emissions with their impact on global warming. However, what must also be factored in are the costs of financial risks inherent in large-scale investment in a nuclear power plant. In addition, the long-term and temporary storage of nuclear waste continues to be problematic.

The MIT study considers that nuclear power can make a contribution to the world's energy needs as long as issues of non-proliferation, safety, economy, and waste are addressed.

The main findings of the MIT study are:

1. Nuclear power is not cost-competitive with coal and natural gas.
2. The cost gap can be reduced by reductions in capital and operational costs.
3. Reactor designs could be implemented to reduce risk of serious accidents.
4. Long-term disposal needs to be addressed.
5. Current international safeguards are not sufficient to meet security challenges.

Finally, the Daiichi nuclear plant at Fukushima, Japan, damaged by the tsunami in 2010 makes the world policymakers pause when thinking about the safety of nuclear power, and the ethical and financial implications when both man-made and natural disasters occur.

The Three Mile Island accident was caused by lack of knowledge of the operators on the ground. Chernobyl was caused by lack of maintenance and investment, whereas Fukushima was caused by nature. Were any or all of these foreseeable and could they have been prevented by proper training, investment, and maintenance, and considered siting and safety?

In 2010, newly installed capacity of wind, biomass, and solar power was greater than newly installed nuclear power capacity. In 2011, more reactors were decommissioned than opened, which reduced the number of operating reactors worldwide. Policymakers are questioning whether licences for longer than 40 years are prudent.[107]

Once again the world must ask itself, 'Can we afford nuclear power?'

7.13 HYDROPOWER AND RENEWABLE SOURCES

The United States is a common law country and derives its basic jurisprudence from English common law. As far as water law is concerned, the common law doctrine of riparian rights prevails on the East Coast where water is plentiful. However, on the West Coast, where water is at a premium, the doctrine of prior appropriation has been adopted. This means that the first user can divert flowing water for its own use and can retain that right so long as that use continues.

Although it was the states who took ownership of these water rights for the production of electricity, with the growth of interstate commercial activity, the need for a more unified federal system became necessary.[108] Federal jurisdiction over hydro projects is usually based on the fact that hydroelectric plants use water from navigable waterways that fall within the federal domain, in addition to the Commerce Clause of the US Constitution, Article 1, section 8, clause 3.

The FERC has the authority to license hydroelectric projects subject to the hydroelectric guidelines, which state that:

1. The project must be within the United States.
2. It must be located on navigable waters.
3. It must use water or water power from a government dam.
4. It must affect interstate commerce.

The federal hydropower authority under the enabling legislation was intended to be comprehensive, but has been limited to some extent by litigation as to jurisdiction.[109]

After the question of jurisdiction has been settled, the FERC can authorise preliminary permits, licences, and new licences. The Federal Power Act requires that the FERC should choose the project 'best adapted' to improve or develop a waterway for commercial development. But it is the Electric Consumers Protection Act 1986 that states that the 'best adapted' requirement should be based on a comprehensive review.[110]

Early in the twenty-first century, when the historic private licences granted for 50 years began to come to the end of their term, there was much discussion. In response, the FERC developed an alternative licensing process,[111] which complemented the traditional system but responded to the need for licensing reform. The final Rule was issued on July 23, 2003.[112] This created an integrated licensing process (ILP), which was intended to streamline the Commission's licensing process by providing:

1. Early issue identification and resolution of studies needed to fill information gaps, avoiding studies needed to fill information gaps

2. Integration of other stakeholder permitting process needs
3. Established time frames to complete process steps for all stakeholders, including the Commission.

All hydroelectric projects are subject to NEPA and so present the classic energy/environment conflict.[113] Nevertheless, hydroelectric power will continue to play a role in the US energy supply as a notable renewable energy resource.

As of 2012, the US Congress has no serious intention of addressing climate change in any meaningful way notwithstanding the number of energy policy studies by various study groups of eminent standing, which all promote a movement to a clean energy economy.[114]

ENDNOTES

1. An excellent primer on US Energy Law is Tomain & Cudahy (2011), *Energy Law in a Nutshell*, West Nutshell Series. West Publishing: USA.
2. Tomain & Cudahy (2011), *Energy Law*. West Publishing: USA.
3. APA §553.
4. The staff normally consists of specialists in various aspects of energy development.
5. The Office of Energy Projects has oversight over hydroelectric facilities and natural gas pipelines and energy projects that are in the public interest.
6. The word 'markets' in the title emphasises the importance of deregulation (Tomain & Cudahy, p. 108).
7. (1956) 350 US 332.
8. (1956) 350 US 348.
9. (2008) 554 US 527.
10. (1877) 94 US 113.
11. Tomain & Cudahy (2011), *Energy Law*, West Nutshell Series.
12. *New England Power Co. v New Hampshire*, 455 US 331 (1982).
13. Tomain & Cudahy, pp. 142–145.
14. *Pacific Gas & Electric Co. v State Energy Resources Conservation and Development Comm'n*, 461 US 190 (1983).
15. The preemption doctrine.
16. *New York Public Service Comm'n v FERC*, 535 US 1 (2002).
17. *Tahoe Sierra Preservation Council, Inc. v Tahoe Regional Planning Agency*, 535 US 302 (2002).
18. *Iowa Coal Mining Co., Inc. v Monroe County*, 555 NW 2nd 418 (Iowa 1966).
19. (pa 1907) 65 A. 801.
20. (1858) Pub. L. no. 85-686, 72 Stat. 673.
21. (1920) Pub. L. no. 66-146, 41 Stat. 437 (as amended).
22. (1987) Pub. L. no. 100-203, 101 Stat. 1330.
23. 43 C.F.R. §3103.2(a).
24. These two areas alone comprise approximately 25 million acres.
25. 43 USC §§1303–1315.
26. As amended in 1978 (OCSLA), 43 USC §§1331–1356.
27. 43 USC §1301(b) (1985).
28. 43 USC §1331(a).
29. 339 US 699 (1950).
30. 339 US 707 (1950).
31. 332 US 19 (1947).

32. 16 USC §1451. Also *Secretary of the Interior v California*, 464 US 312 (1984).
33. 42 USC §4321.
34. Tomain & Cudahy, ibid, p. 257.
35. In the case of *Watt v Energy Action Educ. Foundation*, 454 US 151 (1981), an environmental group sought to halt all auctions until the Interior Department developed regulations requiring the Secretary to use a variety of systems. The Supreme Court held that nothing in the Congressional authorisation required this and that therefore alternative bidding was permitted.
36. Rachel Carson (1962), *Silent Spring*. Houghton Mifflin: USA.
37. Bosselman, Rossi, & Weaver (2000), *Energy Economics and the Environment*, Foundation Press: USA.
38. 594 F2d 872 (1st Cir. 1979).
39. See Note 2.
40. This amounted to approximately 1 year's profit for Exxon.
41. *Exxon Shipping Company v Baker*, 554 US 471 (2008).
42. See *Commonwealth of Puerto Rico v SS Zoe Colocotroni*, 628 F2d 652 (1st Cir. 1980); *United States v Dixie Carriers, Inc.*, 627 F2d 736 (5th Cir. 1980).
43. 486 F. Supp. 332 (DDC 1980).
44. 33 USC §§2701 et seq.
45. 33 USC §2701.
46. A methane gas explosion flashed through Massey Energy's (MEE) Upper Big Branch on April 5, 2010, killing 29 coal miners. Massey is the largest coal producer in the Central Appalachian region, with operations in Kentucky, Virginia, and West Virginia. Chairman and Chief Executive Don Blankenship anticipates longer-term liability costs totalling no more than $150 million related to compensatory benefits paid to the dependents of the dead miners, costs associated with the rescue and recovery efforts, insurance deductibles, and federal/state fines. A review of judicial proceedings involving Massey in the recent years, from alleged unfair labour practices to property damage (including environmental litigation), shows an overwhelming propensity of the courts to rule in favour of the company. But then West Virginia is a coal country.
47. Environmental Protection Agency (2009), 'Inventory of US greenhouse gas emissions and sinks: 1990–2007', Annex I at A-4 tbl, A-I, http://www.epa.gov/climatechange/emissions/dpwnloads09/AnnexI.pdf. Accessed 2012.
48. Powers Melissa (2012), 'The cost of coal: Climate change and the end of coal as a source of *"cheap"* electricity', *University of Pennsylvania Journal of Business Law*, 12:2, pp. 101–130.
49. MLA, 30 USC §801.
50. 30, USC §§201–209.
51. 30, USC §§351 et seq.
52. FLPMA, 43 USC §§1701–1782.
53. Various Department of Labor Reports.
54. 30 USC §§451 et seq.
55. 30 USC §§801 et seq.
56. 30 USC §§901 et seq.
57. 26 USC §§9701–9722.
58. 428 US 1 (1976).
59. The Coal Industry Benefits Act 1992 apportioned liability to the coal companies to fund the benefits.
60. 524 US 498 (1998).
61. *Anker Energy Corp v Consolidated Coal Co.*, 177 F3d 161 (3rd Cir.1999); *National Coal Ass'n v Chater*, 81 F3d 1077 (11th Cir. 1996); *Holland v New Era Coal Co.*, 179 F3d 397 (6th Cir. 1999); *The Pitston Co. v United States*, 199 F3d 694 (4th Cir. 1999).

62. Tomain & Cudahy, p. 331.
63. 65 Fed. Reg. 79, 920-80, 107 (20 December 2000).
64. 292 F3d 849 (DC Cir. 2002).
65. See *Donovan v Dewey*, 452 US 594 (1981).
66. 30 USC §811(a)(6)(A) (1982).
67. Tomain & Cudahy, ibid, p. 332.
68. Partly because the states were protective of their coal industries, which could be the main source of employment.
69. RCRA, 42 USC §§6901–6987.
70. SMCRA, 30 USC §§1201–1328.
71. *Haydo v Amerikohl Mining, Inc.*, 830 F2d 494 (3rd Cir. 1987).
72. 550 F. Supp. 979 (DDC 1982).
73. 725 F2d 1434 (DC Cir. 1987).
74. 72 F Supp. 2d 642 (S.D. W. VA. 1999).
75. *Bragg v West Virginia Coal Ass'n.*, 248 F3d 275 (4th Cir. 2001).
76. 567 F3d 177 (4th Cir. 2009).
77. Other similar cases include *Commonwealth v Rivenburgh*, 317 F3d 425 (4th Cir. 2003); see also Forman (2011), 'The uncertain future of NEPA and mountaintop removal', *Columbia Journal of Environmental Law* 36, 163.
78. Environmental health and engineering, http://www.lungsa.org/assets/documents/healthy-air/coal-fired-plat-hazards.pdf. Accessed April 2011.
79. Environmental health and engineering, http://www.lungusa.org/assets/document/heathy-air/coal-fired-plat-hazards.pdf. Accessed April 2011.
80. See *Ariz. Pub Serv. Co. v EPA*, 562 F3d 936 (9th Cir. 2009); *New Jersey v Reliant Energy Mid-Atlantic Power Holdings*, 2009 WL 3234438 (E.D. Pa. 2009); *United States v Cinergy Corp.*, 618 F. Supp. 942 (S.D. Ind. 2009); *Southern Alliance for Clean Energy, Envt'l Def. Fund v Duke Energy Carolinas, LLC*, 2009 WL 2767128 (W.D.N.C. 2009); *Sierra Club v Duke Energy, Ind., Inc.*, 2009 WL 363174 (S.D. Ind. 2009).
81. 657 F2d 298 (DC Cir. 1981).
82. Reitze (2010), *Air Pollution Control Law: Compliance and Enforcement*, 2nd edition, Environmental Law Institute: USA.
83. National Academy of Science (2009), *America's Energy Future Technology and Transformation.* National Academy of Science: USA.
84. See US DOE (2008), *Program Facts: Clean Coal Power Initiative (CCPI)*, http://fossil.energy.gov/programs/powersytems/cleancoal/ccpi/Prog052.pdf. Accessed February 2012.
85. See National Mining Association, *The US Department of Energy's Clean Coal Technology: From Research to Reality*, http://fossil.energy.gov/aboutus/fe_cleancoal_brochure_web2.pdf. Accessed February 2012.
86. *Lochner v New York*, 198 US 45 (1905).
87. *Jersey Central Power & Light Co. v FERC*, 810 F2d 1168 (DC Cir. 1987).
88. EPA Pub. L no. 109-58, 119 Stat. 594.
89. See Rossi (2009), 'The Trojan horse of electric power transmission line siting', *Environmental Law* 39, 1015; Swanstrom & Jolivert (2009), 'DOE transmission corridor designations & FERC backstop siting authority: Has the Energy Policy Act of 2005 succeeded in stimulating the development of new transmission facilities? *Energy Law Journal*, 30, 415.
90. 350 US 348 (1956) and *FPC v Sierra Pacific Power Co.*, 350 US 348 (1956).
91. Tomain & Cudahy, ibid, p. 422.
92. 554 US 527 (2008).
93. Supreme Court 2010.
94. US Supreme Court 549 US 497 (2007).
95. AEA 1954, 42 USC §§2001 et seq.
96. 42 USC §§2210 et seq.

97. 438 US 59 (1978).
98. Daniel Ford (1982), 'Three Mile Island: Thirty minutes to meltdown, and report of the President's commission on the accident at Three Mile Island' (October 1979), Kemeny Commission Report, Penguin Books.
99. *Susquehanna Val. Alliance v TMI Nuclear Reactor*, 619 F2d 231 (3rd Cir. 1980).
100. *TMI Litigation*, 193 F3d 613 (3rd Cir. 1999).
101. NRC, 10 C.F.R. §50.72, www.nrc.gov/what-we-do/regulatory/event-assess.html. Accessed February 2012.
102. See Note 98; NRC Inquiry Group, TMI Report to the Commissioners and the Public (1980).
103. *Union of Concerned Scientists v United States NRC*, 824 F2d 108 (DC Cir. 1987).
104. Pub. L. no. 102-486, 106 Stat. 2776.
105. Final Rule, *Power Reactor Security Requirements*, 74 Fed. Reg. 13, 926 (27 March 2009).
106. (2003), 'The future of nuclear power: An interdisciplinary MIT study'.
107. Froggatt & Thomas (2011), *The World Nuclear Industry Status Report 2010–2011: Nuclear Power in a Post Fukushima World*, http://www.worldwatch.org/nuclear-power-after-fukushima. Accessed December 2011.
108. Rasband, Salzman, & Squilliance (2009), *Natural Resources Law & Policy*, Chapter 7, University Casebook Series, Foundation Press: USA.
109. *First Iowa Hydro Electric Cooperative v FPC*, 328 US 152 (1946); *California v FERC*, 495 US 490 (1990); *United States v Appalachian Electric Power Co.*, 311 US 377 (1940); *FPC v Oregon*, 349 US 435 (1955); *FPC v Union Elec. Co.*, 381 US 90 (1965); *Fairfax County Water Authority*, 43 FERC, ¶61,062 (FERC 1988); *Habersham Mills v FERC*, 976 F2d 1381 (11th Cir. 1992); *City of Centraia v FERC*, 851 F2d 278 (9th Cir. 1981) et seq.
110. *Scenic Hudson Preservation Conference v FPC*, 354 F2d 608 (2nd Cir. 1965).
111. 18 CFR §4.34(i).
112. 104 FERC, ¶61, 109 (23 July 2003).
113. *Scenic Hudson Preservation Conference v FPC*, 354 F2d 608 (2nd Cir. 1965); *Udall v FPC*, 387 US 428 (1967); *Platte River Whooping Crane Critical Habitat Maintenance Trust v FERC*, 876 F2d 109 (DC Cir. 1989).
114. Tomain (2011), 'The next generation is now', Chapter 3, pp. 65–92; 'Concensus energy policy', Chapter 4, pp. 92–121, *Ending Dirty Energy Policy: Prelude to Climate Change*. Cambridge University Press: UK.

8 Energy Law in the European Union

This chapter describes the development of law and policy in the European Union as the Directives and Regulations that come from the European Union affect all of its Member States.

8.1 AN ENERGY POLICY FOR EUROPE

The purpose of this chapter is to provide an introduction to European Community (EC) law within the European Union as it applies to the energy sector.

8.2 THE HISTORICAL DEVELOPMENT OF EUROPEAN ENERGY LAW

The strategic management of energy resources has, from the early days of the European Economic Community (EEC), been a central part of the creation of an internal market. At present, the European Union is only able to produce 50% of the energy that it consumes. For this reason, the first steps were taken towards the creation of the original EEC, with a trans-European strategic management of the most important energy source and industrial raw material of the 1950s, namely coal and steel. In the post World War II era in Europe, it was proposed by the governments of the six founding states that these two basic materials at the heart of the defence industry should be brought under common control. In 1951, the European Coal and Steel Community was established under the Treaty of Paris (the ECSC Treaty). The objective was to form a common market for coal and steel and to 'ensure the most rational distribution of production at the highest possible level of productivity, while safeguarding continuity of employment and taking care not to provoke fundamental and persistent disturbances in the economics of Member States'.[1] Four principal institutions were created; the Council (representing the Member States), the Commission (a supranational executive, which was originally called the *High Authority*), the Assembly, and the Court. Except for the Assembly, which normally met in Strasbourg, they were all located in Luxemburg.

This Treaty was followed by the Treaty of Rome, which established the EEC, now the EC, which was signed on March 25, 1957. On the same day, the Treaty Establishing the European Atomic Energy Community (EURATOM) Treaty,[2] which created the European Atomic Energy Community, was also signed and they both came into force on January 1, 1958.

The three treaties contain many common features, but there are also important differences. One difference is that the ECSC Treaty is more specific as regards the policy to be pursued; the EC Treaty, on the other hand, is concerned mainly with creating a framework, and leaves the creation of policy to the Community institutions. The consequence of this is that the institutions are required to play a more creative role in the EC. This difference is in turn responsible for another distinction between the ECSC and the EC. Treaties: that is, under the former, the Commission has much more independent power, which results in correspondingly less power for the Council, than in the case under the EC Treaty. The self-financing capacity of the ECSC Treaty, through levies on coal and steel production under Articles 49–53, added to its independence and autonomy. The Assembly had relatively few powers, being mainly advisory and supervisory, although it could, in extreme circumstances, require the resignation of the High Authority; but, significantly, Article 21 of the ECSC Treaty provided for the future possibility of direct elections.

On the legal level the EC and EURATOM Treaties are very similar and many provisions concerning legal remedies are identical, notwithstanding that the provisions under the ECSC Treaty are significantly different.

8.3 SCOPE OF THE ECSC TREATY

The ECSC Treaty established rules governing the single market for primary coal products, which include hard coal, including briquettes and coke, but excluding electrode and petroleum coke; brown coal; and brown coal briquettes (Article 81 and Annex 1). Where the Treaty refers to 'undertakings', these are defined under Article 80 to include any undertaking engaged in production in the coal or the steel industry within the territories to which the Treaty applies. In Article 65 on restrictive agreements and Article 66 on concentrations and in relation to any information that is required to apply these provisions, 'undertakings' also includes any undertaking or agency regularly engaged in distribution, other than sale to domestic customers.

8.4 INCONSISTENCIES BETWEEN THE TREATIES

There is a line of decisions from the European Court of Justice (ECJ), which declares that Article 232(1) of the EC Treaty provides that the provisions of the EC Treaty 'shall not affect the provisions of the Treaty establishing the European Coal and Steel Community'.[3] These decisions confirmed the case of *Gerlach & Co. BV*[4] in which it was stated that 'the rules of the ECSC Treaty and all the provisions adopted to implement that Treaty remain in force as regards the functioning of the Common Market in Coal and Steel, despite the adoption of the EEC Treaty' (para 9). The jurisprudence of the ECJ has indicated that provisions of the EC and ECSC Treaties must be interpreted on their own merits. It has been argued that legal consistency in the Community legal order would require that direct effect be given to Articles 65 and 66(7) of the ECSC Treaty in the same way as it is applied to the equivalent competition provisions of the EC Treaty (Articles 85 and 86). However, if the matters are not subject of any ECSC rule, the EC Treaty will apply.[5]

8.5 THE OBJECTIVES OF THE ECSC TREATY

The main objectives, as laid out by Article 3 of the Treaty, are to ensure an orderly supply to the common market, while taking into account the needs of third parties;

non-discriminatory access to production for consumers; fair and non-discriminatory prices; rational and sustainable production; improved working conditions; international trade; and the orderly expansion and modernisation of industry.

Under the Treaty, the Community institutions have sweeping powers to take both direct and indirect actions to regulate production. As an indirect action the ECSC may cooperate with national governments to regularise or influence general consumption or intervene with regard to prices and commercial policy under Article 57. Should there be a significant demand, the Community can go further and fix production quotas under Article 58. If there is a serious shortage, then the Community may take action to ensure that goods are allocated equitably and may impose fines for non-compliance under Article 58. The provisions of the Treaty also deal with competition law and state aid. These provisions include rules on price quotation and the prohibition of unfair and discriminatory pricing practices under Article 60 and Decision 30/53/ECSC; Article 61 provides the power to fix maximum and minimum prices, whereas Article 63(1) provides powers on systematic discrimination by purchasers. Article 63 further provides for Commission control of agents and intermediaries to ensure compliance with the pricing rules under section 2. Article 64 provides powers to impose fines for infringement, whereas Article 65 covers the prohibition of anti-competitive agreements and concerted practices. The requirements to seek authorisation for certain concentrations are covered under Article 66(1) (6). Article 66(7) addresses the prohibition of abuses of a dominant position and the rules relating to state aid are made under Article 4(c).

The ECSC Treaty expired in the year 2002, and was not renewed. Instead, all issues relating to coal and steel will be governed by the relevant provisions of the EC Treaty. It is expected that further provisions will need to be added to the EC Treaty to cover this sector. In the light of the European Union's proposed territorial enlargement under which a number of prospective Member States have significant coal and lignite reserves and steel industries, it is expected that, in the field of competition law, the Commission's policy will be to try and operate the two regimes as closely as possible within the limits of their legal obligations, in order to facilitate eventual integration.[6]

8.6 THE EUROPEAN ATOMIC ENERGY COMMUNITY TREATY

The law relating to nuclear energy is covered in detail in Chapter 10 and therefore only the European Atomic Energy Community Treaty is addressed here.

When the EURATOM Treaty was signed in 1957, the idea of self-sufficiency in the form of nuclear energy was a very persuasive argument for the development of nuclear energy production in the Community. The main objective of the EURATOM Treaty is to create a common market for nuclear ores and fuels. The main proposals under Article 2 of the Treaty are to promote research and disseminate technical information; to establish uniform safety standards; facilitate investment in developing nuclear energy; ensure that users in the Community receive a regular and equitable supply of ores and nuclear fuels; ensure wide commercial outlets and access to the best technical facilities through a common market in those materials and equipment; to prevent nuclear fuels from being diverted for unintended purposes; and to promote peaceful use of nuclear energy.

As far as jurisdiction and ownership are concerned, EURATOM has exclusive jurisdiction over nuclear supply in both internal and external relations. These rights are provided under Articles 52–76 of the EURATOM Treaty and Article 86 provides that all special fissile materials automatically become the property of the Community. These rights were confirmed in the ECJ Ruling 1/78.[7]

8.7 THE EUROPEAN ENERGY POLICY

As Tudway says, the energy policy of the European Union has been influenced by a number of changes in technology; the fluctuations in demand for different sources of energy, including moves towards a policy for renewable sources; and in particular to environmental concerns. A White Paper *Energy Policy for the European Union*[8] was published in 1995, and was approved by Council Resolution in 1996.[9] All of the general objectives of the Community's policies were included, but in addition there are a number of sector-specific issues that are addressed. These include the safeguarding of supply, any dependence on external sources of supply, but more importantly, the balancing of the pursuit of competitiveness with the aims of environmental protection.

8.7.1 European Energy Policy and the Internal Energy Market

A key objective of the EU policy in relation to the cultivation and maintenance of the international competitiveness of Europe's industry base is the promotion of a competitive energy sector. Recently, the political commitment among the Member States to achieve this has led towards steps being taken to create an internal market for energy. Deregulation and the privatisation of energy utilities in some of the Member States has been a major influence in the development of this commitment. A further influence is the general movement towards removing obstacles to trade of goods and services. Subsequent to the successful negotiation of the liberalisation of the telecommunications market, the negotiators from the electricity and gas sectors relied on the arguments already won in particular areas such as third-party access. One of the strongest arguments in favour of the liberalisation of the energy market was the high, and rising, costs of energy prices and the effect they were having on the high energy-using industries within the European Union. These high energy prices were effectively putting EU industry at a competitive disadvantage with their major trading partners internationally.

In 1996, the EC adopted the Electricity Directive,[10] which establishes common rules for an internal market in electricity. In 1998, the EC continued this initiative by adopting the Gas Directive,[11] the aim of which was to liberalise the gas industry. Further, in 1994, the EC had already adopted a Directive on the conditions for granting and using authorisations for the exploration and production of hydrocarbons.[12]

Prior to these Directives, the EC adopted in 1990 a Directive 'concerning a Community procedure to improve the transparency of gas and electricity prices charged to industrial users'.[13] This was followed by a Directive 'on the transit of electricity through transmission grids'.[14]

To enhance the ability of major industrial users to choose among different sources of energy and different suppliers,[15] the Directive on price transparency was adopted.[16] It also aimed to reduce the unfair discrimination on the part of suppliers. The Directive outlines procedures by which undertakings that supply gas or electricity to industrial users are required to supply to the Statistical Office of the European Communities (SOEC) information concerning the prices and terms of sale of gas and electricity to industrial end users, their current price systems, and a breakdown of the volume of sales to particular customers. The information is to be supplied twice a year and is published in an aggregate form so that individual commercial transactions cannot be identified, so as to preserve commercial confidentiality. The prices of gas and electricity for industrial users for each Member State are then published annually in May and November.

The first stage in the creation of a single electricity market could arguably be identified as the Electricity Transit Directive.[17] This Directive facilitated the interconnection among the national high-voltage networks by setting out the conditions that apply where the operator of an electricity network in one Member State transmits electricity either to another operator or across one or several intermediary national grids. Article 3 provides that conditions of transit must be non-discriminatory and fair for the parties concerned; shall not include unfair clauses or unjustified restrictions; and must not endanger security of supply or quality of service. A conciliation service is available through the Commission in the event of failure to agree. This Directive applies only to network operators.

A Committee of experts on the transit of electricity between grids was established by the Commission Decision 92/167/EEC.[18] The Committee advises the Commission on matters relating to the smooth functioning of the transit and related economic, legal, and social factors. It also assists with writing an annual report on the implementation of the Electricity Transit Directive. Where negotiations over transit among network operators breaks down under Article 3(4), the Committee can also act as a conciliation body.

Similar provisions were adopted concerning the gas industry under the Gas Transit Directive.[19] The Directive applies to transmission carried out by the operators of high-pressure natural gas grids in each Member State where gas is being transmitted to or through another Member State in the European Union. A further Committee of experts on the transit of natural gas was established under Commission Decision 95/539/EC[20] to advise the Commission, to assist with drafting of the annual report, and to act as a conciliation body in the event of disputes over transit among Member States.

8.8 INVESTMENT PLANNING

The Member States are required to regularly report all major new investment projects to the Commission to enable it to carry out its task of monitoring developments in the European Union's petroleum, natural gas, and electricity markets. All projects listed in the Annex to Regulation 2386/96, concerning production, transport, storage, and distribution of petroleum and natural gas, which are due to commence in the next

3 years, and all similar projects in the electricity sector due to commence in the next 5 years, have to be communicated to the Commission. The requirements regarding the type of information that should be supplied are set out in the Regulations.[21] These include information such as the planned capacity or power, date of commissioning, purpose and nature of the investment, and the type of raw materials to be used. The Regulations were last updated in 1996 by Council Regulation 1056/72/EEC[22] and Council Regulation 236/96. These were again updated by Regulation 736/96.[23]

8.9 NON-DISCRIMINATION AND THE LICENSING OF HYDROCARBONS

The European Union is not currently self-sufficient in hydrocarbons, but a major contribution is made from the North Sea. Under international law, a Member State retains sovereignty and sovereign rights over any hydrocarbon resources in its territory. However, the EC adopted Directive 94/22/EC[24] in 1994, which ensured that Member States could not reserve exploration and exploitation rights to their own national undertakings. The Directive essentially establishes the ground rules by which licences may be granted, with the fundamental requirements that the licensing procedure should be transparent, objective, and non-discriminatory.

The principle of non-discrimination also applies to non-EU countries, but is subject to the condition that there should be reciprocity in the countries of those third parties who are applying. Applications from third parties outside the European Union may also be refused on the grounds of national security under Article 2.

Apart from non-discrimination Member States are required to grant authorisation on the basis of objective criteria under Article 5. These include the established technical and financial capability of the undertakings; the programme of work that identifies the proposals for exploration and exploitation of the area; and if the authorisation is by 'bid', then the amount offered.

If all enterprises are of equal merit under the above criteria, then other relevant criteria may be applied as long as they are both objective and non-discriminatory. For example, it is permitted under the Directive for the exploitation of hydrocarbons to be carried out by government-owned enterprises so long as the government keeps separate the activities of its authorisation and regulatory departments.

In some Member States, a single entity was given an exclusive right to be granted authorisation in a specific geographic area. Under Article 7 of the Directive, all of these rights were revoked by January 1, 1997 as being inconsistent with the aims of the Directive. Denmark, however, was permitted a special derogation from this requirement in respect of territory in the Danish North Sea where an exclusive concession had been granted to a consortium in the early 1960s. Under Article 4(a) of the Directive, the area covered should be consistent with both technical and economic best practices. There is also a general requirement that entities should not retain exclusive rights longer than is necessary for the execution of the authorised activities. Article 12 of the Directive provides those authorised under the Directive the possibility of an exemption from the full requirements of the Public Procurement Utilities Directive 93/38/EEC[25] on application by the Member State.

8.10 THE LIBERALISING OF THE ELECTRICITY AND GAS MARKETS

The Electricity Directive 96/92/EC[26] created a framework that merely set common objectives. This reflected the difficulty in creating a single market from the existing disparate regulatory systems that already existed in the fifteen Member States. This meant that any legislation promulgated by the EC had to be flexible enough to incorporate those differences while creating a level playing field. The provisions of the Directive are designed, therefore, to form the basis of an ongoing and progressive opening of the market. The Directive allows Member States to choose among different but equivalent methods of market opening.

8.11 THE EUROPEAN ENERGY CHARTER TREATY

The series of measures discussed were introduced by the European Union with the intention of creating a single market for energy and increasing the level of competition among different energy sources. At the same time, it was recognised by the European Union that these measures are insufficient to reduce the dependency of the European Union on outside sources for its energy requirements. The realisation that any energy policy must effectively secure steady, reliable, and safe sources of energy led the Member States and other energy-dependent nations to sign an international Treaty on energy trade and cross-border investment.

The Energy Charter Treaty[27] was signed in Lisbon on December 17, 1994. The focus of the Treaty is essentially on the European energy market, but is open to non-European states and has been signed by a number of other major trading partners, including the United States, Japan, and Australia. The signing of the Treaty was preceded by the non-binding Energy Charter, which was signed by 48 countries in 1991.

The Treaty itself came into force in April 1998 after ratification by over 30 countries and the European Union itself. Most of the Member States of the European Union and also many of the Central and Eastern Europe (the CEE) countries have ratified the Treaty but, so far, one of the largest energy producers, namely, the Russian Federation, has not yet ratified it.

The Treaty creates the framework for non-discriminatory access for investors, freedom of transit, non-discriminatory pricing, intergovernmental cooperation in areas including environmental protection, and dispute settlement. There are also a number of Protocols including one on Energy Efficiency and Related Environmental Aspects. This Protocol came into force with the main Treaty.

The Treaty establishes rules on the treatment of investors in contracting states and prevents discrimination between domestic and foreign investors. The aim of the Treaty under Article 3 is to promote a free and competitive international market in energy. Under Article 6, Contracting Parties are required to remove barriers to competition and to apply their own national competition laws to address any anti-competitive behaviour within the energy sector. Under Article 6(3), a Contracting Party can initiate appropriate enforcement action to ameliorate the situation if it considers that its interests are being adversely affected in another Contracting State. Some Contracting Parties have better national competition laws and the resources and experience to effectively enforce them than others. Under Article 6(3), therefore, there is a requirement that technical assistance will be provided by Contracting Parties with greater experience in this field.

8.12 ENVIRONMENTAL ASPECTS

A further aim of the Treaty under Article 19 is to ensure that 'sustainable development' takes place and that the environmental impact of pursuing certain types of investment strategies must be taken into consideration. Article 19 provides that sustainable development must be carried out in a way that will 'minimise, in an economically efficient manner, harmful environmental impacts'. This Article recognises the principle of 'the polluter pays'. Essentially, it establishes that the Contracting Party should bear the cost of pollution of its own creation, including any transboundary pollution. Disputes between Contracting Parties over the environmental aspects of the Treaty may be taken to the Charter Conference for review. In addition to the principles established under Article 19, there are also provisions concerning 'energy efficiency and related environmental aspects'. These are set out in the Efficiency Protocol,[28] the purpose of which is to stimulate the Contracting Parties into formulating strategies and national programmes for improving energy efficiency and reducing the environmental impact of the whole cycle of energy production, distribution, use, and disposal.

8.13 THE CONSTITUTIONAL BASIS OF EUROPEAN ENVIRONMENTAL LEGISLATION

Notwithstanding its economic basis, the EC is a major and increasing source of environmental protection law. Although the 1957 Treaty of Rome made no mention of the environment in its provisions, this lack of power did not deter the Community from legislating for the environment. The early measures were based on Article 100, which provides for the approximation of national laws that may obstruct the establishment or functioning of the common market. Alternatively, Article 235, which provided for measures that were necessary to attain one of the objectives of the Community, was used, and over 100 pieces of legislation were put in place under these provisions before the Treaty was amended to put environmental legislation on a firmer footing.

In 1987, the Treaty was amended by the Single European Act, which provided for the objectives of the Community to include 'sustainable and non-inflationary growth respecting the environment' (Article 2). Further, Article 3 was adopted to implement this environmental policy and Article 100 was amended to explicitly take a high level of environmental protection (Article 100a) as a base. Also included were Articles 130r, 130s, and 130t. These Articles provide specific justification for environmental protection laws, even where there is no direct link to the economic aims of the EC.

Article 130r(1) lists as objectives of the EC's environmental activities the preservation, protection, and improvement of the quality of the environment; the protection of human health; and the prudent and rational utilisation of resources. Article 130r(2) (now Article 174) sets out the four central principles of EC environmental policy, which must be taken into account when framing policy and legislation:

1. Preventative action should be preferred to remedial measures.
2. Environmental damage should be rectified at source.
3. The 'polluter pays' principle should be followed.
4. Environmental policies should form a component of the EC's other policies.

Article 130s provides the mechanics by setting out the voting procedures in the Council. Unanimity was originally required. However, the Treaty on European Union, agreed at Maastricht in 1992, relaxed the requirement and now most environmental measures can be agreed to by qualified majority voting. There is a derogation for measures significantly affecting a Member State's choice between different energy sources and the general structure of its energy supply (Article 130s(2) – now Article 175), which retains the unanimity requirement.

Article 130t (now Article 176) specifically provides for the possibility that stricter measures than those agreed under Article 130s may be employed by Member States, as long as they are compatible with the rest of the Treaty.

Article 235 (now Article 6) is a catch-all power that permits the institutions of the EC to take appropriate measures to attain any of the objectives of the EC that cannot be achieved through other powers.

8.14 TYPES OF EUROPEAN ENVIRONMENTAL LEGISLATION

Environmental legislation falls into a number of different categories that are divided by some into anticipatory and continuing controls.[29] Anticipatory controls relate mainly to the planning and consent stage of most operations, whereas the continuing controls relate to the operational phase.

Anticipatory controls are measures in which controls are imposed on an activity at its commencement in order to mitigate any potential environmental problems. This often results in the prevention of any activity unless certain requirements or conditions are met. This category includes a wide range of activities that require an authorisation before their commencement or continuation. These are often complemented by a combination of criminal and/or administrative sanctions if the activity starts without the required authorisation or if any conditions attached are contravened. Examples of anticipatory controls are requirements for a minimum of information prior to an authorisation being granted; consent systems that provide for conditions to be applied to any activity for which consent is required; and designation of certain areas to provide protection for habitats or species.

Continuing controls are measures where the carrying out of an activity is controlled on a continuing basis and typically relates to the way an activity is carried on. These could be described as operational activities and examples of such controls are compliance with emission standards that govern the output from certain processes; and an ongoing duty to comply with the terms of a consent, licence, authorisation, or permission granted by a pollution control authority, which will normally be combined with a range of other regulatory controls relating to monitoring and enforcement.

Inevitably, anticipatory and continuing controls are mutually supportive and most regulatory systems combine the two types of mechanisms.

8.15 DIRECTIVES WITH PARTICULAR RELEVANCE TO THE ENERGY SECTOR

8.15.1 The Environmental Assessment Directive

The EC Directive on the Assessment of the Effects of Certain Private and Public Projects on the Environment[30] requires an environmental impact assessment (EIA) and statement prior to the granting of consent for development. This may affect many areas of the energy production and supply sectors. The aim of the Directive is to identify, by expert analysis, all the potential environmental impacts (EIs) that may be associated with a proposed development and to submit these together with the application for consent to develop, to the relevant decision-making body.

Development projects are categorised into Annex I[31] and Annex II.[32] The Directive also sets out the detailed requirements for an assessment, which should include direct and indirect effects of a project on a variety of factors, including human beings, fauna, flora, the environment and material assets, and the cultural heritage.

The developer must also submit certain specified information to the authority dealing with the application. There is also a requirement for consultation with both the authorities likely to be concerned with the project and members of the general public, although the detailed arrangements for consultation are left to individual Member States.

Under the 1997 amendments, which came into force in March 1999, *all* projects that are likely to give rise to significant environmental effects will be required to obtain a development consent. This now clarifies the position in UK law for projects that are exempt from the need to obtain planning permission either by virtue of not being 'development' or by having permitted development rights.

For all Annex II projects, Member States are required to adopt the criteria for assessment on a case-by-case basis or by means of thresholds and criteria for specific classes of project. Currently, the United Kingdom lays down indicative criteria and leaves the decision as to whether an assessment is required to the discretion of the decision-making body on a case-by-case basis, thus mixing the two approaches. Whichever method is chosen, all Member States must take into account criteria for selecting Annex II projects that are set out in the new Annex IIA. These criteria include the characteristics of the project, the sensitivity in terms of location, and the nature of the impacts involved. Any decision regarding whether or not an assessment is required must be made public.

Member States are also required to introduce measures to ensure that, if a developer makes a request prior to submitting an application, the competent authority is under an obligation to give an opinion on the information that is supplied. The competent authority may also request further information should they feel it is necessary at a later date.

The developer must provide all the main alternatives to the preferred option supplied as part of the assessment, and an indication as to the reasons for selecting the preferred option must be given, taking into account the environmental effects.

It is the duty of the competent authority to make public the results of any consultations with a developer in addition to whether the authority has decided to grant or deny an application. In the case when consent has been granted, the authority is under a duty to give reasons for the decision and a description of the mitigating measures that are proposed. This new transparency will make challenging the decision by way of judicial review much easier and may even facilitate the introduction of third-party rights of appeal against the grant of development consent.

These amendments have extended the scope of the Directive significantly. Although some changes are the result of clarification or rationalisation of existing definitions, there are additional projects in both lists under the old Annex I and Annex II.

Although in the United Kingdom there are distinct and separate procedures for dealing with the granting of development consent and pollution control authorisation, under the amendments to the Directive, Member States may introduce a single procedure to deal with both authorisations.

8.15.2 The Environmental Impact Assessment Directive and Direct Effect

The modern practice as far as the transposition of European Law into state law is concerned is to translate the Directive 'literally' into the legislation, but this was after the Environmental Impact Assessment Directive came into force. Moreover, most Member States failed to implement the Directive within the time laid down, which led to a number of challenges to decisions made on projects that purportedly fell within the scope of the Directive. As far as the UK cases were concerned, the central issue was one of whether the Directive had direct effect within the UK legislation.

Unfortunately, this issue remains unclear. Most of the cases involved projects that fell within one of the Annexes of the Directive and were applied for or published before the date of the implementation of the Directive but received consent afterwards; or which were applied for after implementation of the Directive but before the implementation date of the relevant domestic legislation.[33]

McCullough J. appeared to accept that the Directive passed the test of being 'unconditional and sufficiently precise' in its terms and was therefore capable of having direct effect in relation to all projects under the Directive in the *Twyford Down* case.[34] Subsequently, in the Scottish court of Sessions in the case of *Kincardine and Deeside District Council v Forestry Commissioners*,[35] Lord Couldsfield held that Article 4.2 of the Directive[36] was not 'plain and unconditional' as it allowed the authorities a discretion not as to the means of implementation of the Directive, but as to whether an assessment was required. Therefore, the Directive would not appear to have direct effect in respect of Annex II projects.

Further confusion was created by the decision of Tucker J. in the *Wychavon* case.[37] This was another 'pipeline case' in which the High Court held that the Directive had no direct effect in relation to any Annex I or Annex II projects, and in doing so, Tucker J. found it unnecessary to analyse each Article of the Directive in order to identify whether every one was sufficiently precise. If only one Article were sufficiently imprecise, this offended the principle that lay behind the direct effect doctrine.

Following this decision, there were three separate judicial views on the direct effect of the Directive, none of which referred to the others and one of which was based on the interpretation of European law, which appeared to be contrary to the view of the ECJ. However, Hidden J. preferred to follow the *Wychavon* decision and held that the Directive did not have direct effect.[38] However, in 1998, the Court of Appeal decided in this case that the Directive was directly applicable.[39] The issue before the court was whether the Directive applies to setting conditions by virtue of section 22 of the 1991 Act on Interim Development Orders or whether the section 22 determination entitled the developer to proceed with the project. It was decided that the rationale of the Directive was to subject to EIA the approval of those private and public developments that are likely to have a significant effect on the environment. The primary obligation of Member States is to adopt all measures necessary to ensure that, before consent is given, projects likely to have such effects by virtue inter alia of their nature, size, or location are made subject to an EA (Article 2(1)). The Court, therefore, adopted the purposive approach and concluded that the Directive did apply.

The ECJ has now considered the direct effect of the Directive in *Annemersbedriff P.K. Kraaijeveld BV* v *Gedeputeerde Staten von Zuid-Holland*.[40] The Court was asked to rule specifically on the question of the direct effect of Article 4.2 of the Directive, which provides for the assessment of Annex II projects. The applicants had challenged the modification of a zoning plan that dealt with the reinforcement of dykes, arguing that the works were subject to the requirement of environmental assessment under the terms of the Directive. The project itself fell within Annex II of the Directive but the modifications fell below the threshold as set out in the domestic legislation. However, the ECJ did not deal with the issue specifically, but held that anyone concerned with the implementation of the Directive could raise the issue of an improper use of discretion in selecting the method of transposition.

This does not help with the issue of certainty so far as Annex II procedures having direct effect is concerned, but it does undermine the certainty that the whole Directive does not have direct effect as per the *Wychavon* decision.

8.15.3 Procedures for an Environmental Impact Assessment

Most projects in the energy production and supply sector must have an environmental impact statement prior to the granting of consent for development. The aim of such a statement is to provide information by expert analysis to any decision-making body as part of the application for development consent. As part of the process, there is also an emphasis on consulting the public.

There is no statutory provision as to the form of an environmental statement, but it must contain information as mandated in Schedule 3 of the UK Regulations,[41] which reflect Articles 3 and 5 of Annex III of the Directive.
Factors to be included in the statement are:

- A description of the proposed development including:
  - The site
  - The design;

The size and scale of the development
- All data necessary for the assessment of the main effects that the development may have on the environment
- The impact of the project on:
  - Human beings, fauna, and flora
  - Soil, water, air, climate, and the landscape
  - Material assets and cultural heritage; and
  - Any interaction of the above
- Of prime importance is that where significant adverse effects are identified, a description of the measures proposed to mitigate the effects must be included.

Should the developer wish to include a further supporting statement that could amplify or explain any of the information given then this could include:

1. The physical characteristics of the proposed development
2. The main characteristics of the production processes involved
3. The estimated type and quantity of emissions resulting from the proposed development
4. The main alternatives studied under Best Practicable Environmental Option (BPEO)

The environmental statement must also contain a non-technical summary of the information, which enables non-experts to understand the findings.

Each type of information indicated above must address the construction and commissioning stage as well as the operational stage. The UK Department of the Environment has published a *Guide to Environmental Assessment*, which includes a checklist of items for inclusion in any environmental statement.

Annex II(13) also requires that 'any change or extension of projects listed under Annexes I and II, already authorised, executed or in the process of being executed, which may have significant adverse effects on the environment' are subject to an environmental assessment under Article 4(2).

8.15.4 Strategic Developments

Towards the end of 1996, a draft Directive[42] was proposed to deal with strategic environmental assessment issues. Strategic assessment would operate in a similar way to project-based assessment but would address policy-based plans, the aim being to simplify the procedures for assessing individual projects. The developers would then be able to incorporate any information from the strategic environmental assessment into the environmental statement for the individual project. It is also the intention that Annex II projects would no longer need a specific environmental assessment as sufficient information would be available under the strategic environmental statement. The UK government are of the opinion that under UK planning law there is sufficient strategic assessment provided under the requirements of the current strategic plan system.

8.15.5 The Integrated Pollution Prevention and Control Directive (IPPC)

A further directive with particular relevance for the energy sector is the Integrated Pollution Prevention and Control (IPPC) Directive.[43] This Directive follows a different approach to controlling environmentally harmful activities at the consent stage. It concerns the emissions to air, water, and land from industrial activities and requires that these should be subject to the granting of a permit which identifies certain conditions to prevent or reduce those emissions. The conditions lay down measures designed to 'prevent or, where that is not practicable, to reduce emissions'.[44]

Historically, emissions to these three media were regulated separately from each other, and often controlled by different regulatory bodies. The imposing of emission limits to one medium would often result in the shifting of the polluting emissions to another. The Directive seeks to address this problem by considering all media when granting a permit. Each Member State may define the mechanisms for control, and so it is still possible for different agencies to be involved in the licensing process. Nevertheless, to ensure that a fully integrated approach is taken, Member States must introduce a fully coordinated consent procedure that involves all the relevant regulatory bodies.

The United Kingdom has had an Integrated Pollution Control (IPC) system in place since it was introduced under Part 1 of the Environmental Protection Act 1990 and both this and the new European IPPC Directive are discussed in more detail in Chapter 10.

The types of activities to which the IPPC Directive applies are laid out in Annex I to the Directive and those relevant to the energy sector are:

- Combustion installations with a rated thermal input exceeding 50 MW, or where the sum of all combustion installations on the same site exceeds 50 MW
- Mineral oil and gas refineries
- Coke ovens
- Coal gasification and liquefaction plants

Excluded from the provisions of the Directive are installations or parts thereof that are used for research, development, and the testing of new products and processes. Permits under the IPPC Directive are required for all existing installations no later than October 30, 2007.[45] Some provisions, mainly relating to inspections and exchange of information, must be applied by October 30, 1999 under Article 4, and there are interim measures, which will apply until the respective implementation dates. A new Directive on Industrial Emissions 2010/75/EU was adopted on November 24, 2010. It entered into force on January 6, 2011 and has to be transposed into national law by Member States by January 7, 2013. The IED replaces the IPPC Directive and the sectoral directives as of January 7, 2014, with the exemption of the LCP Directive, which will be repealed with effect from January 1, 2016.

8.16 THE PROTECTION OF HABITATS AND SPECIES

In recent years, the conservation of nature has become a popular subject that has drawn the attention of legislators. This is partly owing to the dramatic growth of interest in all things connected with nature and conservation, but also as a consequence of the rate of decline in and loss of natural environment. There is now an extensive international legal framework through international conventions and European Directives that may have relevance for the energy sector, in that they can have an effect on development.

The Bern Convention on the Conservation of European Wildlife and natural Habitats (Cmd. 8738, 1979), was, arguably, the driving force behind the two main EC Directives discussed later. Additionally, the Ramsar Convention on Wetlands of International Importance (Cmd. 6465, 1971) has played a key role in relation to the protection of one particular habitat that the United Kingdom has in relative abundance. There is also the Bonn Convention on the Conservation of Migratory Species of Wild Animals, 1979, and the Washington Convention on International Trade in Endangered Species of Wild Fauna and Flora (Cmd. 6647, 1973).

However, with regard to the energy sector, the two EC Directives that undoubtedly have the greatest influence are the Wild Birds Directive[46] and the Habitats Directive.[47] The main reason for this relates to the specific requirements laid down by these Directives, which concern the designation of protected areas and their impact on future development.

8.16.1 The Habitats Directive

The full title of this Directive refers to the Conservation of Natural Habitats and of Wild Fauna and Flora. However, although there are some provisions to protect individual animals and plants, its main importance is in the protection of habitats, and as such is usually referred to as the *Habitats Directive.*

The central feature of this Directive is that it provides for the creation of a network of special conservation areas, which will make up a system of European sites known as *Natura 2000.* The network will consist of sites of the type listed in Annex I of the Directive and in addition sites containing the habitats of the species listed in Annex II. Incorporated into the network will be the special protection areas classified under the Wild Birds Directive. The Directive also contains general duties that cover the monitoring of the conservation status of all habitats by the Member States,[48] and a general obligation regarding the management of certain important landscape features.[49]

For sites adopted by the European Commission as sites of Community importance and special protection areas designated under the Wild Birds Directive, any plan or project not directly connected with the management of the site that is likely to have a significant effect on it must be subject to an appropriate assessment of the implications; the competent national authority can then agree to the plan or project only if it will not 'adversely affect the integrity of the site concerned'.[50]

If there are no alternative solutions, a plan or project may be given permission if there are 'imperative reasons of overriding public interest, including those of social or economic nature'. It may well be that projects in the energy sector could be recognised under this exception. In such a case, the Member State has a duty to take compensatory measures to ensure the overall coherence of Natura 2000[51] and must inform the Commission of those measures.

8.16.2 The UK Regulations

Rather than remodel the law on nature conservation, the UK government decided to simply add the additional protections for 'European Sites' as required by the Directive to the existing town-planning procedures and site designations.

European Sites are defined as follows:

1. A special area of conservation[52] (Regulation 8)
2. A site adopted by the Commission as a site of Community importance
3. A special protection area designated under the Wild Birds Directive

Where a special nature conservation order has been made, Regulation 24 sets out the procedures under which an owner or occupier may apply for consent to carry out a plan or project. English Nature must then decide whether the plan or project is likely to have a significant effect on the site, it is then under duty to carry out an appropriate assessment and must refuse consent, unless it is satisfied that the plan or project will not affect the integrity of the site. The owner or occupier may then appeal to the Secretary of State, who is given the power to direct English Nature to grant consent if there is no alternative solution and the plan or project must be carried out 'for imperative reasons of overriding public interest'.[53]

8.16.3 European Marine Sites

European sites that are marine or tidal are less strictly regulated. However, there is a duty on every public body having functions relevant to marine conservation to exercise its functions in order to secure compliance with the requirements of the Directive.[54]

8.17 EUROPEAN LEGISLATION COVERING INDUSTRIAL ACTIVITY IN THE ENERGY SECTOR

Further to the more general European environmental Directives covering habitats and species conservation, there is a body of legislation that covers the operational phase of industrial activities, relating to energy production, processing, transmission, and transport.

Marine oil pollution is addressed extensively under international law (see Chapter 5) and the EC has adopted the 1974 Paris Convention for the Prevention of Marine Pollution from Land-Based Sources.[55] Offshore installations are included in its definition of land-based sources.[56] The Convention requires that the discharge into the sea of those substances identified and placed on a 'black list' must be eliminated. The discharge of those substances identified and placed on the 'grey list' must be licensed and reduced.

Further, the Bonn Agreement for Co-operation in Dealing with Pollution of the North Sea by Oil and Other Harmful Substances was ratified by the Community in 1994.[57] The Bonn Agreement concerns exchange of information and technical assistance in the event of severe marine oil pollution.[58]

8.17.1 Hydrocarbon-Based Motor Fuels

With the primary aim of controlling pollution from such fuels, the use and composition of hydrocarbon-based motor fuels is highly regulated by the Community. This is mainly achieved by laying down standards for the composition, storage, and distribution of fuels marketed for use in vehicles.

The standards for liquid fuels are constantly revised and are expected to become more stringent in the year 2000 and further reductions are projected for 2005. Targets are set long term to allow both the vehicle manufacturers and oil refineries to implement the necessary technologies. Although the expected values for each substance are strict, temporary derogations will be permitted for Member States who can show that social or economic hardship would otherwise result. Directive 98/70/EEC has since been updated by Directive 2000/71/EC and Directive 2003/17/EC. The latest update was Directive 2009/30/EC which came into force on June 25, 2009 and was transposed into national legislation by December 31, 2010. Directive 2009/30/EC also covers the content of lead in petrol.

The *lead* content in petrol for use in vehicles has long been regulated under the Directive 85/210/EEC.[59] This Directive also regulates the benzene content of unleaded petrol, and requires Member States to ensure the availability and balanced distribution of such petrol. If it is considered justified for the protection of human health or the environment, Member States are permitted to prohibit the sale of leaded petrol with a motor octane number of less than 85 and a research octane number of less than 95. However, there is a condition that the Member State promotes the availability and balanced distribution of unleaded petrol. Currently, unleaded petrol must not exceed a lead concentration of 0.013 g Pb per litre, and in leaded petrol the concentration of lead must lie between 0.40 g Pb per litre and 0.015 g Pb per litre. The marketing of leaded petrol has been banned from the year 2000, with a 5-year derogation for Member States who could demonstrate that a ban would cause 'severe socio-economic problems'. The sulphur standards have been further updated by the European Directive 2011/63/EU which was required to be transposed into national law by June 2, 2012.

The purpose of regulating the *sulphur* content in fuels is to reduce emissions of sulphur dioxide that affect the catalytic converters fitted to vehicles. Limits are set on the allowable sulphur content in gas oils that are marketed in the Community.[60] Member States must also prohibit the marketing of diesel fuels with a sulphur content exceeding 0.05% by weight, and all other gas oils with a sulphur content exceeding 0.2% by weight, with the exception of aviation kerosene.

8.17.2 The Storage and Distribution of Petrol

The precise design and operation standards for the storage of petrol and its distribution from terminals to service stations are set out in the Annexes to the Directive covering volatile organic compounds (VOCs).[61] The Directive also lays down specific technical standards and target levels to limit the escape of VOCs into the atmosphere during the course of specified operations. Member States may also lay down alternative requirements if it can be shown that they are equally effective at meeting a target reference value.[62] They may also be more stringent where specific conditions so require for the protection of human health or the environment. The provisions apply to new installations as from December 31, 1995. Provisions for existing installations are phased in by schedule depending on the type and scale of operation. The largest installations must have complied by December 31, 1998, and all installations must comply by December 31, 2004.

The following categories are covered by the Directive:

a. Article 3: Storage installations at terminals
b. Article 4: Loading and unloading of mobile containers at terminals[63]
c. Article 5: Design and operation of mobile containers
d. Article 6: Loading into storage installations at service stations

8.17.3 Emissions from Large Combustion Plants

The EC introduced a general framework Directive[64] to enable the drawing up of a system of limit values for a whole range of emissions. The daughter Directive[65] on Large Combustion Plants (LCPs) was introduced to further the programme by reducing sulphur dioxide (SO_2) and nitrous oxide (NO*x*) emissions.[66] These provisions apply to plants with a rated thermal input of 50 MW or more, but excluding coke battery furnaces and plants powered by gas turbines. However, it is expected that plants powered by gas turbines, with a rated thermal input of 50 MW or more will eventually be brought within the scope of the Directive.

The Directive sets emission limits for each Member State, ending in the year 2003, for the total national emissions from existing large combustion plants.[67] This creates a 'bubble'. for each Member State and so avoids setting uniform limits for plants, but retains an emission level that is progressively reduced. This permits the Member State to offset the emission levels from the more polluting plants against those from less polluting plants. On December 21, 2007 the EC adopted a proposal for a Directive on Industrial Emissions which includes the LCP Directive. This new Directive is still being drafted.

However, each new plant must comply with the emission limits set in the conditions of its licence. These in turn are set out in Annexes III to VII under Article 4. The plants must discharge all waste gases from a stack.[68] Exceptions and derogations from these provisions include inter alia: a limit value of 88 mg/Nm^3 for new plants with over 400 MW capacity, operating for less than 2200 hours per annum;[69] indigenous fuels enjoy less stringent conditions if the nature of the fuel makes it impossible to meet the sulphur dioxide limits,[70] this being subject to Article 6, which provides for the use of 'best available technology not exceeding excessive costs' (BATNEEC). Furthermore, an exemption may be granted for up to 6 months when interruptions in fuel supply make it unrealistic for the emission levels to be met.[71] These derogations must be notified to the Commission when invoked. They must also be monitored at the operator's expense by methods given in Annex IX to the Directive.

8.18 COMMUNITY ACTION: ENERGY TAX AND INDUSTRY

The proposed CO_2/Energy tax[72] is part of the 'climate changes' package assembled by the Commission and unveiled in 1991,[73] aimed at reducing the 'greenhouse effect' by stabilising carbon dioxide emissions at their 1990 levels by the year 2000. Climate changes were one of the main topics at the Rio de Janeiro Conference of the United Nations on the Environment and Development, and as a result the *United Nations Framework Convention on Climate Change*[74] adopted the above objectives for emission reduction.

The proposal for an energy tax ran into immediate and firm opposition from industry. The criticism evolved around many arguments. First, a major price rise in oil, gas, and coal would not necessarily effect a reduction in CO_2 emissions, which should be viewed as a global and not just a Community problem. Secondly, if the CO_2/Energy tax were set at a higher level than corresponding taxes imposed by the main trading partners of the Community, it would only worsen the competitiveness of Community industry, leading perhaps to a shift of production facilities to outside of the Community, which would imply job losses without reduction in worldwide CO_2 emissions.

The Commission, for its part, claimed that the proposal had been specifically structured to protect the competitive position of the European industry. The Commission pointed to the extensive exemptions available for the energy, steel, paper, and cement industries and to the possibility of tax breaks for companies already making efforts to save energy or reduce CO_2 emissions. The Commission also stressed that it had made its proposal dependent upon the willingness of the Community's main trading partners to adopt equivalent measures.

These arguments have so far failed to convince all Member States to support the Commission's proposal.[75] Throughout 1993, the proposal was debated in various meetings of the Council when many Member States were in favour of the proposal forming part of a combined initiative at the Organisation for Economic Co-operation and Development (OECD) level. The United Kingdom opposed the idea of a new tax being imposed by the Community, and France wanted electricity produced from nuclear sources to be excluded. The poorer Member States including Spain, Portugal, Greece, and Ireland requested that they be exempt from imposing the tax owing to their economic situation and their low emission level. A burden-sharing formula was introduced in the course of discussions and, to appease opposition to the tax, it was linked to a reduction in labour charges.[76]

In December 1993, the proposal for the tax was sent back to a Council working group. The mandate of this group was made more precise in March 1994 to include burden sharing; the CO_2 and energy components of a tax; the relationship with excise duties; and the impact on competitiveness. Member States still disagree on the best way to comply with their international duties to reduce CO_2 emissions. However, they are almost unanimous in acknowledging that fiscal measures of some sort will prove necessary. In 1997, the Commission put forward a proposal for a Community framework for taxation of all sources of energy. This was extensively changed before being adopted as Directive 2003/96/EC. The Directive entered into force on January 1, 2004. On April 13, 2011 the European Commission presented its proposal to restructure this Directive in order to promote energy efficiency and consumption of more environmentally friendly products and to avoid distortions of competition in the single market. These proposals are currently being debated.

8.19 ENFORCING EUROPEAN ENVIRONMENTAL LAW

The EC Treaty places the Commission at the heart of the system for the enforcement of Community law. Article 169 provides that

> If the Commission considers that a Member State has failed to fulfill an obligation under this Treaty, it shall deliver a reasoned opinion on the matter after giving the State concerned the opportunity to submit its observations.
>
> If the State concerned does not comply with the opinion within the period laid down by the Commission, the latter may bring the matter before the Court of Justice.

The enforcement actions under Article 169 thus comprise an administrative as well as a judicial phase. The vast majority of proceedings commenced never reached the European Court. In its twelfth report on monitoring the application of Community law,[77] the Commission notes that in 1994, as regards Member State failure to notify measures implementing environmental directives, it issued 42 Article 169 letters, 46 reasoned opinions, and only three referrals to the European Court.[78]

For a long time, criticism of this enforcement mechanism focused upon the declaratory nature of the European Court's judgements, finding a Member State to be in breach of its EC law obligations. On the entry into force of the Maastricht Treaty, the Commission may now institute a new action against any Member State failing to comply with a judgement of the Court. The European Court may, in such circumstances, impose a lump sum or penalty payment upon the offending Member State.[79] The scale of the potential fines involved[80] and the widespread publicity has had the desired effect of 'persuading' recalcitrant states to review their environmental legislation.

The Commission's capacity to perform its enforcement function is, however, contingent upon the quality of the information it receives. In the absence of a Commission environmental inspectorate, complaints by private parties and non-governmental organisations (NGOs) are the main mode in alerting the Commission to possible infringements. It is significant that the Commission stops short of proposing any such formal inspectorate or enforcement function for the European Environment Agency; instead it proposes to consider adopting guidelines establishing minimum criteria for Member State inspections in relation to the monitoring of industrial emissions and environmental quality standards. It is proposed that it may be necessary to combine this with the establishment of 'a limited Community body' with responsibility for auditing Member State fulfilment of their inspection duties.[81]

Further, the Commission also proposes to consider adopting guidelines governing nonjudicial complaint investigation procedures in Member States. This would result in complaints, in the first instance, being investigated by the appropriate body within the Member State.

8.20 THE INTERRELATIONSHIP BETWEEN THE THIRD ENERGY PACKAGE AND THE CLIMATE CHANGE PACKAGE

The EU legislative landscape in the energy sector has seen some significant changes over the last few years inasmuch as in September 2007, the European Commission published a proposal for a Third Energy Package (TEP), which was finally adopted in July 2009 and came into force on September 4, 2009. The main focus of the TEP is to further liberalise and harmonise the European internal energy market. It includes, inter alia provisions as to the unbundling regime, the role of the national regulators and investments by third countries in European transmission systems.[82]

In January 2008, the European Commission had proposed a legislative package on a range of measures designed to shape the European Union's climate change policies and actions, which was adopted under the co-decision procedure. The Council adopted the new legislation in April 2009 and thus passed the Climate Change Package into European Law. The following analyses the European Directives and Regulations of the Two Packages and their effects at European level.

8.20.1 The Policy Context

In 2005, the European Commission undertook an inquiry into competition in the gas and electricity markets under Article 17 of Regulation 1/2003. This regulated the implementation of the EC Treaty rules on competition and was aimed at assessing the prevailing competitive conditions and establishing the causes of the perceived market malfunctioning. This sector enquiry examined eight of the key areas.

10.2 Market concentration
10.3 Vertical foreclosure
10.3 Lack of market integration
10.4 Lack of transparency
10.5 Price formation
10.6 Downstream markets
10.7 Balancing markets
10.8 Liquefied natural gas (LNG)

8.21 THE THIRD ENERGY PACKAGE

On October 6, 2011, the Ministerial Council of the Energy Community of the EU rules on the internal market for electricity and gas (known as the *Third Energy Package*) was adopted. The Energy Community was established on July 1, 2006, and it links the European Union to Albania, Bosnia Hezegovina, Croatia, the Former Yugoslav Republic of Macedonia, Kosovo, Moldovia, Montenegro, Serbia, and Ukraine. The TEP Decision is an attempt to create a pan-European energy policy in the absence of any further enlargement of the European Union in the immediate future. It will affect investors and energy undertakings across South-East Europe as transmission assets will now gradually need to comply with the unbundling regime of the European Union, and non-EU Contracting Parties are likely to transpose the TEP Decision at differing times.

In 2005, the European Commission undertook an inquiry into competition in gas and electricity markets under Article 17 of Regulation 1/2003 on the implementation of the EC Treaty rules on competition. The aim was to assess the prevailing competitive conditions and establish the causes of the perceived market malfunctions. As a result of the inquiry, the Council and Parliament adopted a proposal for the TEP, which came into force in September 2009.

The TEP contains three Regulations and two Directives as follows:

- Directive 2009/72/EC concerning common rules for the internal market in electricity and repealing Directive 2003/54/EC (the New Electricity Directive)
- Directive 2009/73/EC concerning common rules for the internal market in natural gas and repealing Directive 2003/55/ECA Gas Directive, amending and completing the existing Gas Directive 2003/55 (the New Gas Directive)
- Regulation (EC) No 713/2009 establishing an Agency for the Cooperation of Energy Regulators (the ACER Regulation)
- Regulation (EC) No 714/2009 on conditions for access to the network for cross-border exchanges in electricity and repealing Regulation (EC) No 1228/2003 (the New Electricity Regulation)
- Regulation (EC) No 715/2009 on access to the natural gas transmission networks and repealing Regulation (EC) No 1775/2005 (the New Gas Regulation)

With this introduction of unbundling options, the TEP does not, in fact, bring in the unbundling that the Commission originally envisaged, but it does help to clarify a number of issues concerning the existing rules, in particular in respect of third-party access exemptions. The TEP also gives legal status to the Commission Guidelines in addition to expanding the competence of the Commission in respect of the issuing of Guidelines in a number of policy areas.

The harmonisation of the competence of the National Regulatory Agencies (NRAs) and the introduction of ACER has been interpreted as a step further towards a European Regulator. However, some have expressed concerns that the NRAs will be kept busy with an array of monitoring tasks, whereas decision-making power in respect of regulatory policy will be claimed by the ACER. Although the full effect of the TEP will only become apparent in 2013, some are already wishing for a Fourth Energy Package.

8.22 THE EU CLIMATE CHANGE PACKAGE

The Climate Change package contains four Directives, one Regulation, and one Decision as follows:

- Directive 2009/29/EC amends Directive 2003.87/EC with the aim of improving and extending the greenhouse gas emission allowance trading scheme (the New EU Emissions Trading Scheme (ETS) Directive).
- Decision No 406/2009/EC on the effort of Member States to reduce their greenhouse gas emissions to meet the Community's greenhouse gas emission reduction commitments up to 2020 (the Greenhouse Gas [GHG] Reduction Decision).
- Directive 2009/28/EC on the promotion of the use of energy from renewable sources amending and, subsequently, repealing Directives 2001/77/EC and 2003/30/EC (the Renewable Energy Directives).

- Directive 2009/31/EC on the geological storage of carbon dioxide, amending Council Directive 85/337/EEC, Directives 200/60/EC, 2001/80/EC, 2006/12/EC, 2008/1/EC, and Regulation (EC) No 1013/2006 [the carbon capture and storage (CCS) Directive].
- Directive 2009/30/EC as regards the specification of petrol, diesel, and gas-oil and introducing a mechanism to monitor and reducing GHG emissions and amending Council Directive 1999/32/EC as regards the specification of fuel used by inland waterway vessels and repealing Directive 93/12/EEC (the Biofuel Directive).
- Regulation (EC) No 443/2009 setting emission performance standards for new passenger cars as part of the Community's integrated approach to reduce CO_2 emissions from light-duty vehicles (the Emissions Standards Regulation).

8.22.1 The New EU ETS

The new Directive introduces a number of important changes to the EU ETS that will take effect from Phase III in 2013 to 2020, which should provide a clearer sense of the future of the scheme. The main changes include a declining emissions cap with longer trading phases, and an increase in the auctioning of allowances. This increase in auctioning will be introduced more slowly than was originally proposed by the European Commission when it first announced the climate and energy package in 2008.

The new Directive expands the EU ETS to cover both new activities and new gases including carbon dioxide emissions from the petrochemicals, ammonia, and aluminium sectors, in addition to nitrous oxide emissions from the production of nitric, adipic, and glyocalic acid, and perfluorocarbon emissions from the aluminium sector.

The inclusion of these new activities and gases confirm the fact that the EU ETS will continue to be focussed on the large energy-intensive sectors.

One of the main features of the New Directive is the increased harmonisation and centralisation of the operation of the scheme.[83] Owing to this centralisation, the allocation of allowances will be done centrally rather than under Member States National Allocation Plans. This caused some problems in the first and second phases. The National Allocation Plans caused competitive distortions within sectors due to different allocation rules being adopted by Member States. In addition, the administration of the New Entrant Reserve, which is equivalent to 5% of the total annual allowances to be awarded, will also be centralised. There will be a central register rather than one that is held nationally, and the proceeds from the auctioning of 300 million allowances reserved for new entrants to the EU ETS will be used to support renewable energy projects and up to 12 CCS demonstration projects.[84]

Overall, the European Union-wide allowance cap is decreased, and from 2013 the cap will decrease year on year by 1.74% of the Phase II cap from the total amount of 1.974 billion allowances in 2013 to 1720 billion in 2020.[85] After 2020, the cap will continue to decrease by 1.74% per annum, but this rate of reduction may be revised before 2025. Allowances issued from 2013 onwards will be permitted to be banked for use in subsequent phases of the scheme.[86]

A further major change will be the move away from allocation of allowances to operators free of charge to a process of auctioning of allowances. The phasing out of free allocation of allowances to a process involving compulsory auctioning will be progressive. From 2013, all allowances not allocated free of charge in accordance with Article 12 will be auctioned.

With regard to the electricity sector, stricter rules will apply inasmuch as no allowances will be allocated free of charge to electricity generators after 2013. However, there is an exception for district-heating schemes, high-efficiency combined heat and power schemes, and where eligible, Member States have opted to derogate from this rule.

If a global successor agreement to the Kyoto Protocol has been reached, then the limits on the use of joint implementation (JI) and clean development mechanism (CDM) credits will be automatically increased by up to half of the additional reduction effort and operators of participating installations may use certified emission reductions (CERs), emission reduction units (ERUs), or other approved credits from the third countries that have ratified the international agreement on climate change succeeding the Kyoto Protocol.[87] A further change from current practice is the extension to cover the capture, transport, and storage of carbon dioxide.[88]

Member States are required to transpose the New EU ETS Directive into national law by December 31, 2012.

8.22.2 The Greenhouse Gas Reduction Decision

This Decision sets binding targets for individual Member States for sectors of the economy not covered by the EU ETS and provides an indication of the extent to which Member States will be required to address and reduce emissions from sectors such as transport, construction, and agriculture over the next decade.

Member States already monitor and report greenhouse gas emissions annually and the GHG Decision provides that now if a report indicates non-compliance with a limit set for that year, the Member State will have to submit a corrective action plan to the Commission.

8.22.3 The Renewable Energy Directive

The Renewable Energy Directive provides for the promotion of the use of renewable sources for electricity generation and sets a target of 20% of total energy consumption to be generated from renewable sources across the European Union by 2020. The Directive also includes a further target of 10% for energy from renewable sources for each Member State's transport energy consumption.

To achieve these overall targets, the Renewable Energy Directive sets a mandatory national target for each Member State, which states the overall share of gross energy consumption that must come from renewable energy sources. This takes into account the differing levels of progress achieved by Member States to date.[89] The mandatory national targets provide certainty for investors and are aimed at encouraging technological development. Member States are required to follow an indicative trajectory towards achieving their target and each will have to produce a National Action Plan. These Plans will set national targets for the share of energy from renewable sources to be used to meet demands for transport, electricity, heating, and cooking by 2020. Each Member State is free to decide their preferred mix of renewable sources, but they were required to present their National Action Plans, based on an 'indicative trajectory', to the Commission by June 30, 2010.[90] Progress reports are to be submitted to the Commission every 2 years, and the plans must be constructed so that the three sectors are identified separately as electricity, heating and cooling, and transport.

Member States can apply financial support schemes in relation to the mandatory targets but it will not be required to link these with schemes in other Member States.

The Renewable Energy Directive contains a series of interim targets for all Member States to ensure a steady progress towards the final targets, which are as follows[91]:

- 25% of the overall 2020 target to be achieved between 2011 and 2012
- 35% to be achieved between 2013 and 2014
- 45% to be achieved between 2015 and 2016
- 65% of the overall 2020 target to be achieved between 2017 and 2018

However, there are no financial penalties for failing to achieve these interim targets, but the Commission does reserve the right to issue infringement proceedings if Member States do not take 'appropriate measures' to try to meet their targets.

In order to meet their targets, two or more Member States can cooperate on joint projects relating to the production of energy from renewable sources.[92] They can also join with non-EU countries on renewable electricity generation projects.[93] Member States may link their national support schemes to those of other Member States, and may, under certain circumstances, count the import of 'physical' renewable energy from third countries towards their targets. It may even be possible to count 'virtual' imports, based on investments in non-EU countries towards a Member State's national target.[94]

The Renewable Energy Directive states that Guarantees of Origin in relation to the renewable energy may only be used to prove the quantity of energy from the renewable sources in a suppliers energy mix to the final consumer, and Member States must ensure that a Guarantee of Origin is issued in response to a request from a generator of renewable energy, which will be given in relation to each 1 MWH generated.[95]

The Directive also establishes binding criteria to ensure that biofuel and bioliquid production are environmentally sustainable. These criteria relate to biodiversity; the protection of rare, threatened, or endangered species and ecosystems; and greenhouse gas emission savings.

After 2017, the greenhouse gas emissions savings resulting from the use of biofuel production plants must be 50% compared to the emissions from using fossil fuel-powered plants.[96] Those installations that commence production after January 1, 2017 must emit 60% lower greenhouse gas emissions from the use of biofuel than those from fossil fuels. It is thought that the second-generation biofuel does not present the same risks to the security of food supplies as the first generation. The second-generation biofuels are normally produced from wastes, residues, or biomass such as algae, wood residues, or paper waste.

Previously, small producers of renewable electricity argued that there was a lack of transparency and restricted access to electricity grids prevented them from competing in the market. The new Directive now requires Member States to ensure that transmission and distribution system providers give priority access to the grid for all electricity produced from renewable sources. Member States are also required to develop transmission and distribution grid infrastructure, intelligent networks, storage facilities, and systems that can operate safely while accommodating renewable generation.[97]

Member States are required to assess whether there is a need to build new district infrastructure for heating and cooking using energy produced from renewable sources and include any such need in the National Action Plan. This will include large biomass, solar, and geothermal facilities. Local and Regional authorities are to be encouraged to include heating and cooling systems when planning city infrastructures

> . . . to ensure equipment and systems are installed for the use of heating, cooling and electricity from renewable sources, and for district heating and cooling when planning, designing, building and refurbishing industrial or residential areas.[98]

8.22.4 The Carbon Capture and Storage Directive (CCS)

The climate change and renewable energy package includes a directive providing a framework for CCS in the European Union, which supports CCS as an emissions reduction option. The main provisions are as follows:

- The creation of a permit-based CCS storage regime to be administered by Member States and the amendment of existing EU legislation that either prohibits or inhibits CCS.[99]
- The establishment of a regime for operators holding permits to pass long-term liability for leakage from storage sites to the licensing Member State, after certain 'handover' criteria have been met.[100]
- There are requirements for all new combustion plants in the European Union built without CCS to be capable of retrofitting, and by carrying out feasibility studies as to the availability of storage sites.[101]

The Climate Change Package provides that CCS will be financially viable by joining up the funding mechanism under the New EU ETS Directive and the Provisions of the CCS Directive from Phase III (2013–2020) and Member States can opt for the inclusion of CCS in Phase II (2008–2012). There is also an allocation of up to 300 million EU ETS allowances from the new entrant reserve to fund up to 12 CCS demonstration projects.[102]

Under the CCS Directive, carbon dioxide stored in a geological formation will not be classified as 'emitted' for the purpose of the European Union Emissions Trading Scheme (EU ETS); therefore, credit is given to power stations with CCS technology and they will not be required to surrender any allowances for carbon dioxide that is stored.

The CCS Directive provides for two types of permits. The first permit will be an exploration permit for the specified exploration works to be carried out and entitles the holder to explore the area covered by the permit for suitable geological formations on an exclusive basis.[103]

The second permit is a storage permit for the development and utilisation of geological formations within the permit area as storage sites for carbon dioxide and also allows the injection of carbon dioxide.[104]

There are rigorous criteria for the grant of either permit, which involve site characterisation in order to assess the site suitability for permanent storage in addition to technical and financial criteria. The permits will also contain provisions for the identification of delineation of the site, the operation of the facility, and the total amount and composition of the carbon dioxide stream, in addition to an approved monitoring plan.[105] The permits will be issued by the designated competent authority in each Member State, after review by the Commission. Before any permit is awarded, the competent authority must take into consideration any comments by the Commission.[106] The CCS Directive also includes provisions for any liability for damage resulting from leakage of carbon dioxide from a storage site, which includes damage from the local environment and the climate.[107] In respect of the environment, the CCS Directive is linked to the Environmental Liability Directive,[108] whereas liability for climate damage resulting from a leakage will be covered by the inclusion of CCS in the revised EU ETS Directive so that EU ETS allowances will need to be surrendered for leaked emissions.

The storage operator, who must take corrective measures to remedy any leakage, remains responsible for the storage site for as long as it represents a risk, until the site is handed over to the competent authority of the relevant Member State.[109] From this point of handover, the Member State assumes liability.[110]

Where there is a fault on behalf of the operator, including deficiencies in data, concealment of relevant information, negligence, wilful deceit, or a failure to exercise due diligence, the competent authority may recover any costs incurred by the authority, even after the transfer of responsibility has taken place.[111]

Part of the permitting regime is that Member States may require operators to provide financial security before the injection of carbon dioxide into the storage facility commences. In addition, Member States may require a contribution from the operator to cover future liabilities prior to handover of responsibility. This contribution shall be no less than the cost of monitoring the site for 30 years post-closure.[112]

There is no compulsory requirement for new power stations to have CCS initially, however, there is an obligation for a feasibility study as to what suitable storage sites are available and whether transport facilities are technically and economically feasible to retrofit the plant for carbon dioxide capture. There must also be suitable space on-site for the installation of equipment necessary to capture and compress carbon dioxide.[113] By amending the Directives relating to waste- and groundwater to permit the injection of carbon dioxide in storage sites, the Climate Change Package removes a number of the current prohibitions on storage under EU legislation.

Member States must transpose the CCS Directive into law by June 25, 2011 most of which have done so.

8.22.5 The Biofuels Directive

The Biofuels Directive was introduced to give a boost to the European biofuel market. It amends two Directives relating to the quality of petrol and diesel.[114] The changes bring in a mechanism for the reporting of and reduction in the lifecycle of GHG emissions from fuel; enable the more widespread use of ethanol in petrol; and tighten environmental quality standards for specified fuel parameters. The Directive obliges fossil fuel suppliers to reduce GHG emissions from their fuels throughout their life cycle by 6%.[115]

The most significant change brought about by the Biofuel Directive is the increase in the permissible content of biological components of petrol up to 10%. The Directive also provides for changes to the current diesel specifications, and allows a content of fatty acid methyl ester (FAME) of up to 7% by volume. Once implemented, these changes are likely to have a significant impact on fuel suppliers throughout the distribution chain as well as on fuel producers, who will have to adapt to meet the new quality criteria.

8.22.6 The Emissions Standards Regulation

Carbon dioxide emissions from road transport across Europe increased by 26% between the base year of 1990 until 2004, despite improvements in fuel efficiency. They now account for a third of all European Union total emissions and have demonstrated that a voluntary car industry reduction target would not be met. The Commission proposed the Emissions Standards Regulation, which sets the first legally binding standards for CO_2 emissions from passenger cars, which required reductions to a new EU average for new cars of 130 g of CO_2 per kilometre travelled, through the adoption of improvements in motor technology.

Owing to opposition from the car industry to bringing in new and more stringent targets by 2012, the obligations will now be phased in between 2012 and 2015. From 2012, 65% of a manufacturer's newly registered cars must comply with the manufacturer's target, followed by 75% in 2013, 80% in 2014, and 100% from 2015 onwards.[116] Credit will be given for very low emission vehicles, and in certain circumstances, for biofuel capability until 2016. Manufacturers of heavier cars will be allowed higher emissions than those of smaller cars, but will also be required to make larger cuts from current fleet average emission levels.[117]

If manufacturers decide to pool together to meet the emissions target as provided for under Article 7 of the Directive, they must have a nominated pool manager who will be responsible for paying any penalties, and evidence must be provided that it is sufficiently financially robust to pool the emissions. Pools must allow open, transparent, and non-discriminatory participation on commercially reasonable terms in order to discourage cartel behaviour as the usual anti-competition rules will apply.[118]

Small and niche manufacturers will benefit from lower targets, but the targets must still be achieved by 2012 and the same financial penalties will apply to larger manufacturers.[119]

Penalties are on an escalating scale with those who miss the target by a small amount being penalised less severely.[120]

The Emissions Standards Regulation is already in force and is directly applicable in all EU Member States.

8.23 LOOKING FORWARD UNDER THE EUROPEAN CLIMATE CHANGE REGIME

Taken as a whole, the Climate Change Package is the European Union's first attempt to create a comprehensive, legal, Europe-wide regime that covers carbon and the renewable energy sector. This regime will help investors by providing certainty under which they may make their decisions. It attempts to provide a secure future for carbon trading by laying the foundations for future investment in renewable technologies for biofuel and the development of CCS.

The Climate Change Package must be fully transposed into national law by 2012 and so the impact will not be fully known until some time later. However, this is only the start of the European Climate Change project and both the Commission and the European Parliament have made it clear that there will be stricter targets in the future and a greater drive to stimulate technological developments to ensure that industry players, particularly those in energy-intensive industries, will embrace those new technologies.

ENDNOTES

1. Article 2 ECSC Treaty.
2. The reason a separate Treaty was signed governing the non-military use of atomic energy was the fear that the EEC Treaty might be rejected by the French Parliament. It was hoped that, in such an eventuality, at least the EURATOM Treaty could be saved.
3. *Hopkins v National Power plc*, Case C-18/94 (1996) E.C.R. I-2281; *H.J. Banks & Co. Ltd v BCC*, Case (1996) 4 C.M.L.R. 745.
4. *Gerlach & Co. BV*, Case C-239/84 (1986) E.C.R. 3507.
5. *Deutsche Babcock Handel GmbH v*, Case 328/85 (1987) E.C.R. 5119.
6. See Robert Tudway, Editor (1999), *Energy Law and Regulation in the European Union*. Sweet & Maxwell: UK.
7. (1978) E.C.R. 2151 at 2176–2177.
8. COM (95) 682.
9. (1996) O.J. C224/1.
10. Directive 2009/72/EC.
11. Directive 2009/73/EC.
12. Directive 94/22/EC; (1994) O.J. L164/3.
13. Directive 90/377/EEC; (1990) O.J. L185/16.
14. Directive 90/547/EEC; (1990) O.J. L313/30.
15. Many of these were still national state-owned monopolies.
16. Directive 90/377/EEC.
17. Directive 90/547/EEC.
18. (1972) O.J. L74/43, amended (1997) O.J. L272/54.
19. Directive 91/296/EEC; (1991) O.J. L147/37.
20. (1995) O.J. L304/57.
21. Council Regulation 1056/72/EEC; (1972) O.J. L120/7.
22. (1996) O.J. L102/1.
23. (1996) O.J. L326/13.
24. (1994) O.J. L164/3.
25. (1993) O.J. L199/84.
26. (1997) O.J. L27/20.
27. (1998) O.J. L69/1; Trade amendment, Council Decision 98/537/EC; (1998) O.J. L252/21.
28. (1993) O.J. L69/93.
29. See Tudway, *Energy Law and Regulation in the European Union*, p. 1402.
30. Directive 85/337/EEC, as amended by Directive 97/11/EC.
31. It is mandatory for applications for development, which are listed in Annex I of the Directive, to have an accompanying environmental assessment.

32. The decision regarding Annex II projects is within the discretion of the decision-making body as to whether an environmental impact assessment is required.
33. These are generally known as *the pipeline cases.*
34. *Twyford Parish Council v Secretary of State for the Environment* (1993) 3 Env LR 37.
35. *Kincardine and Deeside District Council v Forestry Commissioners* (1993) Env LR 151.
36. This lays down the framework for Annex II projects.
37. *Wychavon District Council v Secretary of State for the Environment* (1994) Env LR 239.
38. See *North Yorkshire County Council, ex parte Brown* (1997) Env LR 391.
39. (1998) *J.E.L.*, 10:2, p. 363.
40. *Kraaivjeveld BV v Gedeputeerde Staten Van Zuid-Holland*, Case C-72/95 (1996) E.C.R. I-5403.
41. Town and Country Planning (Assessment of Environmental Effects) Regulations 1988.
42. (1997) O.J. C129/14.
43. Directive 96/61/EC.
44. Article 1.
45. 'Existing' installations are defined as operating on 30 October 1999, or being the subject of an application for authorisation by 30 October 1999, if it subsequently starts operating before 30 October 2000.
46. EC Wild Birds Directive 79/409.
47. EC Habitats Directive 92/43.
48. Article 11.
49. Article 10.
50. Article 6(3).
51. An alternative habitat may be created in another favoured area to replace the one affected by the project.
52. Once designated, the regulations reproduce the procedures and timetable set out in the Directive.
53. This is defined in Article 6(4) of the Directive and includes the more restrictive test for priority sites.
54. Regulation 3(3).
55. By Decision 75/437/EEC.
56. See Chapter 9.
57. Decision 84/358/EEC as amended by Decision 93/540/EEC.
58. The Agreement includes other substances as well.
59. As amended by Directive 87/416/EEC.
60. Directive 93/12/EEC relating to the sulphur content of certain liquid fuels.
61. Directive 94/63/EEC.
62. Under Article 2(1), 'Target reference values' are a technical measure of the effectiveness of a technique and do not refer to the performance of individual installations.
63. Under the Directive, 'mobile containers' means any tank used for the transfer of petrol transported by road, rail, or water from one terminal to either another terminal or a service station (Article 2e).
64. Directive 84/360/EEC.
65. Directive 88/609/EEC.
66. This Directive has since been amended by Directive 94/66/EC.
67. This includes any plant for which a construction licence or an operating licence was granted before 1 July 1987.
68. Article 10.
69. Article 5(1).
70. Article 5(2).
71. Article 8.

72. Proposal for a Directive introducing a tax on carbon dioxide emissions and energy, (1992) O.J. C196/1.
73. Communication on a Community strategy to limit carbon dioxide emissions and to improve energy efficiency. COM (91) 249 final. The proposal (and overall strategy) was further developed in the Commission Communication on a Community Strategy to limit carbon dioxide emissions and to improve energy efficiency, COM (92) 246 final.
74. (1994) O.J. L33/13.
75. As a tax proposal based on Article 99 EC Treaty, it requires unanimity in the Council.
76. See the White Paper on *Growth, Competitiveness and Employment.*
77. COM (95) 500 final.
78. See Macrory & Purdy (1997), 'The enforcement of EC environmental laws against member states', Holder, Editor, *The Impact of EC Environmental Law in the United Kingdom.* Wiley: Chichester, UK.
79. Article 171 EC.
80. A fine of up to a quarter of a million European currency unit (ECU) per day was imposed in respect of a German failure to respect Community law regulating the environmental quality of groundwater.
81. *Communication on Implementing Community Environmental Law,* COM (96) 500, p. 9.
82. This was very controversial and was the subject of intense negotiations among the Member States and between the Council and the European Parliament.
83. Article 1(12) of the New EU ETS Directive.
84. Ibid.
85. This is equivalent to an overall reduction of 21% in allowances available by 2020 compared to 2005.
86. Article 1(9) of the New EU ETS Directive.
87. Article 1(13) of the New EU ETS Directive.
88. See Section 8.22.4 on the CCS Directive.
89. Annex 1 of the Renewable Energy Directive.
90. Article 4 of the Renewable Energy Directive.
91. Article 3(2) and Annex 1(B) of the Renewable Energy Directive.
92. Article 7 of the Renewable Energy Directive.
93. Article 9, ibid.
94. Article 9 in conjunction with Article 10.
95. Article 15, ibid.
96. Article 17(2) of the Renewable Energy Directive.
97. Article 16, ibid.
98. Article 16 of the Renewable Energy Directive.
99. Articles 5–11 of the CCS Directive.
100. Articles 12–20 of the CCS Directive.
101. Article 33 of the CCS Directive, amending Directive 2001/80/EC.
102. Article 11(12) of the New EU ETS Directive (Article 10a(8) of the amended Directive 2003/87/EC).
103. Article 5 of the CCS Directive.
104. Article 6 of the CCS Directive.
105. Articles 7 and 8.
106. Articles 8(2) and 10 of the CCS Directive.
107. Article 16 of the CCS Directive.
108. 2004/35/EC.
109. Article 16 of the CCS Directive.
110. Ibid.
111. Article 20 of the CCS Directive.
112. Ibid.

113. Article 33 of the CCS Directive (Article 9a of the amended Directive 2001/80/EC).
114. Directives 98/70/EC and 2003/17/EC.
115. The Commission will review this target in 2012 in the light of technological advances, such as the use of electricity in transport, and will reserve the option to increase the reduction target by a further 2%. (Article 1(5) of the Biofuel Directive.)
116. Article 4 of the Emissions Standards Directive.
117. Article 5, ibid.
118. Article 7(5) of the Emissions Standards Directive.
119. Article 11, ibid.
120. Article 9, ibid.

9 Energy Law in the United Kingdom

9.1 ENERGY POLICY

In 2010, no political party achieved an overall majority in the general election and the major Conservative Party formed a coalition government with the Liberal Democrat Party.

On April 8, 2010, the Energy Bill of the previous government received Royal Assent to become the Energy Act 2010, which implements some of the key measures required to deliver Department of Energy and Climate Change (DECC)'s low-carbon agenda. The provisions included in the Act are discussed in the following.

9.1.1 Carbon Capture and Storage (CCS)

This introduces a new carbon capture and storage (CCS) incentive to support the construction of four commercial-scale CCS demonstration projects in the United Kingdom and the retrofit of additional CCS capacity to these projects should it be required at a future point of time. The statute also requires the government to prepare regular reports on the progress made on the decarbonisation of electricity generation in Britain and the development and use of CCS.

9.1.2 Introducing Mandatory Social Price Support

This is aimed at tackling fuel poverty by lowering the energy bills of more of the most vulnerable consumers and giving greater guidance on the types of households eligible for support. This was to be funded by requiring the energy companies to make available at least £300 million per annum by 2013–2014.

9.1.3 Fairness of Energy Markets

The statute makes it clear that the Office of Gas and Electricity Markets (OFGEM), in relation to its principle objective of protecting the interests of existing and future consumers, must include the reduction of carbon emissions and delivery of secure energy supplies in its assessment of the interests of consumers; it should also step in proactively to protect consumer interests as well as consider longer-term actions to promote competition.

Later in the same year, The new coalition government published its own Energy Bill, which had its first reading in the House of Lords in December 2010. The Bill establishes the framework to implement the coalition government's 'Green Deal' plans, intended to improve the energy efficiency of properties in the United Kingdom. The Bill contains provisions that are key to the implementation of the government's wider energy policy as it sets up a framework for the energy efficiency improvements to 'revolutionise the energy efficiency of British properties'. The framework enables private firms to offer consumers energy efficiency improvements to their homes, community spaces, and businesses at no upfront cost to the owner/occupier, and recoup the costs through a charge.

Energy security has always been high on the political agenda, but its importance has risen as the United Kingdom has become increasingly dependent on imported energy, experienced high and volatile oil and gas prices, and addressed the challenge of reducing the carbon dioxide emissions.

DECC needs to define 'energy security' and adopt a more strategic and systematic approach to provide a clear goal for policy interventions, taking a more holistic view in order to ensure that the energy system is resilient to both short-term shocks and longer-term stresses.

9.2 THE PRIMARY ENERGY SUPPLY

It is inevitable that the United Kingdom will become increasingly reliant on energy imports. This is not necessarily incompatible with increasing energy security; it can be maintained by a diverse energy portfolio that does not rely too much on either a single supplier or a single fuel. The decline of the oil reserves on the UK continental shelf (UKCS) is not a major concern in terms of energy security to the Coalition government, but the way in which the £2 billion levy on North Sea producers was announced in Budget 2011 may have undermined investor confidence. The Government needs to work closely with the industry to restore that confidence.

9.3 INFRASTRUCTURE

The government needs to communicate a clear strategy to encourage more gas storage if it is to ensure timely investment. It is only by having sufficient gas storage that the United Kingdom can build broader system resilience. It is also recommended that the government set up an independent stock-holding agency—funded by industry—to manage privately held strategic oil stocks.

The proposals set out in the White Paper on reforms to the electricity market are not convincing that they strike the right balance between encouraging investment in new gas-fired plants in the short term (to fill the gap that will be created by the closure of around 19 GW of nuclear, oil-fired, and coal-fired plants by 2020) and the need to decarbonise the power sector over the course of the 2020s; this will ultimately entail only a very limited role for unabated gas-fired capacity. In particular, the proposed form of the Emissions Performance Standard could risk locking the United Kingdom into a high-carbon electricity system in the future.

Electrification of heat and transport will result in significantly increased loads on the local distribution network. An increase in distributed energy generation combined with greater use of demand side measures could mean that distribution network operators will need to move away from the relatively passive operation model at present towards becoming distribution system operators (DSOs) with responsibility for balancing supply and demand on their network. The government needs to do more work to ensure that distribution network operators are sufficiently prepared for the changes ahead.

9.4 ENERGY USERS

Even though improving energy efficiency will bring benefits for energy security, it is often difficult to deliver in practice. Failure to deliver could have serious consequences for energy security.

New 'smart' technology will provide opportunities for energy users to engage in demand-side response measures, which could play a vital role in ensuring the security of the electricity system. The full potential of such measures to contribute to energy security is not yet known and the government needs to investigate this further.

Although energy users are a key component of the energy system, they are perhaps not as well understood as the technologies that make up the supply side of the system. If the United Kingdom is to make a successful transition to a low-carbon economy, it is essential that the government understands both the social as well as the technical feasibility of new technologies in the energy system.

9.5 THE INTERRELATIONSHIP BETWEEN ENVIRONMENTAL AND ENERGY POLICIES

The energy and environmental policies are very closely linked and any developments in energy policy immediately raise a significant number of environmental considerations. Such impacts may include carbon dioxide emissions, other greenhouse gas emissions, sulphur dioxide emissions, the development of renewables, combined heat and power projects, and any localised environmental impacts. There may also be wider impacts on diversity of fuel mix, sustainablity of energy supply, and social impacts. The new coalition government after much consultation produced an Overarching National Policy Statement for Energy (EN1) in June 2011. This Statement covers the policy context for the development of nationally significant infrastructure. It reflects the commitment in the coalition programme for the government to take forward the Energy National Policy Statements (NPSs). As energy is vital to economic prosperity and social well-being, it is important to ensure that the United Kingdom has secure and affordable energy. To produce the energy the United Kingdom requires and getting it to where it is needed necessitates a significant amount of infrastructure, both large and small scale. The Energy NPSs consider the large-scale infrastructure that plays a vital role in ensuring that the United Kingdom has energy security.

9.5.1 Carbon Dioxide Emissions

Energy policies, which have impacts on power generation or industry, and in particular affect the fuel mix, will in turn affect carbon dioxide emissions from the electricity industry. Electricity generation is responsible for just over one-quarter of the total UK carbon dioxide emissions. Any significant changes in the carbon intensity of generation will, therefore, result in significant changes on the range of measures needed to be deployed to ensure that the United Kingdom meets its climate change targets. The EU Emissions Trading Scheme (EU ETS) has formed the cornerstone of UK action to reduce greenhouse gas emissions from the power sector since 2005. It is expected to deliver reductions from the energy sector of 21% on 2005 levels by 2020, so underpinning the transition to low-carbon electricity generation. The coalition government intends to go beyond the EU ETS and ensure that developers deliver the required levels of investment in low-carbon generation to decarbonise the way in which electricity is produced and so reinforce the security of supply, while retaining efficiency and competitativeness through the Electricity Market Reform Project.

9.5.2 Other Greenhouse Gas Emissions

The scale of UK coal production, and the mix of deep-mined and opencast production, affects emissions of methane, which is the second most important greenhouse gas.[1] Therefore, any developments in the electricity generation sector may affect future methane emissions and in turn have implications for the range of policies and measures necessary for the United Kingdom to meet its climate change commitments. However, as in the case of carbon dioxide emissions, it is not possible to quantify the impact of any new energy policy framework on methane emissions.

9.5.3 Sulphur Emissions

Trends in economy-wide sulphur emissions are heavily dependent on the mix of different fuels in demand to meet energy requirements. The combustion of oil and coal produces significant amounts of sulphur, unless the emissions are abated by an end-of-pipe technology such as flue gas desulphurisation (FGD). Alternatively, other energy sources such as gas, nuclear, and hydro sources do not give rise to any sulphur emissions. The United Kingdom is committed to reducing its sulphur emissions under the Second Sulphur Protocol. The main regulatory instrument used to that end is the requirement placed on the Environment Agency to apply best available techniques not entailing excessive cost (BATNEEC) criterion when granting an authorisation for regulated processes, including oil and coal power stations. These authorisations, combined with commercial decisions by generating companies, have recently led to a substantial switch from oil and coal to gas in power generation. This means that sulphur emissions from the generation of electricity have fallen rapidly.

9.5.4 Renewables

Power generation from renewable sources has a significant contribution to make to sustainability. It reduces harmful emissions and provides more sustainable forms of energy. The government announced the Fifth Non-Fossil Fuel Obligation (NFFO 5) towards the end of 1998; this was the largest obligation so far with the lowest average prices. However, other more sustainable energy technologies need to be encouraged in order to reduce the United Kingdom's dependence on nonrenewable fossil-fuel-based energy sources. The United Kingdom has committed to sourcing 15% of its total energy from renewable sources by 2020[2] and new projects need to continue to come forward urgently to ensure that this target is met. Renewable electricity generation is currently supported in the United Kingdom through the Renewables Obligation (RO), which is a market-based support mechanism to encourage investment. Renewables have the potential to improve security of supply by reducing reliance on the use of coal, oil, and gas supplies to keep the lights on and in power businesses. The United Kingdom has substantial renewable energy resources, for example, the British Isles have 40% of Europe's wind and some of the highest tidal reaches in the world. Unlike other technologies, the cost of renewable energy is in the construction and maintenance alone as the resource itself is usually free, so it helps to protect consumers against the volatile but generally increasing cost of fossil fuels.

9.5.5 Combined Heat and Power

The environmental benefits of combined heat and power projects come from the very high levels of efficiency obtainable from such plants; the transmission losses that are avoided; and, in many cases, the reduced emissions achievable from shutting down previous steam-raising plant. The new government policy is to see an expansion of such projects to increase the efficiency of use of natural resources, and contribute towards achieving emission targets.

9.5.6 Localised Impacts

Energy production and use can have a wide range of environmental impacts; for example, the construction of a new power station may affect the local community through increased traffic levels, noise, dust, and disruption during the construction phase, and visual intrusion once the plant is in place. Any coal production may have similar impacts, which may vary according to the production method, for example opencast mining or deep-mined. The impact of North Sea oil and gas production arises both during production and also from the need to dispose of redundant facilities at the end of their economic life. The impact of emissions from electricity generation on local areas can be considerable, with the main issues being sulphur, nitrous oxide, and particulate emissions.

As seen above, most of these impacts are regulated by local authorities through the regime of planning consents and air pollution control permits[3] and also the Environment Agency and the government through Integrated Pollution Control (IPC) and the consideration of applications for consent.[4]

Some impacts (such as landscape and visual impacts) arise from the development of any type of energy infrastructure. Others (such as air quality impacts) are relevant to all types of energy infrastructure but nevertheless arise in similar ways from the development of types of energy infrastructure covered in at least two of the Energy NPSs. Both these classes of impacts are considered as 'generic impacts'. However, in some cases, the technology-specific NPSs provide detail on the way these impacts arise or are to be considered in the context of applications that are specific to the technology in question.

It is not possible to assess the net effect of any changes in energy policy on these local impacts as yet, as it also depends on how the market responds to the new government policy. However, any new energy policy initiatives must be consistent with the United Kingdom's ability to achieve its environmental commitments.

9.6 UK REGULATORY BODIES

9.6.1 Department for Environment, Food, and Rural Affairs (DEFRA)

In 2001, the majority of the powers and duties of both the Department of the Environment, and the Department of Transport were brought together to form the new Department for Environment, Food, and Rural Affairs (DEFRA). This new ministry was constituted by amalgamating components of the old Department of the Environment, Transport and the Regions (DETR), much of the Ministry of Agriculture, Fisheries and Food (MAFF) and fragments of other departments. Inevitably, there is some crossover between DEFRA and in particular the Department of Trade and Industry (DTI) with regard to environmental control of the energy sector, but mostly the influence of DEFRA is in policymaking. However, inasmuch as any devolved powers involving discretion are normally circumscribed by the need to take central policy into consideration, any central policy announced by DEFRA has a major impact, and of course the Treasury operates its usual 'dead hand' as far as the implementation of any policy is concerned. In October 2008, the climate change team at DEFRA was merged with the energy team from the Department for Business Enterprise and Regulatory Reform (BERR) to create the DECC.

Owing to the nature of the framework for environmental legislation, the Secretary of State enjoys wide powers to create legislation and quasi-legislation. He or she may also issue directions and guidance notes, approve actions of regulatory bodies, and appoint persons to those regulatory bodies.

Under the Environment Act 1995, the main legislative powers have been reserved to the Environment Agency; however, DEFRA and the Treasury control the funding for the independent agency and it is through DEFRA that the agency is accountable to Parliament.

9.6.2 Department for Business

The energy scene in the United Kingdom has been transformed by the rapid introduction of competitive markets over the past few years, but some aspects have not worked properly or effectively and the Labour government committed itself to further initiatives in setting an overall framework of legal rights and obligations, including property rights and contract law, for regulation. It also committed to more specific functions such as licensing of hydrocarbon activities, both offshore and onshore, and the rules for the exploitation of coal resources. The then government promised to set these new rules in such a way as to facilitate the optimum economic exploitation.[5] These themes are developed and discussed in more detail in the energy-source-specific sections in the following.

9.6.3 The Department for Business Innovation and Skills

The Department for Business Innovation and Skills was set up by the new coalition government to bring together all of the levers of the United Kingdom economy in one place with a view to help drive growth in the areas of higher education to innovation and science to business and trade. The work of the Corporate Law and Governance (CLG) Directorate and the Consumer and Competition Policy (CCP) Directorate work is to ensure that businesses operating in the United Kingdom are supported by legal and governance frameworks to promote open and competitive markets. The CLG Directorate is responsible for company and partnership law, corporate governance, audit, accounting, and reporting, while the CCP is responsible for mergers and monopolies, European Commission and international competition policy, and state aid regulations of the Enterprise Act.

9.6.4 The Department of Energy and Climate Change

The DECC was created in October 2008, to bring together the energy and climate change mitigation policies. The DECC takes the lead in tackling the challenges of both energy security and climate change and reflects the fact that climate change and energy policies are inextricably linked—two-thirds of the UK climate change emissions come from the energy used. Decisions in one field cannot be made without considering their impacts on the other. The DECC has to make a statutory declaration to Parliament annually setting the government's targets for the following year and stating the government record for the previous year. This is the first time that the government is required to achieve statutory targets.

One of the DECC's key responsibilities is to ensure the United Kingdom continues to enjoy secure, competitively priced energy. The United Kingdom is importing progressively more of its energy, and competition for energy resources can make prices volatile. In order to meet the UK emission targets, the DECC is charged with accelerating the move to a low-carbon energy supply.

9.6.5 Parliamentary Select Committees

The long-term oversight over both public expenditure and the formulation and implementation of policy is exercised by the Parliamentary Select Committees related to the work of the government departments. These committees have the function of obtaining information, appointing specialist advisers, making reports to the Commons of their conclusions, and making special reports on any matters they think proper to bring to the attention of the House of Commons. Usually, the committees manage to present unanimous conclusions to the Commons, although where a highly controversial and diverse issue is examined, this may lead to a split in a committee and the issue of a minority report. There is no specific time allocated for the debate by Parliament of committee reports, although there is an undertaking by government that it will always respond to a committee report, and thus there will be a reaction to any criticisms made that can be taken up by Members and the media. The chief value of the committees is the production of vast amounts of information for the public.[6]

In any given situation, a committee may be more or less successful depending on the nature of the subject matter, the quality of its specialist advice, and the membership of the committee. However, if committees are too critical of government policy, this may have a disproportionate and unintended effect.[7]

9.6.6 The Environment Agency

9.6.6.1 Background

The Environment Agency (the Agency) was provided for under the Environment Act 1995, and took up its statutory duties at vesting on April 1, 1996. It brings together the functions previously carried out by the National Rivers Authority (NRA), Her Majesty's Inspectorate of Pollution (HMIP), the waste regulatory functions of 83 local authorities (including the London Waste Regulating Authority) and a small number of units from the DETR dealing with aspects of waste regulation and contaminated land.

The bringing of together of these into a single body was to enable management of the environment in a more coherent and integral way, and to provide a more streamlined service for those activities that the Agency regulates.

The Agency has a range of functions to do with regulation, management, monitoring, and understanding of the environment and the processes that affect it. In carrying out its functions, the Agency is required to act in accordance with the government's overall environmental strategy, as set out in the reports on the 1990 White Paper *This Common Inheritance* and in the UK Strategy for Sustainable Development. Reflecting this, the Environment Act defines the principal aim of the Agency as 'to protect or enhance the environment, taken as a whole, as to make the contribution towards attaining the objective of achieving sustainable development that Ministers consider appropriate'.

The declared principal aims of the Agency are designed to build on the work of the Agency's predecessor bodies in protecting and enhancing the environment as a whole, through an integrated approach. This would put the activities of the Agency more firmly in the context of sustainable development, as defined in the 1987 Bruntland report, *Our Common Future*, as 'development that meets the needs of the present without compromising the ability of future generations to meet their own needs'.

To this end, various ministers have issued statutory guidance to the Agency on its contribution to sustainable development and have set seven main objectives governing the manner in which it should carry out its functions.

9.6.6.2 Functions of the Agency

The main statutory responsibilities of the Agency for protecting and improving the environment are to:

1. Regulate over 2000 industrial processes with the greatest pollution potential ensuring that the BATNEEC are used to prevent or minimise pollution to the environment
2. Regulate the disposal of radioactive waste at more than 6000 sites, including nuclear sites

3. Regulate the treating, keeping, movement, and disposal of controlled waste, so as to prevent pollution of the environment or harm to public health, in a manner that is proportionate to the threat posed
4. Preserve or improve the quality of rivers, estuaries and coastal waters through pollution-control powers that include discharge consents and regulation of sewage treatment works
5. Take any necessary action to conserve, redistribute, augment, and secure proper use of water resources through the regulation of water abstraction licences
6. Exercise supervision over all matters relating to flood defence
7. Maintain, improve, and develop salmon, trout, freshwater, and eel fisheries and regulate these through a system of licensing, by-laws, and orders
8. Conserve and enhance the water environment, including areas of outstanding natural beauty (ANOBs) or environmental sensitivity
9. Maintain or improve navigation on waters for which the Agency has responsibility
10. Regulate the remediation of contaminated land designed as special sites
11. Administer registration of businesses and exemption schemes in accordance with regulations on producer responsibility

9.6.6.3 National Standards and Targets and the Agency

The United Kingdom has maintained a long-held stance in Europe over its preference for the use of environmental quality standards and objectives as a basis for pollution control. It is, therefore, surprising that virtually all of the standards that have been incorporated into UK legislation arise from EC directives rather than from national initiatives.

Extensive environmental monitoring programmes are carried out by the Agency to assess compliance with these standards in both fresh and tidal waters and the results reported back to the DTER. Nationally agreed, but non-statutory, standards also exist for air quality, such as those recommended by the government's Expert Panel on Air Quality Standards (EPAQS). Standards based on EPAQS's recommendations for 80 priority pollutants have been incorporated into the government's air quality strategy, which established compliance criteria for achievement by the year 2005. These have now been updated by the Air Quality Standards Regulations 2007 (Statutory Instrument 2007/64).

9.6.6.4 Environmental Surveillance by the Agency

There are several national programmes in existence that have been established to assess trends, or just changes, in environmental quality both over time and across the country, but without reference to specific standards or targets. Although they have no formal statutory basis, some of these are long-standing programmes. In some cases, the Agency contributes to UK-wide programmes, which continue to be maintained by government departments as the principal customers. This is the case, for example, for the national Harmonised Monitoring Scheme (HMS) for rivers set up by the Department of the Environment in 1974 to assess long-term trends in river quality and to provide a basis for estimating the total quantities of materials passing from rivers to estuaries. The Agency also contributes to the government-sponsored national programme for the surveillance of estuarine and coastal water quality—the National Marine Monitoring Plan (NMP)—which was established to rationalise monitoring activities arising from international agreements, such as those related to Oslo and Paris Commissions (OSPARCOM), and national initiatives. The Agency also makes a contribution to various smaller national programmes such as the Environmental Change Network (ECN) managed by the National Environmental Research Council (NERC) and the Acid Water Survey sponsored by the DETR.

Some national programmes previously managed by government departments have been handed over to the Agency to maintain because they fit closely with its functional responsibilities. For example, the long-standing programme for the assessment of the background level of radionuclides in air particles and rainwater has been passed to the Agency as principal customer. Similarly, responsibility for the Radioactive Incident Monitoring Network (RIMNET) set up as part of a national response plan to deal with the consequences for the United Kingdom of overseas nuclear accidents, was also passed to the Agency in 1997.

The Agency has progressively developed a national surveillance programme for coastal waters using a combination of ship-based measurement and aerial surveillance. Other Agency programmes that involve surveillance of the state of the environment are carried out to support its water-management responsibilities. The Agency maintains an extensive network of hydrometric monitoring sites across England and Wales to support its water resource and flood defence management functions. A comprehensive network of gauging stations across the main hydrometric areas of the country provides both daily and monthly data for river flows. Groundwater is also measured and reviews are carried out to examine options for rationalising both groundwater quality and quantity networks to improve national consistency.

9.6.7 Local Authorities

The latest local government reorganisation under the Local Government Act 1992 provided for three main types of local authority. Firstly, there are the single-tier London boroughs and metropolitan districts, which have responsibility for all matters, although some functions are run by joint boards of the metropolitan district councils. Secondly, there are the non-metropolitan areas, which have a two-tier system consisting of the county council and the district councils. These are equal but have different duties. All of these different types of local authority exercise control over the following areas.

9.6.7.1 Town and Country Planning

It is the local authority that is normally the local planning authority, which has responsibility for making development plans and grants planning permission for development that requires it. Additional powers cover responsibility for hazardous substances consents and the control of derelict land, as well as the protection of listed buildings, conservation areas, and tree preservation. Although the county councils are responsible for the drawing up of strategic plans and for national parks and mineral waste disposal, it is the district council that has responsibility for other development control. The areas of responsibility are laid out in the Local Government, Planning and Land Act 1980. The Department for Communities and Local Government directly influences the decisions made by the Planning Authorities by issuing Planning Policy Guidance Statements. Energy planning is influenced by Minerals Policy Statements.

9.6.7.2 Air Pollution

Responsibility for the control of smoke, dust, grit, and fumes has fallen to the local authorities under the Clean Air Acts and related legislation. Under Part I of the Environmental Protection Act (EPA) 1990, the local authorities were given greater powers to control air pollution from plants that do not come under the control of the Environment Agency within its responsibilities for IPC.

9.6.7.3 Contaminated Land

Under the Environment Act 1995, local authorities were charged with the duty to inspect their area and produce a register of contaminated land. Once contaminated land has been identified the local authority has to issue a remediation notice to the 'appropriate person' under the statutory guidelines, which came into force on April 1, 2000. Contaminated land legislation and regulations can be found at www.netregs.co.uk.

9.6.7.4 Noise Control

It is the local authorities that have the primary responsibility for the control of noise from both industrial and domestic premises, with powers to bring an action under section 79 of the EPA 1990 for statutory nuisance. Noise nuisance is covered by Part III of the EPA 1990. This law empowers local authorities to deal with noise from fixed premises, including land, if they consider that the noise amounts to a statutory nuisance. Proceedings may be taken against noise from factories, shops, pubs, houses, and traffic.

9.6.8 The United Kingdom Atomic Energy Authority (UKAEA)

The United Kingdom Atomic Energy Authority (UKAEA) is a statutory corporation established under the Atomic Energy Authority Act 1954. Both the chairman and members are appointed by the Secretary of State. Under the 1971 Atomic Energy Authority Act, the authority's trading enterprises were transferred to other organisations, in particular British National Fuels Ltd. (BNFL), which is a state-owned company. BNFL now provides all aspects of production and reprocessing. The activities of the company have been expanded to include the Thermal Oxide Reprocessing Plant (THORP) at Sellafield. The 1995 Atomic Energy Authority Act provides for the Secretary of State to direct all or any of the authority's property, rights, and liabilities to be transferred to any person or persons. Further discussion on the regulation of nuclear fuels is covered in the following. In addition, there is a safety regulator, the Nuclear Installations Inspectorate (NII), which normally conducts pre-licensing reviews prior to the formal licensing process. The NII would then scrutinise preliminary safety and pre-construction reports to ensure general compliance with UK safety regulations.

9.6.9 The Coal Authority

The Coal Authority was established as a body corporate on the privatisation of the coal industry. Section 1(1) of the Coal Industry Act 1994 sets out the stated purposes:

(a) holding, managing and disposing of interests and rights in or in relation to unworked coal and other property which is transferred to or otherwise acquired by it by or under this Act;
(b) carrying out functions with respect to the licensing of coal mining operations;
(c) carrying out functions with respect to coal mining subsidence and in connection with other matters incidental to the carrying on of any opencast or other coal mining operations;
(d) facilitating the establishment and maintenance of arrangements for the information to which persons are to be entitled under this Act to be made available to them; and
(e) carrying out the other functions conferred on it by virtue of this Act.

9.6.9.1 Functions and Duties

One of the main functions of the Authority relates to surface property and unworked coal and other minerals. Interests in unworked coal and coal mines were passed to the Authority from British Coal under the Act.[8] Surface interests were passed from British Coal under a series of restructuring schemes made by the Secretary of State.[9] The Authority is required to coordinate its practice in relation to relevant property dealings with the carrying out of its licensing functions,[10] and must also have regard to the need to secure the safety of members of the public in carrying out its property functions.[11] The Authority is also required to make available to others, land and other property in which the Authority has no statutory interest, but has a duty to dispose of such land or interests at the best terms reasonably available.[12] Other duties contained within section 3 include the exploitation of coal-bed methane (CBM) and an environmental duty when formulating proposals for works on such of the Authority's land as has been previously used for the carrying on of coal mining operations.

To assist in the carrying out of these duties, the Authority is granted certain powers under section 5, which allow it to acquire land, to carry out works, to develop or improve land, and to acquire land with a view to disposing of it jointly with land already in the Authority's possession.

Part II of the 1994 Act contains the main body of the licensing provisions and, as usual, mandates that no coal mining operations shall be carried out except in accordance with a licence granted under the Act. The Authority is required to carry out its licensing function in order to secure an economically viable coal mining industry,[13] to ensure that licensees are able both to finance the proper carrying on of coal mining operations and to discharge liabilities arising from those operations, and also to ensure that any persons suffering subsidence will not suffer any loss.[14]

As far as subsidence is concerned, the owner of property that is damaged by lawful coal mining operations may have remedies under the Coal Mining Subsidence Act 1991 as amended by the 1994 Act. If the damaged land is not within the 'area of responsibility' of the licence holder as defined in the licence, then the liability for subsidence lies with the Authority.[15] The Authority also has a duty to keep registers of licences, which must include the transfers of licences, revocation of any authorisations, and any enforcement orders made. There is a provision under section 59 for confidentiality concerning certain commercially sensitive information.

9.6.10 The Office of the Gas and Electricity Markets (OFGEM)

OFGEM was set up to protect consumers as their first priority. This is done by promoting competition, wherever appropriate, and regulating the monopoly companies that run the gas and electricity networks. The interests of gas and electricity consumers are their interests taken as a whole, including their interests in the reduction of greenhouse gases and in the security of supply of gas and electricity to them.

Other priorities and influences include helping to secure Britain's energy supplies by promoting competitive gas and electricity markets and regulating them in order that there is adequate investment in the networks. The priorities also include contributing to the drive to curb climate change and other work aimed at sustainable development.

OFGEM is governed by an Authority, consisting of non-executive members and a non-executive chair. The Authority determines strategy, sets policy priorities, and takes decisions on a range of matters, including price controls and enforcement. The powers of the Authority are derived from the Gas Act 1986, the Electricity Act 1989, the Utilities Act 2000, the Competition Act 1998, and the Enterprise Act 2002.

OFGEM recovers its costs from the licensed companies whom it regulates.

9.7 REGULATIONS FOR THE DIFFERENT ENERGY SECTORS

9.7.1 Oil and Gas

9.7.1.1 General Regulatory Framework

The first commercial exploitation of oil on land in the United Kingdom was in East Sussex in 1895. Subsequently, oil and gas production has continued on a relatively small scale in the East Midlands. However, drilling for oil did not take place until after World War I, which had the effect of influencing the regulatory framework in a number of different ways. Firstly, British experience was mostly gained in highly regulated jurisdictions such as the United States. Secondly, the influence of international law was felt in the 1950s because the major finds of oil and gas had been offshore. Thirdly, the policies and legislation of the European Community began to have an impact from the time of the accession of the United Kingdom in 1973. Fourthly, the move from one of participation, under a Labour government, to one of privatisation, under a Conservative government, served to increase the amount of regulation and alter the taxation regime. Fifthly, the influences of the growing awareness of health and safety and environmental issues have served to engender comprehensive legislative reforms in these areas.

The fastest growing influence on UK oil and gas law since the offshore industry began has been legislative regulation for the protection of the health and safety of workers, and of the environment. With the enactment of the Continental Shelf Act in

1964, it was assumed that these concerns could be adequately met by the application offshore, so far as appropriate, of relevant onshore legislation such as the Factories Acts. In addition, there would be an insertion in the licence of provisions that would enable the minister to ensure that the licensee followed good industry practice in oil and gas operations, which would include non-interference with fishing and navigation. Further, to secure the safety, health, and welfare of all employees under the Continental Shelf Act 1964, section 1(4) requires the licence to contain provisions to secure the safety, health, and welfare of persons employed offshore (this was repealed by the Offshore Safety Act 1992, section 7(2)). The relevant clauses, in their most modern form, are to be found in Schedule 4 of the Petroleum (Production) (Seaward Areas) Regulations 1988.

This approach had inherent problems. It was not clear which onshore legislation might be applicable in the specific circumstances of offshore operations, nor to what extent and with what result. As far as the licences were concerned, it was an impracticably blunt regulatory instrument for the control of unfamiliar, high-risk, technically advanced operations. This 'hands-off' approach came to an end with the loss of the *Sea Gem* drilling rig in 1965, which convinced the government that a specialised set of regulations for the construction and operation of offshore installations was necessary. The result was the enactment of the Mineral Workings (Offshore Installations) Act 1971, and the associated Regulations, which was to govern the construction and operation of offshore installations for the next 20 years.

Even as the 1971 Act was being debated in Parliament, the general approach to health and safety at work was the subject of inquiry by the Robens Committee. The report that came out from this inquiry was the foundation of the Health and Safety at Work Act (HSWA) 1974. This statute marked a move away from the prescriptive regulation that was the characteristic of the Factories Act and imposed more general duties on employers, together with the recognition of employee responsibilities in relation to safety. It was the results to be achieved that became the feature of the Regulations. Under arrangement with the Petroleum Engineering Division (PED), which was responsible for offshore operations, the Act was selectively applied to certain offshore operations. This dual regime persisted, despite criticism, until a further disaster involving the *Piper Alpha* oil rig in 1988. The report from the resulting inquiry by Lord Cullen recommended the transfer of regulatory responsibilities from the PED to the Health and Safety Executive (HSE), and the absorption of the 1971 and 1975 Acts, and the Regulations under them, within the general scheme of the 1974 Act. This transferred the burden of safety schemes onto the operators with continuous safety assessment and self-monitoring, and was enacted under the Offshore Safety Act 1992.

There are many similarities between the development of health and safety legislation and the legal protection of the environment against oil pollution. The original licensing regime did not address environmental protection at all other than under the general injunction to carry out procedures within the concept of 'good oilfield practice'. Nevertheless, the legislative development in this field at international level was growing exponentially. Such international law making was well established in the maritime arena by the time the UK offshore industry became established, and this was embraced by the enactment of the Oil Pollution Act in 1971. Since that time, further developments have taken place not only in the international field but also within the jurisdiction of the European Community. Particular significance should be attached to the Community directive requiring an environmental impact assessment (EIA) for major and some less environmentally intrusive developments. The requirement for such an assessment was applied offshore by the implementation of the Offshore Petroleum Production and Pipelines (Assessment of Environmental Effects) Regulations in 1999. It is, however, through the exercise of ministerial powers under the licence that this requirement is to be imposed. It is interesting to note that ministerial powers may, in certain circumstances, be used as an effective method of regulation. In this respect, the minister may impose such conditions precedent to the developments that licensees must comply with if those licence rights are to be a source of profit. This economic instrument rather loses its incentive value when the field is nearing the end of its economic life, and so the environmental effects of the abandonment of offshore installations is addressed in the Petroleum Act 1998 Part IV. The laws in this area are, then, influenced by both international and European Community law.

9.7.1.2 Ownership of Petroleum

Unlike the ownership of petroleum on land, where there is clear sovereignty over all of a state's landmass, there is no ownership by the state of that which constitutes the continental shelf; the state merely has jurisdiction over it.

The Continental Shelf Convention 1958 and the 1982 UN Convention on the Law of the Sea provide that the coastal state has sovereign jurisdiction for purposes of the exploration and exploitation of mineral resources. This is a functional sovereignty. The state has no right of ownership of the resources *in situ* in the continental shelf, but merely sovereignty for the purposes of exploring and exploiting. This means that no one else may explore and exploit without permission from the state. The coastal state may grant licences for that purpose, but it does not itself own the petroleum.

The holder of an offshore licence cannot, therefore, get title to a resource over which the licensing government does not itself have title. What the holder of the licence does get is an entitlement to explore and exploit and to reduce into possession. It is the reduction into possession that gives the licensee title to the petroleum. Thus, in the North Sea, title does not pass with the granting of the licence but at the well head, when the recovered petroleum is reduced into possession.

This has implications for any lending institution. Loans to the licensee cannot be secured against collateral in the form of resources owned by the licensee. The government could always stop the intended reduction into possession through unilateral action if the conditions of the licence were not fully complied with. This has led to a variety of financial techniques, which in turn have led to complicated questions as to whether these various arrangements represent an interest in the petroleum itself.

9.7.1.3 The Legal Basis of Licensing

Most of the licences familiar to those living and working in the United Kingdom may be legally analysed as administratively granted exemptions from that which is prohibited by statute. The purpose of the licensing scheme could be revenue raising or the preservation of standards. Originally, production licences granted under the Petroleum (Production) Act 1918 came in this form, but with the vesting of proprietary rights in petroleum in the Crown under the 1934 Act, the position changed. As owner of the rights, the Crown was protected by the ordinary laws of property, and so there was no longer a need for a prohibition on exploration and exploitation. In addition, as owner, the Crown was free to decide if and how the petroleum was to be exploited. The licensing regime provided for under the 1934 Act was, therefore, based on the property rights of the Crown both on land and under the territorial sea. The Petroleum (Production) Act 1934, section 1, as substituted by the Oil and Gas (Enterprise) Act 1982, section 18, is now contained in the Petroleum Act 1998, section 2 and Schedule 3, para 3. The 1982 amendment makes it clear that the section vests in the Crown petroleum that is *in situ* under the territorial sea as well as that under the land mass. In the case of *Earl of Lonsdale v. Att.-Gen.*,[16] Slade J. had confirmed the view that the Crown's rights on petroleum under the territorial sea were derived from common law, and not from the 1934 Act, which did not extend to the territorial sea; if this were the law, then all grants by the Crown of seabed minerals, if construed to include oil prior to the 1934 Act, would operate as an obstacle to the grant of licences under the 1934 Act.

As far as the licensing of activities on the continental shelf is concerned, this is by virtue of the 1982 Law of the Sea Convention, which provides that the coastal state has a sovereign jurisdiction for the purposes of the exploration and exploitation of the resource. It is a functional sovereignty and states must take care that any legislation they pass that purports to have application on the continental shelf, whether it be criminal, civil, or tax legislation, is limited to matters relating to the exploration and exploitation of shelf resources. The reach of criminal jurisdiction and civil liability has led to interesting case law in a variety of jurisdictions. In addition, the issue of whether a state may tax a company that operates on the continental shelf without having any office within its jurisdiction, and which is not itself engaged in exploring and exploiting, has occasioned particular difficulty in the North Sea. Some service companies, with vessels providing food or entertainment on oil platforms, have vigorously resisted any liability for tax at the hands of the coastal state. Therefore, the coastal state must interpret the tax legislation against the specific jurisdiction authorised under international law.

9.7.1.4 Application of Criminal Law to the UK Continental Shelf

Under section 10 of the Petroleum Act 1998,[17] the government may, by Order in Council, extend the application of the criminal law to any act or omission that takes place on, under, or above an offshore installation within the territorial sea of the UK-designated waters[18] or any waters within 500 m of an installation.[19] Therefore, in principle, any criminal act or omission committed on an offshore installation may be prosecuted in any of the jurisdictions of the United Kingdom. However, it must be noted that under Article 77 of the United Nations Convention on the Law of the Sea (UNCLOS) 82, the UK criminal jurisdiction in designated areas is limited by reference to the nature of the interests to be protected.

9.7.1.5 Application of the Civil Law to the UK Continental Shelf

The Civil Jurisdiction (Offshore Activities) Order 1987 does not state that the civil law should apply in its entirety. Determining the exact scope of the application of civil law offshore is therefore difficult. References to 'acts and omissions', in addition to the case law over the last 40 years, indicate that the general laws of tort apply offshore, whereas the relevant health and safety law is explicitly extended offshore. As far as jurisdiction is concerned, this is specifically conferred 'without prejudice to any jurisdiction exercisable apart from this section by any other court'.[20] Given that a court's jurisdiction is usually founded on the place of domicile or primary residence of the defendant, in many cases, it will be possible to raise an action in more than one forum. A specific example is the litigation that arose out of the *Piper Alpha* disaster. Although the installation was located in the Scottish area of the designated waters, it was owned by Occidental Oil, a company domiciled in the United States. The victims came from a variety of jurisdictions. Therefore, an action could have been raised by an injured party in Scotland, England, or the United States. Although the majority of claims were settled out of court with the size of the awards for damages allocated according to a 'mid-Atlantic' basis, the majority of those cases that did come to court were heard in the English High Courts.[21]

In the area of administrative law, however, it is interesting to note that when Greenpeace attempted to challenge the decision by the Secretary of State for Scotland to allow Shell to dispose of the *Brent Spar* storage buoy by sinking it offshore, the High Court held that the Civil Jurisdiction Order gave exclusive jurisdiction to the Scottish courts on the grounds that the facility was located within Scottish waters.

The difficulties of knowing what law will apply offshore was highlighted in the case of *R v. Secretary of State for Trade and Industry, ex p. Greenpeace Ltd* (*Greenpeace* 2).[22]

The question for the courts in this case was whether the European Habitats Directive[23] on the conservation of natural habitats and wild fauna and flora applied to the UKCS and the waters above, and was not restricted to the territorial waters, which were limited to 12 mi. The Habitats Directive had been transposed into UK law by enacting the Conservation (Natural Habitats, etc.) Regulations,[24] which, by regulation 2(1), applied only to the UK territorial waters. The court decided that this was contrary to the directive and held that the directive did apply to the UKCS and must therefore be taken into account of in offshore licensing and exploration.

The difficulties referred to above do not arise where legislation has been specifically designed, in whole or in part, for offshore operations.

9.7.1.6 Licences

The licensing system in the United Kingdom is administered by the Bank for International Settlements (BIS) under the final discretion of the Secretary of State as provided for in the Petroleum (Production) Act 1934 and amended by the Oil and Gas (Enterprise) Act 1982, the UK Petroleum Act 1987 and the Petroleum Act 1998.

The UK government grants licences for designated blocks or areas on the continental shelf. The normal procedure is for the Secretary of State to issue invitations relating to particular blocks for which potential licensees may apply according to specified criteria. This is known as the *Round System.* There is no fixed timing of the rounds in either timing or number of blocks offered. However, the DTI undertook a review of the licensing system in 1996 and identified some areas for change. This was followed in July 1997 by the DTI announcing outline plans for future oil and gas exploration opportunities on the UKCS, which set out a 5-year programme of licensing rounds to be held under new environmental regulations to implement the EIA Directive.

The licensing provisions are contained in the Petroleum (Production) Regulations 1982 as amended by the Petroleum (Production) (Seaward Areas) Regulations 1995, and the Hydrocarbon Licensing Directive Regulations 1995.

9.7.1.6.1 Background

The government previously awarded licences against its own criteria, which included both experience and financial capacity. Experience in the United Kingdom of exploration and exploitation of oil and gas deposits on the continental shelf was minimal and so it was to the US operators with experience of drilling in the Gulf of Mexico that the government looked. However, technically there was a substantial difference between the hostile waters of the deep North Sea and the calmer and shallower waters of the Gulf of Mexico. Nevertheless, the government of the day was keen to take advantage of the higher prices for oil on the world market, but wanted to keep regulation to a minimum. The government retained control of operations through the British National Oil Company (BNOC), which held a major share in each company awarded a licence. In 1979, with the advent of the new government, the policy of privatisation was begun with the winding up of the BNOC. The 1982 Petroleum (Production) Regulations then became the main regime for regulating oil and gas operations on the UKCS.

9.7.1.6.2 The Model Clauses

The model clauses for incorporation in petroleum production licences were revised in 1995 to amend the principal 1988 Petroleum (Production) (Seaward Areas) Regulations.[25] Amendments to the regulations in 1992 extended the scope of the regulations to areas within the territorial waters adjacent to Northern Ireland. The amendments made to the regulations and model clauses in 1995 were partially to implement the requirements of the Hydrocarbons Directive, but only applied to licences granted after June 30, 1995. The 1996 amendments enabled the Secretary of State to offer for licence, in addition to single blocks, tranches of blocks and so reduce the fee payable upon application for a production licence. The model clauses were also modified to encourage additional and earlier exploration drilling by reducing the initial term from 6 to 3 years and enabled the licence to be continued thereafter for further consecutive terms of 6, 15, and 24 years. Any extension beyond the initial term of 3 years was dependent upon the provision of a further exploration programme. All the original clauses in those licences granted before the amended regulations remained in force.

Under the original 1982 model clauses, the licensee is granted an exclusive licence and liberty during the continuance of the licence to search and bore for, and get, petroleum in the seabed and subsoil under the seaward area as described in the licence. There is an initial term of 6 years for exploration of the block. Before the expiry of the initial 6 years, the licensee must relinquish part of the block. This term was inserted initially to encourage licensees to start exploration as soon as possible rather than 'banking' the authorisation and concentrating on exploring in less politically stable areas of the world. The licensee must, therefore, start exploration to determine which parts of the block to retain and which to relinquish.

Three months before expiration of the 6-year period, the licensee may give notice that it wishes to continue in respect of the continuing part for a term of 30 years after the surrender date. The licence will be renewed if an authorised working programme is adhered to. The licensees must demonstrate the ability to exploit the block and also show that the company is carrying out an authorised programme. If the minister is not satisfied with the detail of the programme, the company can ask for it to be revised; or the minister can also modify the programme. This decision can only be challenged by the licensee on technical grounds and this will thence go to arbitration for resolution. If the arbitration panel finds that the scheme of work is insufficient, compliance is required. In case of breach or non-observance of these provisions, the licence, or part thereof, may be revoked without prejudice to any liabilities already incurred. The minister may also modify the programme on 'national interest' grounds. There is no definition as to the national interest in this respect but it is usually thought to include any rate-of-depletion policy.

During the further 30 years, an authorised production programme must be in place. This may also be amended by the minister. Any programmes must be expeditiously considered by the minister, and the programme may be rejected if it is contrary to good oilfield practice or if it is contrary to the public interest. Any authorisation to commence operations may contain conditions. In the case of a limitation notice, the licensees may make technical and financial representations, and any agreed production levels cannot be increased by the minister without consent, unless there is a national emergency. Once imposed, these conditions may not be changed unilaterally as a licence granted by the Secretary of State constitutes a contract between the Secretary and the licensee. However, the granting of a licence comes under public law regulation and any unilateral change of conditions or revocation would be subject to judicial review under public law provisions.

Where an oilfield crosses from one block to another, the minister may insist on a 'unit development' agreement to be drawn up if it is in the public interest. These are usually very complex agreements of two or more groups of licence holders to develop and manage the blocks as a whole unit, and as such must be drawn up by specialists in the area.

9.7.2 The EC Hydrocarbons Directive

The European Hydrocarbons Directive has had a significant impact on the activities of the oil and gas industry in the procedures adopted and the criteria followed by the licensing authorities in both the granting and regulation of new licences. The Directive lays down common rules that must be followed by licensing authorities when granting licences under their respective regimes. Member States continue to have sovereignty over their hydrocarbon reserves and also retain the right to determine which areas within their territories they will license.

Limits to authorisation under the Directive are stated in the recitals of the Directive. The objective of limiting a single entity's retention of exclusive rights over areas that could be better developed by several entities is extended to both the extent of areas of authorisation and the duration of any authorisation. The area of authorisation must be limited on objective, technical, and economic criteria, and the duration of an authorisation may only be limited to that necessary to carry out the authorised activities. An authorisation may be extended if it is 'insufficient' and provided activities have been performed in accordance with the authorisation.

The Hydrocarbons Licensing Directive was implemented in the United Kingdom on June 30, 1995 by the 1995 regulation. These regulations set out the criteria that the Secretary of State may take into consideration when awarding licences. The Directive forms part of the European Community's drive to make the hydrocarbons market more accessible and provide for equal access to prospecting, exploring for, and producing oil and natural gas.

Every application for a licence must be determined on the basis of the following criteria:

1. The technical and financial capability of the applicant
2. The way in which the applicant proposes to carry out the relevant activities
3. Where tenders are invited, the price the applicant is prepared to pay in order to obtain the licence
4. In cases where there are existing licence holders, any lack of efficiency and responsibility so far displayed by the applicant in operations carried out under that licence

Where two or more applicants show equal merit, when assessed according to the specified criteria, the Secretary of State may apply additional criteria that must be relevant and applied in a non-discriminatory manner. If a licence application is unsuccessful, the applicant may request to be given the reasons for the decision. Good reasons for refusal include both on the grounds of national security, or if the applicant is effectively controlled by another state or by nationals of another state that is not a member of the European Economic Area (EEA). Member States must inform the Commission of any difficulties encountered in applying for licences, prospecting, exploration, or production activities of third countries. In such cases, the Commission may negotiate or states may be ordered by the EU Council to retaliate by refusing authorisation to entities controlled by the third country.

Invitations to apply for production licences must be published in the *Official Journal of the European Communities*, which must be accompanied by the criteria to be applied in determining the application.

The regulations implementing the Directive in the United Kingdom identify the scope and conditions for granting the licences. Under the regulations, such terms and conditions must be justified exclusively for the purposes of ensuring good performance and providing for payment; and also on the grounds of national security, public safety, and public health; security of transport and the protection of the environment, particularly in respect of the protection of biological resources and of national treasures possessing artistic, historic, or archeological value; the safety of installations and workers; and the planned management of hydrocarbon resources and the need to secure tax revenues.

Under the Directive, Member State participation may continue with the provision that a 'Chinese wall' exists between the Member State as the licensing authority and the Member State as the participating licensee. Any flow of information between these two entities is subject to appropriate limitations, and arrangement and guarantees must be in place. Any policy or regulatory decisions must be seen to be at 'arm's length' from the participating entities, which must be allowed to exercise management control and make commercial decisions much as a non-state participant would do.

The European Union is not self-sufficient in hydrocarbons and yet many areas of the Community are not explored. The Commission decided that encouragement was needed and, to this end, barriers should be removed. The Licensing Directive is intended to make Europe a more attractive venue for oil and gas field development by adopting a 'hands-off' regime. The purpose of the Directive is to open up the hydrocarbons market to greater competition, to guard against excessive state participation and lighten the burden of regulation.

9.7.3 Types of Licence

The mode of allocation of licences, and their content and effect, have varied according to the nature of the area they cover and the types of activity they permit. The areas are separated into landward and seaward areas. The landward areas include the landmass of Great Britain down to low watermark, in addition to inland waters, which include waters on the landward side of the baselines from which the territorial sea is measured (off the west coast of Scotland the baseline is drawn to the west of the Outer Hebrides and therefore large areas of sea, including the Miches, are landward areas), and the areas of territorial sea around Orkney and Shetland. Seaward areas comprise the remainder of the territorial sea. In 1987, this was extended by the Petroleum Act 1987, section 19(1), to include the territorial sea adjacent to Northern Ireland, together with areas designated under the Continental Shelf Act 1964 and certain small islands within these areas.

Originally, the landward licensing regime was based on a three-tier system. This required separate licences for exploration and appraisal, and a further one for exploitation. Early in the 1990s, the consensus was that there should be a single licensing system. This came into force with the Petroleum (Production) (Landward Areas) Regulations 1995. These regulations introduced for the first time a single landward licence to replace the three-tier system. The new Petroleum Exploration and Development Licence (PEDL) is similar, in fact, to the Seaward Production Licence and was first used in the Seventh Landward Licensing Round in 1996.

The model clauses for these licences are contained in the Petroleum (Production) (Landward Areas) Regulations.[26] The Model Causes have been updated under the Petroleum Act 1998 and can be found under Statutory Instrument 1999 No. 160. As far as the environmental obligations are concerned, they include the requirement for the avoidance of harmful methods of working,[27] including the prevention of the escape of oil obtained from the licensed area. This particularly concerns the pollution of controlled waters and after April 1, 2000 are also subject to the Contaminated Land Regulations. The flaring of gas, other than with the consent of the minister, is also prohibited. Unjustifiable interference with the conservation of the living resources of any controlled waters in, or in the vicinity of the licensed area, is prohibited, in addition to interference with fishing or navigation.[28] All onshore exploration and production (E&P) activities require all the relevant consents under the legislation, in particular under the planning system for development control.

While the landward licence confers an exclusive right to explore and produce petroleum from such designated blocks, it does not grant either any right of entry on to the land or any planning permission for the development of the site in order to exercise the licensed right to explore for and produce petroleum.

The grant of the permission to enter onto the land and the grant of planning permission to develop the land must, therefore, be obtained separately.

9.7.4 Government Policy on Minerals

In recognition of the fact that the long-term national demands for minerals must compete with growing demands for environmental protection, the Mineral Planning Guidance Note (MPG) 1 reiterates that the many unique characteristics of mineral development inevitably result in environmental harm. A unique problem is that minerals can only be worked where they are found, and this is often beneath high-grade agricultural land, or land that is designated as a site of special scientific interest or under the European Natura designation. Alternatively, investment in mineral exploitation also has a positive economic effect both at local and national levels. There is, therefore, a presumption in favour of development and, to this end, mineral planning authorities must have a strong disposition to grant permission unless there are very strong environmental objections or there is no real economic need.[29] However, this is now subject to the emergence of the principle of sustainable development as a central strand of planning policy.

Successive governments have encouraged, and even funded and managed, mineral exploration and exploitation, including oil and gas. The strategic and commercial arguments in favour of such a policy include the inevitable conclusion regarding the security of home-produced oil and gas as opposed to the need for imported supplies. Therefore, notwithstanding the principle of sustainable development, national policy still favours the development of onshore oil and gas. It is only when, exceptionally, the environmental implications are so great that the proposed development cannot be permitted on a particular site, that environmental concerns will take precedence over the national interest in developing indigenous onshore reserves.[30]

Nevertheless, even the commercial and economic considerations that underlie particular investments must, in the interests of sustainable development, be subject to the environmental acceptability of specific projects as determined by the land-use planning system and other environmental controls. Many potential oil and gas reserves lie under high-grade agricultural land and applications in these areas are subject to the government policy, which aims to ensure that no more than the essential minimum of agricultural land is diverted to development. In particular, where agricultural land is taken for oil and gas development the question of the restoration to its original usage should be considered before the land is released.[31]

Any oil and gas development is also subject to the substantial body of legislation relating to water abstraction and drainage under national as well as European legislation. Air pollution controls on gas flaring and the problems of noise during the drilling stage are all subject to regulation. If the development is hazardous[32] then the HSE must be consulted prior to any determination regarding the grant of planning permission.[33]

The new coalition government has produced six new NPSs relating to the energy sector that attempt to bring them into line with the new Electricity Market Reform changes. These were expected to come into force by the end of 2011 but are yet to be finally agreed.

9.7.5 Pollution Control

9.7.5.1 The IPPC Directive and Offshore Installations

Consequent upon the adoption of the Integrated Pollution Prevention and Control (IPPC) Directive in 1996. The purpose of the Directive is much wider than the UK IPC regime provided for under the EPA 1990. Whereas IPC sought to prevent or reduce harm to the environment, IPPC seeks to prevent or reduce emissions, the amount of waste produced, and the conservation of resources.

Article 3 of the Directive places a duty on Member States to require competent authorities to ensure that installations are operated in accordance with certain principles, including taking measures to prevent pollution; avoiding the production of waste in accordance with the Waste Recovery and Safe Disposal Directive 75/442; the implementation of safety measures to prevent accidents and limit their consequences; energy efficiency; to take measures to avoid pollution risk, which would include the cessation of certain activities; and, where necessary, restoring the site to a satisfactory state. The definition of IPPC covers not only the release of substances but also the effects of noise, heat, and vibration, which may be harmful to human health or the environment. IPPC permits will need to include conditions to ensure compliance with these general principles and so will have to be considered as part of the best available techniques (BAT) requirements.

9.7.5.2 The Environmental Assessment Directive

The Council Directive on the Assessment of the Effects of Certain Public and Private Projects on the Environment 85/337/EEC as amended by Council Directive 97/11/EC, was incorporated into UK law by the Offshore Petroleum Production and Pipelines (Assessment of Environmental Effects) Regulations 1998 (SI 1998/968) as amended in 1999. The purpose of the regulations is to require the Secretary of State to take into consideration environmental information before making decisions on whether or not to authorise various offshore projects.

The main change in the amended regulations was the inclusion of four classes of projects that no longer require a Petroleum Operations Notice (PON15) seeking dispensation or an EIA. These cover renewals of consents, drilling of a well at same entry point of previous consent, installation of a new pipe within 500 m from a well, and sixth or later well from the same platform.

The assessment itself must be just that—an assessment. It must not be raw data but a discussion of the likely significant effects supported by what the licensee considers to be 'likely' and 'significant'. Where significant adverse environmental impacts are identified, the licensees must explain what the alternatives are and why they were not being adopted. It is also important that any incremental impact due to the project is addressed and any enforceable procedures undertaken to mitigate the impact must be identified. Should the minister consider that the statement is unsatisfactory, consent will not be granted.

The process not only involves a group of statutory consultees but there is also an obligation for public consultation by public notice. The final decision whether or not to grant consent made by the minister is, of course, subject to judicial review in the usual manner.

9.7.5.3 Decommissioning

Just as decisions on exploration for oil and gas and on design and construction of installations for their exploitation have been important in the development of the United Kingdom's hydrocarbon resources on the UKCS, the decisions on the decommissioning of those installations are equally important. The Petroleum Act 1998 contains the basic statutory framework for those decisions, supplemented by the regulatory regimes for the chosen course.

Although the DTI is responsible for the implementation and development of UK government policy on decommissioning offshore oil and gas installations and submarine pipelines, the formulation of recommendations to ministers is a collective process that involves not only government departments with direct statutory responsibilities but also those representing wider interests, including those relating to the environment, fishing and navigation, and health and safety risks. The objective in reaching decisions is to identify the best practicable environmental option (BPEO). Those involved in the decision-making process included the MAFF, which was responsible for Part II of the Food and Environment Protection Act 1985, as amended by the EPA 1990 and the Waste Management Licensing Regulations 1994. MAFF was also responsible for the development and implementation of domestic and international policies to protect fisheries and the marine environment from deposits of waste and other materials at sea. MAFF also carried out an extensive programme of aquatic environmental monitoring.

The Hydrographic Office of the Ministry of Defence is responsible for maintaining Admiralty charts on which installations and pipelines are marked, and the DETR is responsible for coordinating government policy on the marine environment.

The role of the Foreign and Commonwealth Office is to work in liaison with other departments to ensure the full discharge of the United Kingdom's international obligations in the implementation of the decommissioning programme. The role of the Department of Transport is to allow safe navigation and regulate the shipping and the placing of offshore installations and pipelines. It has statutory control over the placement of installations on the UKCS, which may obstruct or endanger navigation. This control is exercised through consents issued under section 24 of the

Coast Protection Act 1949.[34] This control is used to determine where installations are placed and how they are marked. The same control also enables the Department to ensure that the decommissioning of installations does not create danger or obstruction to navigation.

The Environment Agency is responsible for the enforcement in England and Wales of certain legislation for the prevention of pollution from major industrial processes. The EPA 1990 regulates onshore waste generation and disposal. Any onshore facilities used in the decommissioning of installations requires a consent under the Control of Pollution Act (COPA) 1974 and the EPA 1990. Additionally, the Agency enforces the Radioactive Substances Act 1993, which regulates the use of radioactive materials and the disposal and accumulation of radioactive wastes. The Act applies to installations operating in English and Welsh waters and to a specified area around such installations. Anyone wishing to keep or use radioactive materials must register with the Agency; likewise, anyone wishing to dispose of waste on or from any premises must be granted an authorisation from the Agency to do so. The terms and conditions of registration and authorisation are set out in certificates issued by the Agency and they are enforced by members of the Agency who make regular visits to these premises, including offshore installations.

The HSE plays a role in decommissioning by its enforcement of the HSWA 1974, which applies to offshore installations. Specific regulations made under the Act include the Offshore Installations (Safety Case) Regulations 1992, which require operators of fixed installations to submit an abandonment safety case to the HSE at least 6 months before the planned start of decommissioning. The safety case must show that the methods and procedures to be used will ensure that risks to persons engaged in abandonment operations are adequately controlled. The safety case has to be assessed and formally accepted by the HSE before decommissioning can begin. The HSWA will continue to apply to installations left *in situ* after decommissioning, if worker intervention should become necessary.

The 1998 Act requires the owners of offshore installations to submit a proposed abandonment programme to the Secretary of State, who may require an abandonment programme for each offshore installation. If the programme proposes that the installation, or any part thereof, should be left in place, the programme must include any actions for maintenance. The abandonment programme must also include the costs of the abandonment and how these are to be met, and the timing of the operation.

While the installations remain *in situ*, the Offshore Installations (Safety Zones) (No. 2) Order 1998[35] establishes safety zones with a radius of 500 m around each installation that is specified. Vessels are prohibited from entering these safety zones without the consent of the HSE or in accordance with the Offshore Installations (Safety Zones) Regulations 1987.[36]

The Petroleum Act 1998 section 30(1) requires that a notice to submit a programme for abandonment may be given to a specified group of people. The Secretary of State may then choose the person, from the specified group, who must submit a programme and who may also be required to provide the names and address of every other person specified. The Secretary of State may 'pierce the corporate veil' under section 30(1)(e) of the 1998 Act in order to pursue entities beyond those contracting within the joint operating agreement (JOA). This provision is to protect the public purse against any liquidation of subsidiaries by the parent company to avoid the costs of abandonment.

The 1997 Environmental Impact Directive was extended offshore in 1998[37] and came into force that year. The programme must, therefore, include a plan to remedy the effects of the abandonment programme so as to 'avoid, reduce, and remedy' adverse effects on the environment and this plan must be available prior to consent being granted.

On approval of the abandonment programme, there is a duty imposed on the party who submitted the programme to ensure that it is carried out and that any conditions attached are complied with. Liability is joint and severable under section 36 of the Act. Under section 34 of the Act, the Secretary of State may hold previous owners liable in specific circumstances.

Liability for damage caused by any part of a partially removed installation remains with the owners of the remains. Should any of the corporate owners of the JOA be liquidated, then the remaining owners would be held liable and may not be able to claim from the liquidated owners unless previous provisions have been made.

9.7.5.4 The *Brent Spar* Case Study

Shell owned a 140-m tall floating storage platform, which was originally anchored to the seabed 160 nautical miles north-east of Lerwick in the Shetland Islands. It consisted of a helideck, accommodation block, a diving deck, and six storage tanks that held a total of 30,000 barrels of crude oil.[38] The installation had been decommissioned in 1991. The international law at the time required that a permit be granted by the coastal state should a company wish to dispose of a platform at sea. It is expressly stated in the London Dumping Convention that any disposal of waste and other matter directly arising from or related to the exploration and exploitation and associated processing of seabed mineral resources is not covered. Nevertheless, the definition of dumping includes any deliberate disposal at sea of platforms or other man-made structures. Article 19 of the Oslo Convention defines dumping as any deliberate disposal of substances and materials into the sea by or from ships or aircraft.

Article 210(5) of UNCLOS provides that dumping within the territorial sea and exclusive economic zone (EEZ) or onto the continental shelf should not be carried out without the express approval of the coastal state. Under Part I of the UK Petroleum Act 1987, oil companies must prepare abandonment programmes. The Scottish Office, as the relevant authority, granted a permit under the 1985 Food and Environment Protection Act for the *Brent Spar* to be discarded in water over 2000 m deep in the North Feni Ridge situated 150 mi north-west of the Outer Hebrides in the North-East Atlantic. This was within the UK 200-mi EEZ.[39] On application for consent, Shell provided a detailed site survey and a scientific study, which concluded that marine disposal was the BPEO. Permission was granted in December 1994.

The environmental group Greenpeace, fearing that such a disposal would create a precedent, applied for judicial review of the DTI's decision to grant an authorisation. Greenpeace also occupied the *Brent Spar* and started a campaign to mobilise public opinion. The occupation of *Brent Spar* ended by the removal of the protesters by the Grampian police and the action for judicial review was rejected on the grounds that Article 19(ii) of the Treaty of Rome prohibits a Scottish minister from being tried before an English court.[40] However, the public concern was translated into political pressure on the UK government from other European Member States and commercial pressure being exerted on Shell. Public hostility culminated in a Shell petrol station in Germany being firebombed which resulted in the death of an employee.

The disposal plan was abandoned when the UK government authorised an alternative disposal plan for dismantling the *Brent Spar* and using sections to support a quay extension near Stavanger.

The *Brent Spar* affair raised a number of questions regarding the legal and moral position of disposing of offshore installations at sea. This debate led directly to the Oslo/Paris (OSPAR) Decision 98/3.

9.8 DECOMMISSIONING AND DISPOSAL

9.8.1 The OSPAR Commission

In 1992 a new convention, the Convention on the Protection of the Marine Environment of the North-East Atlantic (the OSPAR Convention) was agreed. This is a regional convention covering specific areas of the North-East Atlantic, including the North Sea and parts of the Arctic Ocean. This new convention replaced and updated both the 1972 Oslo Dumping Convention concerning the protection of the marine environment from dumping from ships and aircraft and the 1974 Paris Convention on the prevention of marine pollution from land-based sources. The OSPAR Convention came into force in 1998.

At its first ministerial meeting, the OSPAR Commission adopted a binding decision[41] to ban the disposal of offshore installations at sea.[42]

Decision 98/3, which entered into force on February 9, 1999, prohibits the dumping and leaving wholly or partly in place of offshore installations. Under the terms of the decision, the topsides of all installations must be returned to shore, and there is a presumption in favour of land disposal. All installations with a jacket weight less than 10,000 tonnes must be completely removed to shore:

1. Only the footings or part thereof may be left in place.
2. Minimum water clearance of 55 m is required above any partially removed installation that does not project above the surface of the sea.

The decision recognises that there may be difficulty in removing the footings of steel jackets weighing more than 10,000 tonnes; consequently, the decision permits derogations from the main rule. However, such derogations will only be granted if the assessment and consultation procedure shows that there are significant reasons why an alternative disposal option is preferable to reuse or recycling or final disposal on land, and, likewise, for the removal of concrete installations.

9.8.2 Disposal on Land

Under the waste management licensing system, established under Part II of the EPA 1990, anyone who in the course of a business, or in any other way for profit, transports controlled waste within Great Britain is required to register as a waste carrier. Relevant legislation is the Control of Pollution (Amendment) Act 1989 and the Controlled Waste (Registration of Carriers and seizure of Vehicles) Regulations 1991. Registration of establishments or undertakings that transport waste on a professional basis is also a requirement of the EC Framework Directive on waste.

The common European definition of waste[43] has been transposed into domestic legislation by Regulation 1 of the Waste Management Licensing Regulations 1994. The legal definition of waste is 'any substance or object which the producer or the person in possession of it discards or intends or is required to discard'. Whether or not a substance is waste is determined on the facts and this area is rich in case law.[44] In addition to the Waste Management Licensing Regulations 1994, depending on the nature and composition of the waste, under the Control of Pollution (Special Waste) Regulations 1996, it may be defined as special waste[45] or radioactive waste under the Radioactive Substances Act 1993. These are potentially the most difficult and dangerous waste materials. The regulations require all movement of special waste to be tracked from the moment it is produced until it reaches a waste management facility by way of consignment. Under this system, the regulating body must be pre-notified of any movements of hazardous waste so that it reaches an appropriate facility and the correct records are kept.

Once the material has been classified as special waste, the International Convention for the Prevention of Pollution from Ships 1973, as amended by the 1978 Protocol, would require that waste reception facilities are provided.[46] In addition, the Dangerous Substances in Harbour Areas Regulations 1987 as amended and the Waste Management Licensing Regulations 1994 control the carriage, loading, unloading, and storage of all classes of dangerous substances in port areas.

Underpinning all of these regulations is section 89 of the Water Resources Act 1991, which identifies as an offence in England and Wales to cause or knowingly permit any poisonous, noxious, or polluting matter to enter any 'controlled waters'.[47] Controlled waters extend to 3 mi from a defined baseline in England and Wales,[48] and in Scotland, it is 3 nautical miles from the baseline within the meaning of section 30A of the 1974 COPA. In addition, there are other named activities under Crown control, as outlined in the Continental Shelf Act 1964.

If the remains are taken to landfill for disposal, then each landfill site within the United Kingdom must be licensed by the appropriate waste regulatory authority (WRA). Under the licensing procedures, each site is approved to take particular wastes and the wastes that a site is specifically permitted to take and that it is prohibited from taking are identified. In addition, under section 34 of the EPA 1990, a duty of care applies to anyone who 'produces or imports, keeps or stores, transports, treats, recycles or disposes of' controlled waste. This duty also applies to anyone acting as a broker to arrange any of these activities. Those subject to the 'duty of care' are required to take all reasonable steps to keep the controlled waste, for which they are responsible, safe throughout the whole of the waste cycle. If the duty holder complies with all of the Environment Protection (Duty of Care) Regulations 1991, then the Environment Agency is likely to conclude that under 'due diligence' the duty holder has taken all 'reasonable steps'.

Where installations, pipelines, and/or waste is brought to land for disposal, then the operation will also be subject to all appropriate health and safety legislation.

9.8.3 Post-Decommissioning Surveys and Monitoring

Once the decommissioning operations are complete, a survey of the site to a radius of 500 m around the location of the installation, must be carried out. Any remaining debris must be removed and an independent verification certificate submitted to the DTI's Offshore Decommissioning Unit. In addition to this, a post-decommissioning environmental seabed sampling survey must be undertaken to monitor levels of hydrocarbons, heavy metals, and other contaminants in sediments and biota.

If it is agreed that a concrete installation or the footings of a steel installation should be left in place, the condition of the remains will have to be monitored in accordance with Annex A of the OSPAR decision. A post-disposal report must be submitted to the DTI's Decommissioning Unit as required under paragraph 10 of the OSPAR decision.

The operator has a duty to mark the remains with (and maintain) any necessary navigational aids and notify mariners and the appropriate hydrographic services of the change of status of decommissioned installations and pipelines. Details of the action to be taken to advise mariners and mark any remains should be included in the decommissioning programme.

9.8.4 Residual Liability

The residual liability arising from any remains or residues left on the continental shelf resides with the persons who own an installation or pipeline at the time of its decommissioning. Any residual liability remains with the owners in perpetuity. Any claims arising from damage caused by such remains will be a matter for the owners and the affected third parties and will be governed by the general law. Owners/operators must post a financial instrument prior to decommissioning in respect of any residual liability.

9.9 ONSHORE OIL AND GAS

The first commercial exploitation of oil on land was in East Sussex in 1895. Subsequently, oil and gas production has continued on a relatively small scale in the East Midlands. However, it has only been in the last 20 years that onshore fields have become a feature. This renewed interest has not passed unchallenged in areas where there is a strong middle-class opposition to environmentally intrusive works, which serves to highlight the potential conflicts between environmental protection and the development of onshore oil and gas. This section looks at the licensing system and the uneasy relationship between planning consent and pollution control, vulnerable sites including the coastal strip, and the use of environmental management systems (EMS).

9.9.1 The Licensing System

Rights to explore and exploit oil and gas reserves in Great Britain are vested in the Crown in Parliament. Licences for developments are awarded by the DTI while planning decisions on oil and gas matters are the responsibility of the relevant minerals planning authority. Two related systems have been developed to regulate activities on land (landward) and offshore (seaward), reflecting the jurisdictional limits of mineral planning authorities who are responsible for the former. The situation is complicated, however, by the fact that the landward regulations also cover estuaries and the waters around the Scottish islands, which lie outside local authority control. Thus, a third set of procedures is required to deal with activities in the landward marine zone.

Both landward and seaward licences grant exclusive rights to the licensees to explore and produce petroleum in one or more particular blocks. Such licences may only be applied for in response to an invitation from the Secretary of State as part of a licensing.

The Energy Minister announced the ninth round of onshore licences in February 2000. These were to include licences to enable companies to explore for and exploit the petroleum resources, including conventional oil and gas, mine vent gas, and coal-bed methane trapped in coal seams. The inclusion of coal-bed methane gas (CMG) was continuing the Government's policy on cleaner coal technology and its wish to encourage projects that aim to develop better methods of mechanically enhancing coal permeability and achieving commercial CBM production.[49] Following a suggestion from the Oil and Gas Industry Task Force, and in keeping with the Government's own declared policy of improving transparency of the decision-making process, the DTI has signalled its intention to discuss the possible options with the onshore industry during the ninth round of negotiations. All the areas on offer were above the mean high watermark.

9.9.1.1 Coal-Bed Methane Gas (CBM)

Methane can be extracted by drilling directly into coal seams, reducing the pressure in the target area by a dewatering process, thus causing the gas to flow out of the coal.[50] The fact that unworked coal was not vested in the Crown under the 1934 Act[51] but in the National Coal Board[52] under the Coal Industry Nationalisation Act 1946 created difficulties in the exploration for and exploitation of such methane. All rights in offshore coal were similarly reserved to the National Coal Board under the Continental Shelf Act 1964, section 1(2). This was an exception to the general grant of all rights on the continental shelf to the Crown. The DTI, as petroleum licensing authority, took the view that CBM was petroleum and granted licences in the early 1990s to explore for it under the landward licensing regulations. Methane drainage licences were limited to the purpose of keeping mines safe. British Coal, however, considered CBM as part of the coal excluded from the ambit of the 1934 Act, and in any event controlled access to CBM that could only be obtained by drilling through coal seams, which were incontestably the property of British Coal. Therefore, the permission of British Coal was needed to explore for CBM under section 10(3) of the 1934 Act[53] (now section 9(2) of the 1998 Act). British Coal demanded both a degree of operational control, in addition to significant payments in respect of any drilling for CBM. Inevitably, the licensees were reluctant to accede to these demands. It was not until the Coal Industry Act 1994 that these conflicts were resolved. The 1994 Statute sold into private ownership the state interest in coal mining and vested coal deposits in a new public body, the Coal Authority. Under section 9(1) of the Act, the Coal Authority has no right or interest in CBM, which is vested in the Crown and is subject to the general petroleum-licensing regime. The statute preserves the right of the Authority to control access to CBM under section 9(4), but provides that in managing its property, the Authority must have regard to the desirability of exploiting CBM where it is economic to do so. The Authority must also settle terms with the Secretary of State for Trade and Industry under sections 3(5) and 3(6).[54] The DTI does not expect the Authority to grant CBM operators access to coal deposits that are being worked or are likely to be worked.

9.9.2 Planning Controls

It is not within the remit of this book to give an exposition of the planning laws of the United Kingdom as there are many and varied excellent books on the subject for both practitioners and academics.[55] However, a brief description of the planning system and how it may affect onshore development of oil and gas projects in respect of the environmental controls should be of interest to the reader.

Since its inception in 1909, planning legislation has given to local authorities direct power and responsibility for carrying on the daily administration of land use control. The role of central government has always been the supervision and coordination of the way in which those powers are exercised. Subsequent upon the Stevens Committee on Planning Control over Mineral Working Report 1976, the government implemented many of the recommendations in the Town and Country Planning (Minerals) Act 1981, the main features of which have been incorporated into the 1990 Act. The 1981 Act established Mineral Planning Authorities (MPAs) to be responsible for all planning control over mineral working. Since the winning and working of minerals is a 'county matter' the MPA will be the county planning authority.[56] The 1981 Act also authorises the MPA to impose both restoration and aftercare conditions.

9.9.2.1 The Planning System

Mineral planning authorities have a duty to produce mineral structure plans for their area, which must include statements of policy for the development of oil and gas. For the purposes of information for both developers and those who sought to object to any development, the policy should contain any presumption against exploration and development in particular areas that may be particularly environmentally sensitive.

Conversely, should a developer wish to submit a planning proposal, it would be in the interests of all concerned if an open dialogue were conducted between the developers, the MPA, and the general public. Local issues may then be discussed and actions taken to mitigate any environmental concerns; these may then be addressed in an EIA and statement. Oil and gas proposals are unlikely to raise wholly new issues of principle. In fact, their impact on the environment can be considerably less than many forms of mineral working.

The seismic investigations during the exploration phase of hydrocarbon operations would normally have little environmental effect as the investigations are mostly temporary. Nevertheless, this phase could involve the use of the road network through sensitive areas that may involve historic buildings, and the abstraction of water from vulnerable sources.

Although at the moment, the act of drilling deep boreholes for exploration purposes requires the grant of planning permission, consultation has taken place regarding the proposed introduction of a new class to the Town and Country Planning General Development Order 1977[57] (the GDO). This would give permitted development rights in respect of certain exploratory operations undertaken preparatory to the exploitation of a mineral. It is anticipated that the proposals will apply to seismic surveys where these are so as to constitute development within the terms of the TCPA 1971 as amended. However, even if the GDO is amended in this manner, all exploration and appraisal boreholes for oil and gas will continue to require permission from the MPA who will give careful consideration to the environmental effects of such operations. These may well include such factors as visual intrusion, vehicular routing, and the control of the disposal of mud and other drilling residue. In many cases, the regulatory authority will attach conditions to the licence as part of an arrangement under section 106 of the TCPA 1990. Otherwise, conditions[58] may be attached to the planning permission to ensure the mitigation of any adverse impacts on either the local population or the environment.

The appraisal of a field will often involve the drilling of further exploratory wells in order to delineate the extent of the field. It is not until this has been established that a true assessment of the environmental effects of the commercial exploitation of the whole field can be made.

It is however, the production and distribution phase that may raise the most controversial environmental issues. The size of the gathering stations for the separation, purification, and treating of the raw material may well be visually intrusive, but there are limited options as to the siting of these facilities. Again, these facilities may be subject to conditions attached to the grant of planning permission. For example, conditions may include the timing and method of gas flaring; the routing of vehicles to and from the facility; the minimisation of noise by the use of special equipment, and the specification of the means of disposal of any unwanted gas. Should the facility require the laying of pipelines for the transportation of the end product, then these too will be subject to environmental appraisal and, maybe, the attachment of conditions.

The restoration of sites that are unsuccessful following exploratory drilling provide few problems. Those sites that have contained processing facilities may present aftercare difficulties unless these are adequately addressed during the design stage.

9.9.2.2 Planning Conditions

Almost all planning permissions that are granted are subject to conditions. It is often the case when considering planning applications for the development of onshore oil and gas that the question is not whether development should be permitted at all, but on what terms the development should be permitted. Conditions will normally be imposed, therefore, to ameliorate any adverse effects that might otherwise flow from the development. It is not uncommon for the grant of planning permission for the exploration for and exploitation of minerals to contain more than 50 conditions.[59]

Section 70(1) of the TCPA 1990 provides a general power to impose such conditions as the authority think fit. However, this general power is not unlimited. The common law progressively restricted until the House of Lords, when required to consider the validity of a condition imposed under the general power in *Fawcett Properties*

Ltd v. Buckingham County Council [1961] AC 636, held that the conditions must comply with the following tests:

1. They must be imposed for a planning purpose and not for an ulteria one.
2. They must fairly and reasonably relate to the development permitted.
3. They must not be so unreasonable that no reasonable authority could have imposed them.

Although the legal tests for the validity of a condition were laid down in the *Newbury* case, they have been reinforced by Circular 11/95, which sets out a sixfold test that conditions should meet.

Planning conditions were again raised in the *Grampian Regional Council v. City of Aberdeen District Council* case[60] when the determination of a planning application was made by a reporter (Inspector). The application was for the change of use from agricultural land to industrial use. The reporter considered that traffic to and from the site would constitute a hazard. However, he determined that the hazard would be removed if a particular road were closed, but concluded that it would not be competent to grant planning permission subject to a negative condition requiring the closure of the road, as this was not within the power of the respondent. On final appeal to the House of Lords, it was decided that as the reporter had determined that the development would be in the public interest, it would not be unreasonable, but, in fact, highly appropriate, to grant planning permission subject to the condition in question.

Following the *Grampian* case, the imposition of negative conditions[61] has risen in popularity, in particular where there is a public interest in the development going ahead, as may be the case in respect of the development of an onshore oil field.

9.9.2.3 Planning Agreements and Obligations

As mentioned above, it is not unusual for the grant of planning permission to be subject to planning agreements under section 52 of the 1971 Act and the old section 106 of the 1990 Act which provided:

1. A local planning authority may enter into an agreement with any person interested in land in their area for the purpose of restricting or regulating the development or use of the land, either permanently or during such period as may be prescribed by the agreement.
2. Any such agreement may contain such incidental and consequential provisions (provisions of a financial character) as appear to the local planning authority to be necessary or expedient for the purposes of the agreement.
3. An agreement made under this section with any person interested in land may be enforced by the local planning authority against persons deriving title under that person in respect of that land as if the local planning authority were possessed of adjacent land and as if the agreement had been expressed to be made for the benefit of such land.

There were, however, problems with the old planning agreements, one of which was that only the developer and local planning authorities could be party to it. The 1991 Planning and Compensation Act amended the old section 106 by enabling any person interested in land in the area of a local planning authority to, by agreement or otherwise, enter into an obligation. Such an obligation may restrict the development or use of land in some specified way, or require specified operations or activities to be carried out. These planning obligations may then be registered as a local land charge under the Local Land Charges Act 1975 and the obligation then becomes binding upon any future purchaser of the land.

This was the first time that it was accepted that a developer could provide an enhancement to the social infrastructure or that it could be a legitimate requirement of any decision to grant planning permission for development.

However, it fell to the House of Lords to clarify the extent to which a local planning authority can require a developer to provide community benefits beyond the development itself, following a number of conflicting cases decided in the Court of Appeal.[62] In the case of *Tesco Stores Ltd v. Secretary of State for the Environment*,[63] it was held that the test to be applied in determining whether a planning obligation is a material consideration is whether it has a connection with the proposed development which is not *de minimus*. If the connection can be established, then the planning obligation must be considered by the planning authority. In the *Tesco* case, it was established that the Secretary of State had taken the planning obligation into account but had given it little weight, and this the Secretary was entitled to do. The latest guidance on this matter is Circular 1/97 issued in January 1997.

9.9.2.4 Appeals

Most planning decisions are subject to an appeal to the Secretary of State, and from thereon to the courts on a point of law. The right of appeal is under section 78 of the 1990 Act and Circular 15/96 sets out best practice regarding the procedures for determining the appeals.

9.9.2.5 The Scope of Public Involvement

Most oil and gas reserves on land are in the midlands and the south. Given the considerable public concerns regarding the possible effects on the countryside and on the local communities in these areas, the public are taking up their rights in greater numbers to be involved in any decision-making process. The balance to be struck is a difficult one that may include national interest, the interests of the developer to take up his rights under the grant of any licence to explore and exploit any discovered petroleum reserves, the rights of the local authority to consider the employment position in the area, and also the rights of those who are concerned about the effects of any such developments on the environment. All parties need to be vigilant within the terms of the principle of sustainable development.

9.9.2.6 Access to Environmental Information

Before any member of the public can enforce their rights, they need to know what their rights are and how the environment may be affected by any given operation. However, as Stuart Bell[64] so eloquently describes, the roots of secrecy in the environmental field can be traced back to the mid-nineteenth century and the Alkali Inspectorate.

The regulatory bodies hold information regarding the dangers and consequences of particular pollutants emitted into the various receiving environmental media, and yet historically there were statutory obligations placed upon the agencies prohibiting disclosure of information relating to environmental discharges, even if there were no need for commercial confidentiality.

Commercial confidentiality is an important issue as far as many regulated companies are concerned. The belief that competitors may be able to take advantage of information placed on public registers to gain access to secrets continues, as applications to exclude information from the registers remain high.

A further justification for industry to maintain secrecy is the belief that open access would lead to unacceptable interference by environmental groups.

This view persisted and was played out in the debates during the passage of the Environmental Protection Bill through Parliament. Claims were made that the chemicals industry would disappear from the United Kingdom as environmental groups used the information to bring prosecutions for 'media' purposes rather than to protect the environment. The counter-argument was that if the information on a public register shows that a company is not in compliance with the conditions of their licence, and the regulatory agency has seen fit not to bring a prosecution, a question arises concerning the policy of pursuing a non-confrontational enforcement strategy. As one junior minister put it, 'The commitment of industry to pollution controls is the surest defence against such litigation'.

In the event, the fears of industry have proved to be unfounded. The proportion of private prosecutions is minute, notwithstanding the continued reluctance of the enforcement agencies to litigate rather than rely on conciliation and persuasion.

The Royal Commission on Environmental Pollution considered that the public were entitled to know of the risks that they faced from environmental pollution and that the only way to restore public confidence in the regulatory system was to allow the public to take enforcement action themselves, based on public access to the information that would allow them to do so. The Royal Commission concluded that the guiding principle behind all legislative and administrative controls relating to the environment should be a presumption in favour of unrestricted access to information gathered by the regulatory bodies; and also that any restriction on that access to the information must be '. . . only in those circumstances where a genuine case can be substantiated'.

The Royal Commission continued to argue for openness and public access to environmental information. In their Eleventh Report, the Commission commented on the need for such information to enable informed discussion about the risks to the public and the environment in order to evaluate any mitigating actions that might be taken to minimise those risks. The initial reporting of polluting incidents is often left to the general public once they become aware that an incident has taken place. Access to the information would enable anyone who recognised that an incident had taken place to assess the seriousness of such an incident.

Despite this long history of Recommendations from the Royal Commission, it was not until the introduction of the EPA 1990 followed by the Environmental Information Regulations 1992, which were to transpose the European Directive into UK law, that the disclosure of environmental information became a reality.

9.9.2.7 The European Directive

The European Directive on Free Access to Environmental Information[65] establishes an access right independent of showing a legal interest. The Directive defines what this is and identifies the exceptions to the right of access. It also requires that the right is legally protected when transposed into domestic law. As the Directive is based on Article 130s of the EC Treaty, Article 130t, which allows the Member State to go further, is applicable. This allows the national legislators to open up even more information to the general public than what the Directive rather narrowly defines as 'environmental information'. However, the obverse is also true by the introduction of the range of exemptions, which include the protection of trade secrets, privacy, and governmental deliberations.

The exemption protecting the 'decision-making process of the administrative agency' includes raw technical data that have not yet been processed.[66] For example, emission data may be retained as long as it is raw, uncompiled, and unanalysed. However, once the data have been analysed and the document completed, the information is then open to access under the Directive. Further, the exemption includes the 'substance of the decision-making'. This involves any information that may affect the 'confidentiality of the deliberations of the agencies'[67] and any 'internal communications'[68]; moreover, these deliberations and communications are expressly not made accessible once the decision has been taken.

As far as the exemption to access to trade secrets[69] is concerned, the Directive does not define the term, and so the historic problems of interpretation of the term remain. Gerd Winter[70] suggests that the information should be subject to four filters through which the information must pass before it will qualify as a protected secret. Firstly, the information should relate directly to the technical process; secondly, the information must be known only by a small number of people; thirdly, the firm or person should be required to claim secrecy when the information is submitted to the agency; and fourthly, the interests of secrecy must be weighed against the interests of disclosure on a case-by-case basis. For instance, the public interest in relation to public health, public safety, or protection of the environment must be weighed against the person or firm concerned suffering loss if the information were to be disclosed.

The final exemption from the right to access to the information concerns the 'protection of privacy'. Information concerning an incident may also involve information about an individual private person. This may include individuals who have given an expert opinion; the author of a third-party intervention, and so on. It may be that the individual may wish to remain anonymous as to their involvement in the supplying of the information. Again, there must be a balance achieved between the free access to the information and the protection of privacy. What should the substantive criteria be for the resolution of this conflict of interests? Some jurisdictions define a narrow sphere of private life that should be protected, while others use a broader definition by drawing up lists of what may and may not be protected. A third system involves a balancing-of-interests test, which in itself creates its own problems, that is, the devising of such a test.

9.9.2.8 The Environmental Information Regulations, 1992

The implementation of the Directive into domestic legislation was through the Environmental Information Regulations.[71] Under the Regulations, any public body with responsibility for the environment must provide access to such information in response to a request from any person. The term *Environmental Information* includes not only emissions levels as recorded on the registers, but also the quality and state of the air, water, soil, flora, fauna, and the state of habitats including designated sites and others.

In respect of habitats and the protection of the countryside in the United Kingdom, there are many private bodies that carry out public functions. Since the privatisation of the utilities, the water service companies and the waste disposal companies carry out roles that are semi-public. Inevitably, this creates an area of uncertainty as to the obligation to provide environmental information held by such bodies.

The Regulation lists categories of information that are exempt from the regulations, and these are divided into those subject to mandatory confidentiality and those subject to discretionary confidentiality. The mandatory categories cover information that comes under the protection of privacy and includes personal information, which the individual has not given consent to disclosure; any information disclosed voluntarily by a third party; and information which, if disclosed, may result in damage to the environment, such as the location of certain rare species of flora or fauna.

Categories of information that are covered by the discretionary exemption include information affecting national security, defence, or international relations; material from internal communications or unprocessed data; and information that may be considered to be covered by commercial confidentiality.

There is no right to appeal against the refusal of a public body to give accesses to the information other than an application for judicial review. However, without the information sought, it is difficult consider when an individual might bring such an action.

A review of the Directive and the regulations was carried out by the House of Lords Select Committee on the European Communities in 1996. The subsequent report concluded that the actual take-up of the right of access to environmental information was low. This was due to a number of factors, including the problem with the definition of 'information relating to the environment'. A further problem is one of costs. The Regulation permits public bodies to make an 'administrative charge' for the provision of the information but gives no guidance as to the level of such a charge. Some authorities make only a nominal charge or none at all, whereas others quote a prohibitive charge which, more often than not, results in the applicant failing to pursue the application. Friends of the Earth gave evidence to the Select Committee concerning the discretionary exemptions and the large amount of data that were not on the public register because they were in the 'operational' category and so were exempt.

9.9.2.9 Environmental Protest

Recent years have seen a renewed industrial interest in onshore oil and gas development in the south and east of England. The public concern about the possible effects on the countryside has become more overt, with some environmental groups not being content to express those concerns by peaceful means. There have recently been a number of high-profile protests organised by special interest groups. Although these protests have covered a number of diverse operations, energy companies have not escaped the attention of organised protestors, both on the mainland and offshore.

Should a site be invaded by uninvited guests it is not always possible to rely on the police to help remove them.[72] An injunction may be the more appropriate remedy whether or not there is any trespass to the land involved. However, this also raises practical difficulties in cases where the names of the protestors are not known. Under the Order 113 procedures, this is expressly catered for where the originating summons may be against 'persons unknown', or the defendants may be named as 'Susanna', 'Sue', or 'Womble'.[73]

A further consideration is the possible effects of the Human Rights Act (HRA) 1998, which came into force in October 2000. If protesters have established residence on the land they may seek to rely on the European Convention on Human Rights Article 8, which provides a right to respect for family and private life. The right to freedom of expression[74] may be relied upon, or the freedom of assembly and association.[75] The issue of freedom of expression would be of particular relevance should a court be considering whether to grant relief or not.

However, all of the rights and freedoms are subject to the public interest and the balancing of the interest in public safety, the prevention of public disorder, and the balancing of the rights of others under the Convention. The issue is to what extent the protest creates danger to others or risk to public order.[76]

With the advent of the HRA 1998 any company wishing to develop an onshore oilfield would do well to give as much information as possible on the envisaged project and the efforts made to mitigate any possible environmental effects. To include environmental groups and the local population in the decision-making process may well carry the day.

With the coming into force of the new Localism Act 2011, developers must be very mindful of the powerful lobbying of those who oppose development of any kind and say 'not in my back yard' (NIMBYs).

9.9.3 Vulnerable Sites

Conservation and the protection of vulnerable sites in the United Kingdom has traditionally been subdivided into landscape, nature, and archaeological areas of interest. This is a distinction which reflects the historical development of environmental and preservation interests by different disciplines, that is, planning, biological sciences, and archaeology. In addition, the conservation of areas of architectural or historic interest is an integral part of the planning system. This separation has gradually become blurred as the environmental issues promote a more holistic approach to sustainable development. However, the traditional divisions are reflected in the different legislative and administrative frameworks that have been developed to achieve the various conservation objectives.

Nevertheless, it would be wrong to imply that conservation is merely concerned with the designation of important sites. In fact, a number of important bodies and authorities, particularly those with a keen interest in the coastal zone, have formal environmental duties that require them to make provision for conservation interests when carrying out their responsibilities. Numerous voluntary bodies, operating at both national and local levels, have become particularly important in protecting the countryside. Of particular significance is the involvement of a wide range of non-governmental organisations (NGOs) such as the World Wide Fund (WWF) for Nature, The Marine Conservation Society, the Royal Society for the Protection of

Birds (RSPB), the Royal Society for Nature Conservation, and the National Trust. All these organisations also have a significant involvement in active conservation management.

9.9.3.1 Protection for Designated Sites

The first positive legislation was introduced in 1949 as the National Parks and Access to the Countryside Act. This statute made provision for the designation of National Parks and ANOBs. Although the protection of the coastline was not the focus of these designations, 5 of the 10 national parks and 20 of the 38 ANOBs have coastal frontages. The statutory protection afforded to these areas by restrictions on permitted development rights under the GDO and by-law-making provisions have ensured that the designations are considered by many bodies as an effective approach to conservation.[77]

Under the 1949 Act,[78] the Nature Conservancy Council (NCC) and its successors[79] has the power to enter into agreements for the purpose of establishing a National Nature Reserve (NNR). These Reserves are managed by the introduction of by-laws restricting the use of the land by the public.[80] This system was given statutory backing by the 1981 Act, which also extended the designation to areas where there was no agreement with the relevant Country Conservation Council (CCC) or land not in CCC ownership.[81]

NNRs serve a variety of purposes including the conservation and preservation of the sites, the maintenance of sites for research and study, the provision of advice on and the demonstration of conservation management, and for educational purposes. As these reserves are managed specifically for nature conservation and are owned or leased by the relevant CCC, or are managed on their behalf, they aim to provide a powerful safeguard in the protection of important habitats.

Sites of Special Scientific Interest (SSSIs) are widely considered to be the cornerstone of nature conservation in the United Kingdom. Even the European and international obligations are normally implemented through the SSSI system.

Under the 1981 Act as amended, English Nature and its Scottish and Welsh counterparts, are required to notify the Secretary of State, the local planning authority, and the relevant landowner and/or occupier that a site is of special interest. The notification must also specify those operations likely to damage the special interest, which could include the development of any petrochemical reserves. However, such designation does not mean that no development may go ahead, merely that, once notified, owners and occupiers must give the relevant CCC written notice before carrying out a potentially damaging operation (PDO). The operation can be carried out if the relevant CCC give written consent or if it is carried out in accordance with a management agreement with the CCC, or after 3 months have elapsed from the giving of the notice of the operation.[82] If the CCC consider the proposed operation unsuitable for the designated site, they may enter negotiations for a management agreement if they fail to persuade the owner or occupier not to proceed. If these negotiations should fail, then the CCC may apply to the secretary of State for a Nature Conservancy Order.[83] However, there are two exceptions to the requirement for consultation, which are in the case of emergency or where the operation already has planning permission.

Further to the general Guidelines for the selection of SSSI sites, are those that concern earth science SSSIs. These are selected on the basis of national or international importance to research and may well have important relevance to the development of onshore oil and gas reserves. The Geological Conservation Review has selected sites by scientific discipline producing a national network of sites that represent the geology and geomorphology of Great Britain. This system is complemented by Regionally Important Geological and Geomorphological Sites (RIGS). These, by contrast, are selected locally and are based on their educational, research, and historical importance.

9.9.3.2 The Coastal Zone

The coastal zone extends both seaward and landward of the coastline itself and provides important locational benefits for heavy industry and energy-generating facilities. Coastal waters are used as coolants in nuclear and thermal power stations and the coast itself is viewed as a potential source of energy through wave and power.[84] A high percentage of known oil and gas reserves are situated in the coastal zone and these were included in the ninth round of landward licensing.[85] However, the coastal zone supports a great diversity of plant and animal communities, and the consequences of not taking into account the nature of the physical environment are most acute because the coastal landscape cannot be considered to be stable.

9.9.3.3 Landward-Marine Areas

Licenses are awarded by the DTI under the Petroleum (Production) (Landward Areas) Regulations 1991. For these areas, the DTI serves as both the licensing authority and, *de facto*, the MPA. Although estuaries and certain other inshore waters are not covered by the planning system, they are affected by the landward licensing system.

Licences for landward areas have undergone a series of reorganisations with the 1984 Petroleum (Production) (Landward Areas) Regulations being replaced by the 1991 Regulations with the same title, which were in turn replaced by the 1995 Regulations. The 1984 Regulations provided for three forms of licence: the development licence, the appraisal licence, and the exploration licence. The 1991 Regulations added a further type with the supplementary seismic survey licence. The development licence corresponded to the seaward production licence, with the appraisal licence being designed to permit the appraisal of the field with the formulation and approval of development plans and the obtaining of any necessary planning consents, in advance of the grant of a development licence. The landward exploration licences for areas above low watermark differed from seaward exploration licences inasmuch as the former were exclusive and covered a particular area of land and permitted deep drilling. The supplementary seismic survey licence is non-exclusive and only lasts for 1 year. It may be granted to the holder of any of the other three licences in respect of an area contiguous to the principal licence area and not extending more than 1 km beyond it. The aim is to facilitate the surveying of deposits that may extend beyond the confines of the licensed area.

The 1995 Regulations aimed to simplify the petroleum exploration and development licences with effect from June 30, 1995 by dropping the separate exploration, appraisal, and development licences, but retaining the supplementary seismic survey licence as a separate licence. The consolidated Landward Exploration and Development Licence is an exclusive licence that confers the same rights to search, bore for, and get petroleum as does the seaward production licence. Further, they are now awarded by invitation in licensing rounds, for territory organised according to a grid pattern of blocks 10 km square. Given the environmental conflicts that may arise in these areas in respect of the development of hydrocarbon resources, and the conservation of wildlife habitats, a Government View Procedure is requested by the DTI from the DETR or the Scottish Office should an operator wish to produce from a well sited in an estuary. As part of the procedure consultations are carried out with other government departments who may have an interest: the Department of Transport (navigation, ports, oil spill contingencies), the Ministry of Defence, and the Crown Estates Commissioners, as well as the mineral planning, pollution, and coastal protection authorities will all have a view as to how the proposed development may affect any interest for which they have responsibility. Likewise, bodies that represent environmental interests such as the NCC and the Countryside Commission will be consulted. This consultation will take place prior to the award of a development licence and oil contingency plans are required at exploration drilling, appraisal, and production stages. When an application is submitted for a proposed development in an estuary, developers are advised to prepare an environmental assessment as they would for an onshore proposal if it were situated in a sensitive area.

9.10 ENVIRONMENTAL MANAGEMENT SYSTEMS

Growing emphasis is being placed on EMS by organisations of all kinds as a means to achieve and demonstrate sound environmental performance. With the entry into force of the European Eco-Management and Audit Scheme (EMAS) and the adoption of the International Organisation for Standardisation (ISO) 14001 series, interest in EMS certification is set to expand. EMSs are also beginning to assume an important role in the E&P sector. Large companies such as Shell have even adopted their own E&P specific EMS. Additionally, the E&P Forum has adopted E&P-specific guidelines on EMS.[86] The regulatory bodies give credit to sites that are accredited under one of the recognised EMSs, and visit them less frequently for monitoring purposes.[87]

Oil and gas companies are responding to these trends and the regulatory imperatives by setting up environmental, health, and safety departments and by changing their operational practices. More generally, they are adopting corporate EMS designed to deal with the varied environmental regulations.[88]

9.11 THE GAS SUPPLY INDUSTRY

The gas supply industry within the United Kingdom was fundamentally changed by the moving of the former publicly owned British Gas Corporation into the private sector as British Gas plc under the 1986 Gas Act.[89] It was the ideological belief of the then Conservative government that introducing the potential for competition into the system created an incentive similar to that of true competition.[90] Unfortunately, the gas supply industry remained a natural monopoly with new participants required to provide a significant investment in infrastructure, which in turn created a significant barrier to entry into the market. Further, the time factor involved for licences to be applied for and granted created a further impediment. The outcome was that the transfer of the gas supply industry from the public to the private sector did not achieve the aim of true competition. The structure created by the 1986 Act did not lead to self-sustaining competition, but with regulatory intervention a competitive market has since evolved.

The Gas Act rejected the accepted view that the best way to guarantee a reliable energy supply was through state-owned enterprises, and demonstrated the belief that the private sector was equally capable of ensuring a guaranteed energy supply with the additional benefit to consumers of lower prices brought about by competition. It was argued that in a state monopoly, the state sets prices, which in turn are influenced by a service mentality rather than a profit motive.

Prior to the 1986 Act, the state monopoly only related to the transmission and distribution of gas rather than its production. The onshore pipeline network was owned by the state-owned British Gas Corporation, which was solely responsible for the transmission of gas to consumers. In addition, British Gas owned all the gas within its pipelines as its privileges included a monopoly purchase right for all gas produced on the UKCS. The existence of these monopoly purchase rights effectively enabled the government to determine not only the number of offshore gas developments[91] but also the rate of depletion.[92] When the government transferred ownership of the gas supply industry to the private sector, its ability to control the details of offshore gas development was, therefore, largely unaffected.

9.11.1 Regulation of the Gas Supply Industry

The Gas Act provided for a Director General (DG), who was to be an independent regulator. The functions of the gas regulator have now been merged with the electricity regulator to form one DG of Gas and Electricity Markets who is supported by OFGEM. The DG has a number of powers and duties that include the overseeing of the development of competition. However, the DG is only concerned with competition within the gas supply industry and has no powers as a general competition watchdog.[93] As the gas pipeline network is a natural monopoly, it will always require some form of price regulation.

9.11.1.1 Duties

The duties of the DG[94] are mainly aimed at ensuring that his functions are carried out in a manner that he considers is best calculated to secure that all reasonable demands for gas through pipes is satisfied within economic imperatives.[95]

In addition, the DG is under a duty to carry out his functions

1. In a manner that is best calculated to protect the interests of consumers in respect of prices charged and terms of supply
2. To promote efficiency and economy on the part of suppliers
3. To protect the public from the dangers associated with gas supply
4. Have particular regard to the interests of the disabled and those of pensionable age

In the event of these duties conflicting, the DG must balance the interests of consumers with the interests of shareholders. For example, the DG must secure that suppliers can finance their activities, which implies that a high price is desirable, and yet the DG also has a duty to protect the interests of consumers, which would imply a low price, at least in the short term. However, one of the primary duties of the DG is to secure effective competition. These duties were also modified under the Gas Act 1995 to take into account during any decision-making process the effect that decision might have on the environment.

9.11.1.2 The Duty to Secure Competition

9.11.1.2.1 Statutory Third-Party Access to Onshore Pipelines

The Gas Act provided a mechanism whereby any supplier could acquire third-party access rights in the face of opposition from the pipeline company.[96] The statute provides that if the parties cannot come to a negotiated agreement, then one will be imposed under statutory authority. The companies recognise that a negotiated agreement is always preferable to one that is imposed and so this provision encourages a negotiated agreement.

Under the statute, a supplier can gain the right to use a pipeline belonging to the former monopoly but not access to offshore pipes, or to privately owned pipes.[97] The application for access rights is made to the DG, after a period of negotiation. The DG may then give directions as to terms and conditions of access, provided that he is satisfied that such directions would not prejudice the existing operations of the pipeline.[98] However, problems may arise when trying to establish when a pipeline is contractually full as opposed to physically full.[99] It is, therefore, the duty of the DG to satisfy himself that there is available capacity.

When the DG is satisfied with regard to capacity, directions may be given as to the terms on which the supplier and the pipeline company must enter an agreement. Such terms may include:

1. The right to convey gas during a specified period
2. The right to convey a specified quantity
3. Securing that the exercise of the right is not impeded
4. Securing supplies of backup gas in the event of the supplier's failure to input gas into the system
5. Securing the supplier's right to connect their own pipeline to the network

The Gas Act section 19 lays down guidance as to the matters that the DG may take into consideration in determining the tariff.[100] Although the statute clearly envisages that the pipeline company is entitled to make a profit from third-party access agreements,[101] it is the level of the profit that continues to be contentious.

9.11.2 The Licensing Regime for the Gas Supply Industry

As with most licensing regimes within the energy sector, the justification for a licensing system is partly strategic and partly administrative. The selling of the public assets of the gas supply industry into private ownership did not diminish the strategic value of the sector. Given this strategic value, the licensing system is a method of retaining state control and ensuring that not only will 'appropriate persons' be permitted to participate in the industry, but that the state can identify which parties will be using the pipelines.[102]

The requirement that participants should hold a licence is not only a convenient way of identifying participants but it also provides a system for ensuring that all participants are subject to the same rules.[103] Further, conditions within the licence ensure that service standards are maintained within a competitive environment.[104]

Section 7 of the 1986 Act provided for British Gas plc to become the public gas supplier[105] to mainland Britain. While section 8 permitted other suppliers to sell to consumers with a demand level in excess of the monopoly threshold, this was, as such, a restricted right.

9.11.2.1 Types of Licence

The Gas Act 1986 has since been amended in respect of the licensing regime, by the 1995 Act, which provides that it is an offence for any person to transport gas through pipes, or arranges for gas to be transported through pipes, or to supply gas without a licence. This brings the gas supply industry with the United Kingdom into a similar position as the other previously publicly owned utility companies. In addition, the 1995 Act creates new categories of licences for the sector.

Under the 1995 Act, the DG has the power to issue three separate types of licence, which reflects the changed characteristics of the sector. The licensing system is now extended to those involved in arranging transport of gas with the pipeline company. According to Stephen Dow,[106] the clause is clearly aimed at intermediary gas marketers, and recognises that future development of the UK gas market may involve new participants in this role. For example, after full liberalisation of the market, there is no reason why banks or supermarkets should not become involved in the gas industry.[107]

Under section 7(3) of the 1995 Act, a person cannot hold both a transport licence and a supply licence. This is in recognition of the fact that the industry operates with a clear distinction between supply and transportation. It also ensures that suppliers compete on an equal basis as far as transmission costs are concerned.

Supply is the activity of contracting with the consumer to supply the product, and as such has been the most easily opened up to competition. Many companies have entered this market with its relatively low threshold of investment. Although the applicant has to satisfy the DG that he is an 'appropriate person', in fact this merely requires an absence of a history of financial irregularities. If, in addition, the applicant can demonstrate the ability to perform the activity, he is then likely to be granted a licence. With the lowering of the threshold, there is now an increasing need for greater scrutiny of applicants in order to protect consumers. Initially it was the large consumers[108] who were able to choose their supplier. Although the Gas Code[109] requires suppliers to provide some additional degree of consumer protection as well as offering existing suppliers a degree of protection from cut-price but incompetent entrants to the market, there is growing concern about such new entrants carrying out 'sharp practices' with regard to individual consumers.

The third type of licence is the shipping licence. It is expected that most suppliers will also wish to hold a shipping licence as this permits those who arrange the transport of gas with the public gas transporter. However, if the supplier has no experience of the gas industry, he may choose to contract this service to a specialised company.

9.11.2.2 The Network Code

The Network Code sets out the rules for the use of the transmission network. It regulates the relationship between the shippers and suppliers on the one hand with the transmission company on the other. The Code[110] came into force in April 1996 and represents a fundamental industry change in terms of the responsibilities for balancing the system. Under the state monopoly, the balancing of supply and demand was easily carried out, as there was only one supplier. This balancing of supply and demand became more difficult when more suppliers entered the market following liberalisation. One of the major purposes of the Network Code is to identify which supplier is to be paid by which consumer.[111] The payment of suppliers is dependent on the accuracy of metering that was taken over by the suppliers. Metering is also fundamental to the operation of the Network Code in respect of the daily balancing.

How does it work? Each day, each supplier must input an equivalent amount of gas to the system as his consumers collectively withdraw, in order that the system remains in balance. This responsibility for balancing has now been taken away from the pipeline company and the burden placed on the supplier, which essentially requires metering.

A supplier who is not in balance is subject to a financial penalty. As the pipeline company[112] is no longer primarily responsible for balancing the system it remains financially neutral.[113] Given the financial penalties involved, there is a strong incentive for the supplier to stay in balance, the most obvious of which is demand control. If all customers took a specific amount of gas each day, there would be no problem balancing the supply to fit the demand. Unfortunately, this is rarely the case, and demand fluctuates according to many variables, with the weather and temperature being two such factors. An alternative control mechanism would be to cap the demand by either charging a higher price for gas over a certain limit, or by interruptible contracts.

Another way of balancing would be in the direct control of the supplier rather than depending on the cooperation of the consumers. The Code provides for a degree of flexibility inasmuch as the supplier, if it is short of gas, is provided with a number of options. It can either buy in more gas from another supplier who has excess, which would be cheaper than paying the penalty, or it can purchase gas on the spot market.

Alternatively, the supplier may take gas out of storage, which may be in disused or depleted offshore fields, or a small amount may be kept in the pipelines. It is essential that all suppliers have some gas held in storage to assist with the balancing requirements. All such storage arrangements are prescribed in the Network Code.

If the supplier finds he has an excess of gas to be in balance, he may seek to alter the amount purchased from the offshore supplier. However, this will again involve an additional cost depending on the contract, but that cost may be preferable to paying the penalty.

If all efforts fail to keep the supplier in balance, then the pipeline operator will balance the system upon payment of the penalty. The imposition of a large penalty will provide a very strong incentive to keep the suppliers in balance. The problem would be if the threat of the penalty made suppliers unwilling to take on certain categories of consumer. However, experience to date shows that companies learn to use the various options to balance even on those days of extreme demand fluctuations.

9.11.3 The Competitive Market for Gas

9.11.3.1 Competition Policy

The government has emphasised the distinction between ownership and operational control with regard to the relationship between the pipeline operator and the former monopolist supplier. Given the voluntary demerger, there is now a complete separation of ownership of the respective businesses as well as operational control. There is now a general acceptance that the independence of management and accounting is sufficient to separate, first by creating 'Chinese walls' between the pipeline operation and the supply operation, which culminated in the demerging of the businesses.

The government has recently accepted a number of important recommendations consequent upon competition investigations. New regulatory action included the placing of an obligation on the former monopoly to publish firm price lists. This provided a benchmark for consumers to negotiate prices with other suppliers. New suppliers could, in theory, undercut the former monopoly prices and so win consumers. In addition to the requirement to publish tariffs, there was a requirement for publication of tariffs for third-party access. This enabled the supplier to establish a cost base in negotiations with potential consumers.

A further measure that had the effect of developing competition in the gas market[114] was the gas release programme. This was developed to circumvent the problem that the former monopoly was contractually entitled to all gas produced from the UK sector of the North Sea. The gas release programme ensured that this provision did not extend to new fields. The aim was to establish the long-term development of the gas market, by providing gas for new entrants to sell.

Unfortunately, the programme did not address the immediate needs of the new suppliers. Allowing the development of new fields solved the immediate need. However, although it was thought that the market for gas would increase with the expected demand from gas-fired power generation plant, in addition to the new interconnector allowing export to the western European gas market, and the increase of the use of natural gas as a fuel for the transport sector, these new markets have been slow to evolve. In addition, the government's 3-year moratorium on the building of gas-fired power stations held back the expected increase in demand from that quarter. The expected increase in demand was, therefore, much less than anticipated and the unfulfilled expectation also contributed to an over-supply of gas.

9.11.3.2 Oversupply and the Incentive for a Spot Market

The oversupply of gas within the United Kingdom finally reached the point when gas was being produced with no contracted buyer. This, in turn, stimulated the growing spot market.[115] These sales of excess production of gas caused further problems for the former monopoly by establishing a spot price for gas. With the oversupply of gas, the spot price was inevitably lower than the price paid under the existing long-term contracts with the producers. There then became an incentive to break the contracts, much to the producer's concern. However, as in most cases, the financial penalty was equivalent to the contract price, so there was little incentive to buy gas more cheaply on the spot market, but where the penalty was lower, there might be a commercial incentive to break the contract and turn to the spot market.

Given the exchange-based trading developments within the United Kingdom for greater volumes of gas, there becomes less incentive to enter into long-term contracts. According to Stephen Dow, as long as the standard terms and conditions of the exchange are acceptable to the industry, and workable in practice, the trading of gas is likely to become more like the trading of oil. However, given that gas is pipeline constrained, it can never have a world market. This development may, in turn, lead to difficulties in the financing of new gas fields, with lenders used to seeing gas sold before production. Given the lack of a world market, it may be difficult for gas producers to achieve any level of trust and acceptance by lenders.

9.12 COAL INDUSTRY

9.12.1 Historical Development

While mining once enjoyed the status of a 'preferred' land use and economic activity with a presumption in favour of development, public awareness of its negative impacts has intensified. While the majority of mining impacts are local from an environmental perspective, some would consider that mineral development is capable of wider, even trans-boundary, impacts.[116]

In the United Kingdom, coal mining has occurred from the thirteenth century. During the Industrial Revolution, coal was the main source of energy and so production was extended and increased to supply the growing demand; however, the early legislation related mainly to welfare, safety, wages, and hours of work.[117]

9.12.1.1 Ownership

Under the common law, a landowner had an exclusive right to win minerals from beneath his land, including coal, but with the exceptions of gold and silver.[118] The landowner could also grant a lease for the right to extract minerals to others. The lease could be independent of the surface owner or occupier, so long as it included sufficient area of the surface land for the sinking of a shaft to exploit the underground workings.

When mines were in private ownership, they suffered from the lack of investment by many private owners. Compliance with the basic regulations under the legislation was sufficient and there was a great need for modernisation of the industry as a whole.

9.12.2 Coal Production and the Environment

No form of power generation is free from environmental consequences. The principal consequences of burning of fossil fuels, including coal, which need to be considered, are the emissions from power stations. Issues that feature prominently include 'acid rain', to which emissions of sulphur dioxide and oxides of nitrogen contribute, and the threat of global warming, arising from man-made emissions of greenhouse gases, the most important of which is carbon dioxide. Electricity generation is not by any means the only source of these pollutants. It is, however, the major source of sulphur dioxide and carbon dioxide emissions and second only to transport as a source of nitrous oxides. In general, the international commitments into which the United Kingdom has entered on both these issues are recognised as constraining the use of fossil fuels.

9.12.2.1 Acid Rain

As previously stated, the combustion of fossil fuels may cause the emission of oxides of sulphur and nitrogen into the air. These gases add to the natural sources of acidity of rain and to acid deposition in general. This, in turn, increases the rate of damage to buildings and other materials exposed to the weather throughout the United Kingdom, and affects ecosystems in geologically sensitive soils and freshwater over much of upland north and west Britain. Acid emissions from the United Kingdom also affect sensitive areas of the Continent, and vice versa.

The United Kingdom was required to reduce its levels of sulphur dioxide and nitrous oxide under the EC Large Combustion Plants Directive,[119] and is also bound by the commitments entered into under the Long-Range Transboundary Air Pollution Convention.[120] Power stations in Britain are also subject to IPC and now IPPC under Part I of the EPA 1990. These processes require authorisation from the Environment Agency and include conditions to ensure that BATNEEC are used to minimise harmful releases to the environment.

The sulphur content of oil can be removed prior to combustion, or from coal-fired flue gases after combustion, but at a cost in both financial and energy terms. The usual form of removal, where necessary, is FGD. There are many technologies for this, but by far the most common worldwide are the wet-lime or limestone–gypsum processes. These consume limestone and water and produce gypsum (calcium sulphate) and carbon dioxide. This both reduces electricity output and increases the carbon dioxide emissions.

9.12.2.2 Global Warming

The burning of fossil fuels also produces carbon dioxide. This is a 'greenhouse gas' which helps to trap the sun's heat, warming the earth's surface. Greenhouse gases are naturally present in the atmosphere, and without them the world would not be habitable. However, over the last 200 years, man-made emissions have substantially increased the atmospheric concentrations of carbon dioxide and other greenhouse gases. While there are many uncertainties, the scientific consensus in 1993 was that without special action to limit these emissions the global mean temperature would increase by about 0.3°C per decade, leading to the melting of the polar ice-caps and subsequent sea-level rises of about 6 cm per decade over the next century.[121]

In June 1992, over 150 countries signed the UN Framework Convention on Climate Change. The ultimate objective of the Convention is to stabilise greenhouse gas concentrations in the atmosphere at a level that would prevent dangerous interference by man with the climate system. The aim of the Convention is that the ultimate objective is carried out within a timescale sufficient to allow ecosystems to adapt naturally to climate change, ensure that food production is not threatened, and enable economic development to proceed in a sustainable manner.

Under the Framework Convention, the United Kingdom has to take measures aimed at returning emissions of greenhouse gases to 1990 levels by the year 2000.

The UK Climate Change Act 2008 commits the UK to reduce it's GHG emissions by up to 80% by the year 2050 with an interim target of 34% by 2020. In the longer term, the Kyoto Protocol commits developed countries to legally binding targets for reducing a basket of six greenhouse gases to an average of 5.2% below levels over the commitment period of 2008–2012. The European Union agreed to an 8% reduction, of which the UK share is 12.5%.[122]

9.12.2.3 Sustainable Development and Coal

The concept of sustainable development does not require that the United Kingdom should abandon fossil fuels overnight, or compromise today's energy security for climate change reasons. However, it does require the factoring in of economic, resource management, and social considerations as well as environmental ones. Equally, however, sustainable development, and hence the UK climate change targets, clearly require the reduction of greenhouse gas emissions.

In 1994, the government set out the key issues for sustainability and minerals,[123] which were as follows: firstly, the need to encourage prudent stewardship of mineral resources while maintaining necessary supplies; and secondly, provisions for the reduction of environmental impacts of minerals both during extraction and when restoration has been achieved. A new MPG was issued in 1994,[124] which sets out the basic controls under planning law. The Note stated that because minerals can only be worked where they are found, there is still a strong presumption in favour of granting permission unless there are very strong environmental objections or there is no real economic need. This was confirming the common law ruling on the subject.[125]

9.12.3 From Public to Private Ownership

In line with the trend towards privatisation, the rights of the BCC were gradually reduced. The Coal Industry Act 1992 removed the conditions on hours of work. This was followed in 1993 by the British Coal and British Rail (Transfer Proposals) Act 1993, which paved the way for full privatisation.

The Coal Industry Act 1994 completed the privatisation of the coal industry by establishing a new regulatory body, the Coal Authority.[126] The duties of the new Authority included the transfer of the property of the BCC to the private sector; and the dissolution of the BCC, including the Domestic Coal Consumers Council. The ownership of all unworked coal and coal mines, which had previously been vested in British Coal, were now vested in the Authority, although it is prohibited from actually carrying on any coal-mining operations for commercial purposes in its own right.

The intention of the government was that the Authority should be a purely regulatory body that facilitates coal-mining operations by the licensing of those operations. Part I of the Act generally provides for the reorganisation of the coal industry but section 3 imposes various environmental duties on the operators.

Part II of the Act provides for the licensing of coal-mining operations and prescribes the operations for which licences are required, the provisions for the granting of licences, any conditions in a grant, provisions for enforcement, and the requirements for publicity. Any operator wishing to carry on coal-mining operations in respect of coal that is not already under lease or licence must apply to the Authority for a licence. Operators must also secure any access rights needed to take advantage of their licence from the Authority, from the surface owner. It is also for the operator to secure all planning consents needed and any other operating consents from other regulatory bodies.

Under Part III, the areas of responsibility designated define the geographical extent of the responsibility for subsidence damage of a person who is the holder of a licence under Part II of the Act. The licences for opencast operations, as well as those for underground operations, will designate areas of responsibility.[127] In the revised model licensing documents, both the model underground operating licence and the model opencast operating licence designate areas of responsibility. Under this part of the Act, sections 38–40[128] confer on the corporation a general entitlement to withdraw support to enable coal to be worked. However, it is not entitled to withdraw support in any particular instance unless it has given public notice indicating its intention, and it is required to take remedial action under Part II of the 1991 Act in respect of any subsidence damage that is caused.

9.12.4 Liability for Subsidence

The 1991 Coal Mining Subsidence Act continues in force but as amended by the 1994 Coal Industry Act. The definition of subsidence under section 1 of the statute is damage to land, and buildings, structures or works on, in, or over land, which is caused by the removal of support consequent on lawful coal mining. However, mere alterations of gradients or levels of land not otherwise damaged, which do not affect its fitness for the purpose for which it has been used, fall outside the definition of 'subsidence damage'. The person liable in respect of such damage is either:

1. The holder of a licence under Part II of the 1994 Act, that is, the licensed operator, if the damaged land is within the holder's area of responsibility as defined in the licence
2. In any other case, the Authority[129]

To facilitate the transfer of licences from one operator to another, or the extension of an area of responsibility, the liability is that of the person so designated as above, whether the subsidence occurred before or after the time when he became the responsible person in respect of the land in question.[130]

If subsidence is likely to occur, the Authority may include restrictions on coal-mining operations to limit or avoid such damage.[131] The 1994 Act also provides for an independent 'subsidence adviser' to help and advise those who have been damaged by subsidence and wish to claim against the person responsible. The primary functions of the adviser are:

1. To provide advice and assistance to claimants
2. To make recommendations to persons responsible for subsidence damage as to the manner in which they conduct themselves
3. To make reports relating to the above

9.12.5 Protection of the Environment

Section 53 applies in the case of proposals formulated for inclusion in an application for planning permission as related to any of the following:

1. The carrying on of any coal-mining operations
2. The restoration of land used in connection with the carrying on of any coal-mining operations
3. The carrying on of any other operations incidental to any coal-mining operations or to the restoration of land that has been so used

Where planning authorities consider any coal-mining proposals included in such an application, they must have regard to the desirability of the preservation of natural beauty, of the conservation of flora and fauna and geological or physiographical features of special interest, and of the protection of sites, buildings, structures, and objects of architectural, historic, or archaeological interest.

Section 54 provides for the Secretary of State to amend the existing planning permission for coal mines in operation before July 1, 1948 so as to impose site restoration conditions with regard to the demolition or removal of any buildings, plant, machinery, structures, or erections used at any time for or in connection with any previous coal-mining operations.

9.12.6 Mining as an Act of Development

Mining constitutes an act of development under section 55 of the 1990 Town and Country Planning Act and as such is subject to the planning legislation. The government's consultation paper concerning a revised strategy for 'sustainable development'[132] identified a number of key objectives that are fundamental to the sustainable development of the minerals industry including the following: the fact that social progress recognises the needs of everyone; the effective protection of the environment; the prudent use of natural resources; and the maintenance of high and stable levels of economic growth and employment. These objectives mainly reiterated those from previous documents and recognised that although the use of minerals benefits the economy, there will almost inevitably be conflicts between the working of mineral resources and environmental aims. It is unlikely that the United Kingdom will experience the problems of exhaustion of mineral resources in absolute terms. The problem is more likely that it will be increasingly difficult to find sites that are capable of commercial exploitation without damaging the environment to the extent that local communities find unacceptable.

The government, therefore, published a Guidance Note[133] to address such issues. Under the Note, the objectives of sustainable development were laid out to be:

1. To conserve minerals as far as possible, while ensuring an adequate supply to meet the needs of society for minerals
2. To minimise the production of waste and to encourage efficient use of materials, including appropriate use of high-quality materials, and recycling of wastes

3. To encourage sensitive working practices during minerals extraction and to preserve or enhance the overall quality of the environment once extraction has ceased
4. To protect areas of designated landscape or nature conservation from development, other than in exceptional circumstances where it has been demonstrated that development is in the public interest
5. To minimise the impacts from the transport of minerals

Special problems are associated with opencast coal workings because of the large amount of overburden that has to be removed and stored. The necessity for a large engineering plant at the site and the possible transportation of overburden away from the site also creates its own environmental problems. However, some see the removal of such large amounts of overburden as beneficial to the re-scaping of the original landforms on restoration of the site and as providing an opportunity to remediate old mineral workings and so create a better environment. This can only take place after the period of the opencast working, and must be weighed against the damage to the environment and the loss of any amenity value to the population while the workings are in progress.

Following the consultation paper *Opportunities for Change*, the government issued a new MPG in 1999, which reverses the presumption in favour of mineral development and now claims that with the application of the principles of sustainable development to coal extraction, whether opencast or deep mine, there should now normally be a presumption against development unless the proposal would meet the following tests:

(i) Is the proposal environmentally acceptable, or can it be made so by planning conditions or obligations?
(ii) If not, does it provide local or community benefits which clearly outweigh the likely impacts to justify the grant of planning permission?
(iii) In national parks and areas of outstanding natural beauty (AONBs), proposals must also meet the additional tests for these areas.
(iv) Proposals within or likely to affect SSSIs and NNRs must meet the additional tests for these areas.
(v) Proposals within the Green Belt must also meet an additional test.

Nevertheless, MPAs must not allow any other development to take place that will unnecessarily sterilise any coal resources, nor must they allow other development to encroach on existing mineral operations and thus increase any environmental impacts to an unacceptable level. To this extent, the Coal Authority is charged with liaising with both the operators and the MPAs with regard to a long-term dialogue and the production of forward programmes. It is hoped that such plans will provide continuity for the industry, with a degree of certainty for the community, and should, in turn, avoid the problems of the cumulative effects of piecemeal applications.

9.15.7 The Historic Environment

It is not only the natural environment that may be at risk during mineral exploration and working. The remains of sites of historic or archaeological interest may also be at risk from such operations. The MPAs are charged with taking into consideration the desirability of preserving historic buildings as well as landscapes, conservation areas, ancient monuments, and their settings under the Guidance Note *Archaeology and Planning*,[134] as well as the Confederation of British Industry (CBI) Minerals Environment Charter. Both of these documents emphasise the importance of archaeological sites and how they may be affected by the proposed development. These remains may be scheduled under the provisions of the Ancient Monuments and Archaeological Areas Act 1979, and if so, the developer must seek the consent of the Secretary of State for Culture, Media and Sport, before any development may proceed. Should a listed building need to be demolished either totally or partially, again, such works may only be carried out with the minister's consent. Advice on the protection of the historic environment, world heritage sites, and listed buildings is given in the Guidance Note *Planning and the Historic Environment*.[135] Should the site be listed as a world heritage site, the fact will then become a material consideration to be taken into account by planning authorities when determining applications for development of any mineral deposits

Should a proposal for a mineral development be likely to have a significant effect on the environment, then the developer should prepare an EIA under the Town and Country Planning (EIA) (England and Wales) Regulations 1999 and an Environmental Statement (ES) and submit them with the planning application. For any application made after March 14, 1999, it is mandatory to supply an EIA for all proposals for opencast mining where the surface of the site exceeds 25 ha. Even if the surface area is less than 25 ha, any proposal for a new site, and modifications to existing sites, will require an EIA if there is likely to be a significant environmental effect. In any event, an EIA will be required for mineral applications in national parks, SSSIs, and AONBs.

9.12.8 The Environmental Duty

Section 53 of the Coal Industry Act 1994 imposes an environmental duty on the coal industry. In order to discharge this duty when formulating proposals requiring planning permission, the operators are required to consider the desirability of the preservation of not only the natural beauty and the conservation of flora and fauna in the area but also the geological and physiographical features of special interest. The protection of sites, buildings, structures, and objects of architectural, historic, or archaeological interest is also part of the duty. The operator must then formulate proposals for the adoption of measures to mitigate any adverse effects of the development. The MPA must take into consideration the extent to which the operator has fulfilled the duty in this respect, when the application is being considered. To encourage 'good practice' in this area, the Planning Officers Society is proposing a scheme[136] to encourage and reward excellence and good site management.

9.12.8.1 Periodic Reviews

Although the grant of planning permission to carry out mineral workings is essentially a temporary one, the permission may sometimes last for many years. The operation of a site can change significantly in its impact on the environment over its lifetime. Further, the standards of society can also change. To reflect this concern, the power to review old mining permissions was provided for in the Town and Country Planning (Minerals) Act 1981.[137] The Act placed a duty on MPAs to periodically review all mineral sites within their area, and to update any planning permissions they thought necessary. Any compensation sought by the operator would be reduced under certain conditions. In this way, it was expected that the industry itself should bear the reasonable costs arising from the modernisation of these old mineral planning permissions.

Unfortunately, the 1981 Act did not provide the sea change in approach to minerals planning that was expected. The government therefore undertook a review of the operation of the legislation to consider how it could be improved. The 1991 Planning and Compensation Act dealt with the oldest extant mineral consents, Interim Development Order (IDO) permissions, referred to in the 1991 Act as 'old mining permissions'. Holders of the old IDO permissions had to register them with the MPA and submit an operating and restoration scheme, with conditions, to the authority for its approval. Although there was a right of appeal under the statute, there was no entitlement to compensation for the cost of compliance. A distinction was made between conditions that dealt with the environmental and amenity aspects of working the site, which should not affect the asset value, and conditions that would have a fundamental effect on the economic structure of the whole commercial operation. Section 96 and Schedules 13 and 14 of the Environment Act 1995 made provisions for the future periodic review of all mineral permissions thereafter.

It is to ensure that the conditions attached to mineral permissions do not become outdated that the periodic reviews are carried out. Such reviews apply to all mining sites, including sites with IDO permissions, except for mineral permissions granted by the General Permitted Development Order. However, there is no distinction between those periodic review sites that are working and those that are not. Further, should an MPA determine conditions different from those submitted by the applicant, and the effect of those conditions is to restrict working rights further than was the case prior to the review, a liability for compensation will arise, the exception being if the conditions are for restoration or aftercare. To monitor the new system, the government is to engage the Minerals Valuation Agency and the County Planning Officers' Society to prepare an annual report.

9.12.9 Abandoned Coal Mines

As we have seen, in recent years, the British mining industry has contracted substantially. Of the 169 operational collieries which survived the miners' disputes in the 1980s only 17 were still operational by 1994. The environmental impact of these closures was considered by the NRA who reported that approximately 100 discharges from underground workings were causing concern and that 200 km of controlled waters were affected by pollution from abandoned mine workings.[138] The conclusions of the NRA were that the legislation should be tightened as regards to liability for pollution from abandoned mines and that the responsibility should rest firmly with the landowners.

9.12.9.1 Liability for Pollution

Under the Water Resources Act 1991, water discharges from mines are controlled by consents from the Environment Agency. If there is no consent, or consents are exceeded, the owner or occupier of the mine is liable to prosecution under section 85. However, although mine owners or former operators of abandoned mines can be prosecuted for *causing* water pollution under the 1991 Act, the legislation provides an exemption from prosecution under section 89(3), if they *permit* pollution. This exemption was confirmed by the court when the Anglers Cooperative Association brought a private prosecution in which they contended that the closure of the Britannia Colliery in 1990 and the cessation of pumping activities resulted in the pollution of the Rhymney River. The Anglers Cooperative claimed that the switching off of the pumps constituted causation under section 85. The court found in favour of British Coal but it was stated *obiter dicta*, that if the prosecution had argued that once the water was not draining into the specially constructed duct and the defendant noticed this, then they would have been guilty of 'knowingly permitting' pollution.[139] However, this decision is inconsistent with the current trend in relation to interpreting cause, whereby establishing a system by which pollution is likely is sufficient to prove the offence without establishing a positive act.[140] In addition, once a pumping system has been established, the failure to maintain such a system constitutes a positive activity and causes pollution.[141]

The exemption is not new. There has been an exemption for such pollution since the Rivers Pollution Prevention Act of 1876. After the nationalisation of the coal mines, the government was not inclined to give up the exemption and so it was continued under the Rivers Prevention of Pollution Act 1951. Discharges from mines were first regulated under the COPA 1974, now section 89(3) of the Water Resources Act 1991 and the Water Act 1989. Such discharges have also been exempt from the Environment Agency's powers under section 161A of the 1991 Act, inserted by the 1995 Environment Act, to carry out and recover from the discharger the costs of any remedial or preventative work undertaken by the Agency, but this exemption ceased after December 31, 1999.

Following a review by the previous government of the framework of legal responsibility for water pollution from abandoned mines, provisions under the 1995 Act amended sections 89 and 161 of the 1991 Act and inserted new sections 3A–C, 4A, and 4B. These changes removed the defence and exemptions in respect of the owners and former operators of mines that were abandoned after December 31, 1999.

9.12.9.2 Definitions

In addition, section 58 of the 1995 Act inserted Chapter IIA (Abandoned Mines) (sections 91A and 91B) into the 1991 Act.

Section 91A of the Act defines 'abandonment' for the purposes of Chapter IIA and also provides that the term 'mine' has the same meaning as in the Mines and Quarries Act 1954, and as such is:

> An excavation or system of excavations, including all such excavations to which a common system of ventilation is provided, made for the purpose of, or in connection with, the getting, wholly or substantially by means involving the employment of persons below ground, of minerals (whether in their natural state or in solution or suspension) or products of minerals.

Under section 91A(1), abandonment includes:

(i) the discontinuance of any or all of the operations for the removal of water from the mine;
(ii) the cessation of working of any relevant seam, vein, or vein-system;
(iii) the cessation of use of any shaft or outlet of the mine.

In the cases where mines are used for activities other than mineral extraction[142] where it is proposed to cease some or all of those other activities, and this action is to be accompanied by any substantial change in the operations for the removal of water from the mine, then this would also be regarded as an act of abandonment.

The list of circumstances that constitute an act of abandonment as set out in the regulations[143] is not exclusive, and it is stated that only the courts can decide what constitutes an act of abandonment in certain circumstances.

Section 181(1) of the 1954 Act defines the 'owner' as the person for the time being entitled to work the mine. That definition was not imported into the 1991 Act as it was revised by the 1995 Act. In the absence of a statutory definition of 'operator' in the 1991 Act, the word is expected to mean the person actually carrying out the mining operations.

The offences are defined by the new sections 91A and 91B of the 1991 Act, as inserted by section 58 of the 1995 Act. Under these sections, it is an offence to fail to give notice to the Agency of any proposed abandonment at least 6 months before abandonment takes effect. The notice must contain information prescribed in the regulations and include the consequences of the proposed abandonment. However, no offence is committed if the abandonment happens in an emergency in order to avoid danger to life or health, and notice of the abandonment is given as soon as is reasonably practicable after the abandonment has occurred.

Nevertheless, the new section 91B fails to give effect to one of the NRA's main recommendations, which was the imposition of a statutory duty on a mine operator to prepare a complete mine abandonment programme.

The owners and operators of all mines abandoned after 1999 are subject to the offence of causing or knowingly permitting pollution of controlled waters contrary to section 85(1) of the 1991 Act. Not only that, but after April 1, 2000, they are subject to the new Statutory Guidance on Contaminated Land.

9.12.10 Contaminated Land

Part IIA of the EPA 1990, which was inserted by section 57 of the Environment Act 1995, provides a new regulatory system for the identification and remediation of contaminated land. Under the regulations, each local authority has a duty to 'cause its area to be inspected from time to time for the purpose ... of identifying contaminated land'.[144] Under the 'polluter pays' principle, the costs of carrying out any remediation needed fall on those responsible for the contamination.

Potentially, both deep coal mines and strip mines may be designated as contaminated land, either during the operational phase or on abandonment. Operators of mineral extraction sites need to be aware of these new regulations with regard to their land and to make sure that remedial action is taken should the land fall within the definition of 'contaminated land'.

9.12.10.1 Designation

'Contaminated land' is any land that appears to the local authority in whose area it is situated to be in such a condition, by reason of substances in, on or under the land, that significant harm is being caused or there is a significant possibility of such harm being caused, or any pollution of controlled waters is being, or is likely to be caused. This is, therefore a pre-emptive piece of legislation in line with the precautionary principle.

To be designated as contaminated, the land must be the 'source' of some potentially polluting material. There must also be a potential or actual 'receptor' that is capable of being caused 'significant harm' by the polluting material present at the 'source'. Thirdly, there must be a potential or actual 'pathway' between the 'source' and the 'receptor'. If there is no means of the potentially polluting material escaping from the site and reaching a potential receptor or target, then the site would escape designation. This is an important distinction as, once designated as contaminated land by the local authority, there is no means of appeal against the designation. The 'appropriate person' would then be liable for the costs of remediation, which could be extensive.

One question that arises is 'What constitutes 'significant harm'? The draft guidance paragraph 25 advises local authorities to disregard any harm or interference other than in the case of harm to humans, death, serious injury, or clinical toxicity. Should the harm be to ecosystems then it must constitute a significant change in the functioning of the ecosystems in protected areas, such as SSSIs, Ramsar sites, or European sites under the Conservation (Natural Habitats) Regulations. In the case of harm to property, then any physical damage that is continuing and cannot be put right without substantial works. Finally, in the case of animals or crops, any disease or other physical damage causing loss in value.

9.12.10.2 Liability

Once the land has been designated under the criteria, then the local authority has a duty to issue a remediation notice to 'the appropriate person'. There are two classes of 'appropriate person':

- **Class A—the polluter:**
  ...any person, or any of the persons, who caused or knowingly permitted the substances, or any of the substances, by reason of which the contaminated land in question is such land to be in, on or under the land is an appropriate person.[145]
- **Class B—the owner or occupier:**
  If no polluter can be found by reasonable enquiry then the liability falls to the current owner or occupier.

Once the 'appropriate person' has been identified by the local authority, the local authority will serve the remediation notice. The notice will set out any remediating actions that the appropriate person must take, and the timescale in which these actions must be taken.

There is increased importance of the voluntary participation of the person who will be potentially liable for the remediation in the processes of determining whether land is contaminated and deciding what remediation action is to be taken. An agreement to undertake remediation is preferred to the service of a notice requiring this to be done. To encourage this, an exemption from liability to pay landfill tax will be available where contaminated material is being removed from the land in order to remediate it. However, the exemption will not be available where a remediation notice has already been served.

An increased focus is also placed on the consideration of the environmental impact on other sites in consequence of undertaking remediation. The guidance suggests that, in the majority of cases, whether this occurs will be dealt with in the context of the grant of specific pollution control permits. Where this is not the case, local authorities should consider whether an alternative approach or precautionary measures could be adopted.

The guidance also seeks to clarify the procedure for allocating liability for remediation where two or more people are responsible for causing the contamination. This could occur where a succession of occupiers has resulted in the land being deemed as contaminated. The local authority first determines who is responsible for the presence of each contaminant. Once this has been done, the local authority must turn its attention as to whether any action is needed with respect to one or more contaminants and, should there be more than one contaminant involved, whether there is any one action which may be able to address all of the contaminants on the site.

9.12.11 Contamination, Successors in Title, and Statutory Successors

The practical issues that arise as a result of long sequences of occupation and ownership of contaminated sites further complicate the interpretation of 'causing' and 'knowingly permitting'. These sequences may have involved transfers of ownership within company structures, changes of corporate identities, and companies that are in liquidation or have been wound up. Public bodies may have undertaken activities that caused the contamination, but may have reorganised, been nationalised, or privatised. Under company law, each new corporate identity is seen as a different 'person' with potentially different identities and liabilities. However, the purpose of the legislation would be undermined if the creation of new 'persons' meant that the statutory successors avoided liability. This situation was considered in 2008[146] when the Environment Agency sought to identify contaminated land on a housing development on a former gasworks site. The National Grid Company (NGC) was a statutory successor to a series of gas companies both public and private, and was considered to be a 'causer' of contamination on the basis of being a successor to the bodies that has caused the original contamination. The NGC argued that it was not a 'causer' and that, at the time that the liabilities were transferred between previous bodies, there had been no liability in existence. The High Court held that Pt 2A should be interpreted purposively. It had been Parliament's intention to allocate primary responsibility for the remediation of contaminated land on the original polluters in the form of causers and knowing permitters, rather than 'innocent' owners or the public purse. The only way of giving effect to this intention was to include statutory successors, as well as the original polluter, within Class A polluters.

The NGC appealed directly to the House of Lords, which accepted the appeal as being on a point of public importance. The Lords overturned the High Court decision, finding that it was impossible to construe the statutory definitions in a way that would make the NGC a 'polluter'. In doing so the Lords took a very narrow interpretation of who was a 'causer or knowing permitter'. The underlying issue of the unfairness of retrospective liability seems to have played a part in the decision with the Lords. They held that very careful statutory language would be needed to impose any liability to clean up contaminated land on an 'innocent' company that had never owned nor had an interest in the land in question. The Lords considered that it was 'extraordinary' that a public body such as the Environment Agency should seek to impose a liability on a private company and thereby reduce the value of the investment held by its shareholders. This decision ignores the fact that, in construing Pt 2A so narrowly, there is no recognition that in the absence of an original polluter, another party will have to pick up the costs of clean-up, which could well turn out to be the taxpayer by way of the public purse.

While this decision is restricted to the specific statutory liability transfer regime applicable to the gas industry, the construction of the identity of the polluter might be significant in other determinations.

9.12.11.1 Excluded Activities and Lenders' Liability

The primary legislation does not expressly exclude those who provide financial services. Bankers, mortgage lenders, and insurers may, under certain circumstances, be considered as 'knowing permitters'.[147] Attempts to minimise this liability were made by including such persons in the first 'exclusion test'. This means that if there is more than one appropriate person, the providers of financial services are the first to be excluded from liability. However, a lender may find itself liable for remediation should it become a 'mortgagee in possession', where the borrower has returned the property to the bank and the bank becomes responsible for on-site security and is technically 'the occupier'. In practice, it is envisaged that a financial institution would only be found liable if it exercised decision-making control over the environmental performance of the contaminated site.

Test three under the exclusion from liability addresses a polluter who has sold a contaminated site, and has given the purchaser 'information that would reasonably allow that particular purchaser to be aware of the presence on the land of the pollutant . . . and the broad measure of that presence'. If the purchaser then fails to deal with the contamination, the possibility arises that they will be deemed to be 'knowingly permitting' and therefore a Class 'A appropriate person. The purchaser then 'steps into the shoes' of the vendor with respect to liabilities relating to the significant pollutant linkage. The decision making control being the second test.

9.12.11.2 Orphan Sites

If no 'appropriate person' can be found, the local authority has the power to carry out the remediation itself. In cases where the polluter or landowner fails to carry out remediation as required in a remediation notice, the local authority can recover the remediation costs from that person.

9.13 ELECTRICITY SUPPLY INDUSTRY

9.13.1 The Structure of the Industry

9.13.1.1 The Participants

It was the Electricity Act 1989 that covered the restructuring of the electricity supply industry within England, Wales, and Scotland, and came into effect on March 31, 1990. The restructuring of the industry in Northern Ireland was provided for under the Electricity (Northern Ireland) Order 1992. The structure of the industry differs in these parts of the United Kingdom inasmuch as in England and Wales vertical separation of the industry was practical, whereas in Scotland it was not.

9.13.1.1.1 England and Wales

The Electricity Act 1989 provided for the following structure.

National Power plc was a generating company, which acquired 50% of the non-nuclear generating business of the Central Electricity Generating Board, a state-owned statutory corporation.

PowerGen plc was also a generating company, which acquired 30% of the non-nuclear generating business of the Central Electricity Generating Board.

Nuclear Electric plc was the owner and operator of the principal nuclear power stations in England and Wales. In the reorganisation in 1996, Nuclear Electric's five advanced gas-cooled reactors and the pressurised water reactor generators were transferred to a new company by the name of Nuclear Electric. The elderly Magnox reactors remained in public ownership under a renamed company Magnox Electric plc. The ownership of both Nuclear Electric and Scottish Nuclear plc were then transferred to a new company named British Energy plc. This new company British Energy plc was then privatised in 1996 with the consequence that only the old Magnox stations remained in public ownership.

Then 12 regional electricity companies were set up as the successor companies to the electricity distribution and supply part of the former 12 area boards. The successor companies operated in the same geographical areas as the area boards did previously and are the public suppliers for their area.

The NGC is a transmission company which was formed out of the transmission division of the old Central Electricity Generating Board. This company also took ownership of the pumped storage generating stations, which have since been disposed of. Under reorganisation in 1990, the company was owned by the 12 regional electricity companies, but in 1995, the NGC was floated on the Stock Exchange. Not only does the NGC have responsibility for the transmission but it also provides services for the operation of the 'pool', which is the market for the bulk trading of electricity in England and Wales.

Although the nuclear stations were not part of the original vesting process under the Electricity Act, British Nuclear Fuels plc is the owner of two prototype Magnox stations and became owner of the share capital of Magnox Electric on the reorganisation of Nuclear Electric.

Other than the 12 companies that are public electricity suppliers (PESs), there is a second tier of suppliers. The 12 regional electricity companies hold public electricity supply licences that authorise and require them to supply electricity within their franchised area. The second-tier suppliers are others who supply electricity under the authority of second-tier supply licences. The 12 regional companies may also hold second-tier supply licences in order to supply outside their own authorised areas. In addition, the privatised generators in England, Wales, and Scotland, Electricité de France (EdF) and a number of independent suppliers are also second-tier suppliers.

9.13.1.1.2 Scotland

In Scotland, the Electricity Act provided for the retention of two vertically integrated structures. The areas were divided geographically with Scottish Power plc inheriting the assets of the South of Scotland Electricity Board and Scottish Hydro Electric plc inheriting the assets of the North of Scotland Hydro-Electricity Board. The virtual monopoly positions in transmission and distribution of the publicly owned institutions were retained. However, both the privatised companies must allow access to competing suppliers on non-discriminatory terms. In addition, the third participant in the Scottish Electricity Supply Industry is Scottish Nuclear plc. Scottish Nuclear plc retains two advanced gas-cooled reactor nuclear stations, with the old Magnox stations having been transferred to Magnox Electric plc in 1996. Scottish Nuclear is now part of British Energy plc.

9.13.1.1.3 Northern Ireland

In Northern Ireland, the electricity supply industry was restructured under the Electricity (Northern Ireland) Order 1992. Previously, the responsibility for generation, transmission, distribution and supply of electricity was vested in a statutory corporation called *Northern Ireland Electricity*. On restructuring, Northern Ireland's four principal power stations were sold to the private sector, with Northern Ireland Electricity plc (NIE) retaining the other functions of the old statutory corporation, principally as operator of the transmission system in addition to being the sole PES in Northern Ireland. On privatisation of the company in 1993, the company also became responsible for power procurement. The electricity supply industry in Northern Ireland is regulated by the DG for Electricity Supply, with a single Office for the Regulation of Electricity and Gas supply known as *OFREG*.

9.13.2 The Market Structure

9.13.2.1 England and Wales

Under sections 4–6 of the Electricity Act, the generation, transmission, and supply of electricity is required to be licensed and (subject to certain exceptions) sold through the pool. The principal exception is electricity that is produced by an unlicensed generator on the basis that the generating capacity does not exceed 100 MW, and not more than 50 MW is exported off-site. In addition, generators producing more than 100 MW are exempt only if they export less than 10 MW.[148] A condition of the licence provides that all licensed generators and suppliers must be members of the pool and must be signatories to the Pooling and Settlement Agreement under which the pool was formed and is operated.

Generators and suppliers will normally contract a proportion of their generation capacity and supply requirements under contracts for differences (CFDs). These contracts are financial instruments that are used to reduce the exposure of generators and suppliers to the volatility of pool prices and normally relate to specified amounts of electricity at specified times over a specific period covered by the contract. The effect of the CFDs is to fix the price that the generator is paid and the supplier pays for electricity traded through the pool. Basically, they balance the payments to be made by the supplier to the generator or the reverse if the pool price for the relevant time is lower or higher than the 'strike price' set out in the contract.

Sections 32 and 33 of the Electricity Act (as amended by the Fossil Fuel Levy Act 1998) provide the framework for a separate market for the generation of energy from non-fossil fuel sources. In effect, an obligation is placed upon the regional electricity suppliers to contract for a specific quantity of non-fossil fuel generation capacity under Non-Fossil Fuel Orders (NFOs), made by the Secretary of State. Section 33 of the Act established a fossil fuel levy, under which the additional costs incurred by the PESs associated with complying with the NFOs can be recovered by a levy on all leviable electricity suppliers. The bulk of the levy was used to constitute a fund for the ultimate decommissioning costs of the nuclear plant. However, following the flotation of British Energy in 1996, the levy rate has been successfully reduced from the original 10% to the current 0.9%. A further proportion of the levy has been used to finance the higher generation costs of power generation from renewable sources. The output from renewable energy generation plants is purchased by the Non-Fossil Purchasing Agency Limited under long-term contracts.[149] These contracts are normally fixed-price arrangements subject to indexation.

9.13.2.2 Scotland

As stated previously, the two Scottish power companies are integrated vertically, encompassing the functions of transmission, distribution, and supply in addition to generation. Each company holds a generation licence, a transmission licence, and a public electricity supply licence. The pool does not operate in Scotland; however, competition for supply is inherent under the conditions of the licences. The two companies are under a duty to offer access to their transmission and distribution systems to competing electricity suppliers and generators on a non-discriminatory basis. In Scotland, the majority of the second-tier suppliers purchase their power from one or both of the two companies, with the price paid subject to the agreement of the DG of Electricity Supply. The current price under the Scottish Trading Arrangements is the pool selling price plus an uplift, minus 1%.

Under section 32 of the Electricity Act, three orders have been made known as *the Scottish Renewable Obligation*, and require the two companies to make arrangements to secure a specified amount of generation capacity from non-fossil fuel sources. The two companies are compensated for the costs of compliance by the introduction of the fossil fuel levy, which is currently levied at the rate of 0.7%.

9.13.2.3 Northern Ireland

All licensed generators are currently required to sell their output to the power procurement business of NIE under the Supply Competition Code. In turn, under the bulk supply tariff, NIE sells electricity to its own supply business to second-tier suppliers and to large customers with over 1 MW of consumption. The power procurement business of NIE purchases electricity from the four privatised Northern Ireland generators under long-term purchase agreements. The transmission and distribution business of NIE charges suppliers for the use of the system. The second-tier suppliers are permitted to compete with NIE in supplying customers across the full spectrum of the supply market. However, they are not permitted to contract directly with electricity generators, which reduces the scope for competition.

The Department of Economic Development for Northern Ireland has the power to make NFOs under the Electricity Order. These orders oblige NIE to secure for itself amounts of generation capacity fuelled by renewable sources of energy.

9.13.2.4 The Interconnectors

There are interconnectors between the transmission systems of the NGC and EdF, between the transmission systems of the NGC and Scotland, and between the transmission systems of the NGC and NIE. An interconnector with a capacity of 250 MW is being planned between Scotland and Northern Ireland. The interconnector between Great Britain and France has a capacity of 2000 MW. EdF is a pool member and is able to bid into the pool in addition to purchasing electricity from the pool, although this is a rare event. The capacity of the interconnector with Scotland is 1600 MW. The two Scottish companies sell electricity in the pool in England and Wales and may also purchase from the pool, but again this is a rare event as Scotland is self-sufficient in power generation.

9.13.2.5 Generation

In 1990, at the time of privatisation, there were three major power producers in England and Wales—National Power (now Innogy), Powergen (now Powergen/EON), and Nuclear Electric.[150] Since privatisation, the generation market has changed from a highly concentrated market with a few portfolio generators, to a market with many diverse generating companies, including independent power producers. The reorganisation of the industry, new entrants into the generation market, and enforced and voluntary divestment of capacity by Innogy (a division of the RWE company since 2002) and Powergen (who were bought by the German company EON in 1991) has significantly reduced the concentration of generation capacity ownership and brought changes to companies' market shares.

There are now 39 generating companies in the United Kingdom.[151]

9.13.2.6 Transmission

The transmission system in England and Wales is the NGC, which is effectively a monopoly. As such, it is closely regulated by the Electricity Act 1989 (as amended), its transmission licence, and the grid code.

NGC principally has two roles, that of transmission asset owner (TO) and that of system operator (SO). As SO, the NGC is required to balance generation with demand in real time to maintain system security. Arrangements for the definition and settlement of balancing energy are set down in the Balancing and Settlement Code (BSC).[152]

Under its transmission licence, NGC is obliged to offer non-discriminatory terms for connection to and use of its transmission system. Arrangements for connection to and use of NGC's transmission system are set down in and governed by the Connection and Use of System Code (CUSC). This code, and its associated schedules and exhibits, represent the contractual relationship between NGC and users of the transmission system in England and Wales.

9.13.2.7 Distribution

Distribution remains a monopoly business and, under the Utilities Act 2000, it has become a separately licensable activity. There are currently nine distribution companies operating in 12 authorised distribution areas. Distribution companies hold separate operating licences in respect of each area and are governed by the terms of their distribution licences.

9.13.3 Regulation of the Industry

9.13.3.1 The Statutory Structure

The statutory framework for the supply of electricity in England, Wales, and Scotland is provided for in the Electricity Act 1989, and for Northern Ireland in the Electricity (Northern Ireland) Order 1992.

The Electricity Act is divided into three parts.

Part I

Part I establishes the new regulatory structure for England and Wales under sections 1–64 and Schedules 1–9. The DG of Electricity Supply oversees the regulation of the industry and exercises his authority through the Office of Electricity Regulation (OFFER). The main functions of the DG include price regulation in respect of distribution and supply obligations of the PESs, the transmission obligations of the licensed companies, and the standards of service provided by the PESs, in addition to consumer protection. The DG exercises substantial power over the wholesale market for electricity and the trading activities of the electricity generators under the conditions he imposes in the licences and his power to modify them. The DG also has the authority to adjudicate in disputes. The Act provides for the appointment of the DG who then has the power to appoint consumer committees and a National Consumers' Consultative Committee. Part I of the statute establishes the principal duties both of the Secretary of State, the government office in charge of the electricity supply industry, and the Director, in addition to the statutory framework within which all other players must operate.

The licensing system is based on the grant of statutory authorisations for the generation, transmission, and supply of electricity on conditions as set out in the licences. Enforcement provisions are set out in the Electricity Act to ensure compliance by the licence holder with the conditions of the licence. Provision is also made in this part of the Act for the imposition of the Non-Fossil Fuel Obligation and the imposition of the Non-Fossil

Fuel Levy. After consultation, the office of the regulator OFFER has been merged with that of the gas regulator, OFGAS. The new regulatory body is headed by an executive board rather than a single individual. In March 1999, the Director announced a new corporate structure to operate the combined offices, with three management groupings under the headings of customers, supply chains, and regulation and price control, each of which operates under a Deputy DG. A management committee was constituted of the DG, the three Deputies, and a Chief Operating Officer whose responsibilities include the carrying forward of the merger of the two organisations. A board of advisors was constituted to advise on major policy issues and strategy.

Part II

This part is concerned with the structure of the industry and the privatisation of its constituent parts. Sections 65–95 and Schedules 10 and 11 provide for the transfer of all the property, rights, and liabilities of the area boards, the Central Electricity Generating Board, and the Electricity Council to the successor bodies.

Part III

The final part of the Electricity Act provides for the Secretary of State to give, to persons authorised by licence or exemption to generate, transmit, or supply electricity, directions for the purpose of preserving the security of electricity installations and in relation to civil emergencies. These provisions are contained in sections 96–113.

Under the Utilities Act 2000, the electricity supply sector was substantially reformed both in regard to the institutional framework for the regulation of the electricity industry, and the legislative parameters dictating the structure of the industry. The Act provided for:

- The replacement of an individual regulator, the director general of electricity supply, with a regulatory board, the Gas and Electricity Market Authority;
- The merging of the regulatory offices for the gas and electricity sectors into a single regulatory office, the Office of Gas and Electricity Markets (Ofgem)
- The replacement of the former electricity consumer council, known as *energywatch*
- Changes in the primary duties of the Secretary of State for Trade and Industry and the regulatory authority and a shift in responsibilities between the secretary of state and Ofgem
- New powers for the regulatory authority and the secretary of state.

These changes were effected, in the main, through changes to the Electricity Act 1989. The Utilities Act 2000 significantly, also abolished the concept of the PES and introduced a single Great Britain–wide licence for all suppliers, thus putting all suppliers on the same legal footing. The Act also introduced a legal requirement for the separation of the former PES electricity supply and electricity distribution businesses, and introduced a statutory requirement for electricity distribution to become a separately licensable activity. These changes have, and will continue to have, a substantial impact on ownership structures in the industry.

9.13.4 Regulatory Bodies for the Electricity Supply Industry

9.13.4.1 The Secretary of State

The Secretary of State for Trade and Industry enjoys substantial powers to give directions in the interests of national security or civil emergency. He may intervene in the operation of generating stations, to give directions for the maintenance of fuel stocks, direct specified levels of output and specified use of fuel. He may do so without reference to Parliament or to the Director 'if he is of the opinion that disclosure of the direction is against the interests of national security or the commercial interest of any person'.[153] In cases of civil emergency, the Secretary of State may give general directions if they are expedient[154] even if this act results in the licensee being in breach of its licence.

The Secretary of State may also classify electricity generators and suppliers who may be exempt from licensing. However, in this respect, the Director must be consulted as to the form and extent of the exemption orders made by the Secretary of State.[155] In addition, in order to protect the consumer, the Secretary of State's consent is required prior to the Director being able to make regulations setting individual standards of performance for PESs.[156]

The Secretary of State also has direct control over some operational activities, notably in relation to the construction of power stations and overhead lines. These operations are exempt from the usual planning procedures but are subject to a consent procedure,[157] which includes consultation with local planning authorities. The Secretary of State must also be notified, or his consent obtained, for the use of gas as a fuel source in power stations of 10 MW or more of output.[158]

The Secretary of State also enjoys wide powers to obtain access to information about the industry, by requesting statistical information directly from the licence holders or those authorised by exemptions,[159] in addition to the Director's reporting obligations to him. In addition, the Secretary of State enjoys extensive powers to make regulations to implement specific provisions of the statute, with significant discretion as to the exercise of those powers.

9.13.4.2 The Secretary of State and the Director

Section 1 of the Electricity Act requires the Secretary of State to appoint the Director for a term not exceeding 5 years. However, the Director will be eligible for reappointment.

9.13.4.3 General Duties of the Secretary of State and the Director

Under section 3 of the Electricity Act, a framework is set out regarding a series of duties placed upon the Secretary of State and the Director, which fall into four main categories.

9.13.4.3.1 The Supply of Electricity

A principal duty of both the Secretary of State and the Director is to ensure that every person who reasonably demands a supply of electricity can obtain it. The customer must, in return, pay any connection costs and comply with any other statutory requirements as set out in the Electricity Act. Under sections 3(1)(a) and 16(1), the PESs are under an obligation to provide such a supply within their areas.

9.13.4.3.2 Financial Matters

The Director and the Secretary of State have a duty to establish that the licence holder, whether as a generator, operator of a transmission system, or a supplier, has the financial capacity to carry out the functions authorised by the licence. Section 3(1)(b) of the Act imposes a specific duty upon the Secretary of State and the Director to establish that licence holders are able to finance the carrying out the activities for which they are authorised by their licences.

9.13.4.3.3 Promoting Competition

The Secretary of State and the Director are also under a duty to promote competition in the generation and supply of electricity. This duty has taken up a large amount of the Director's time since the Electricity Act came into force. It was hoped that the restructuring of the electricity generating sector into, initially, three principal generators with the provision for further independent power producers, would create a genuine competitive market based on the pool. A notable agreement between National Power and Powergen in 1994 was to cap their bidding into the pool and to dispose of 6000 MW of mid-merit generation capacity. The duty to promote competition has caused the progressive lowering of the franchise limit for supply. In April 1994, the threshold above which second-tier suppliers could supply consumers was lowered to an average monthly consumption of 100 KW. Between 1998 and 1999, access to second-tier suppliers was made available by stages to over 21 million, with the balance of 5 million being brought in May 1999. Theoretically at least, this served to eliminate any monopoly of PESs over any consumer section of the supply market.

9.13.4.3.4 General Considerations

The general considerations that the Secretary of State and the Director are required to follow are provided for under section 3(3) of the Electricity Act, which obliges them to exercise their functions in a manner best calculated to:

1. Protect the interests of consumers of electricity supplied by licensees in respect of the following:
   a. The prices charged and the other terms of supply
   b. The continuity of supply
   c. The quality of the electricity supply services provided
2. Promote efficiency and economy on the part of persons authorised by licences to supply or transmit electricity and the efficient use of electricity supplied to consumers
3. Protect the public from dangers arising from the generation, transmission, or supply of electricity
4. Promote research into and the development and use of new techniques by or on behalf of persons authorised by a licence to generate, transmit, or supply electricity

5. Secure the establishment and maintenance of machinery for promoting the health and safety of persons employed in the generation, transmission, or supply of electricity

While exercising these functions, both the Secretary of State and the Director must have regard to the effects on the physical environment of the activities connected with the generation, transmission, or supply of electricity. Particular regard must be paid to the interests of consumers in rural areas under section 3(4), and the disabled and those of pensionable age under section 3(5).

Note should be taken of the general duties imposed under section 3 of the Electricity Act, which do not control decisions made by the Director in granting or refusing consents for generation stations or overhead power lines under sections 36 and 37, or control decisions to be made by him in the context of the powers he has under the Competition Acts.

Scotland

The Secretary of State and the Director are under special duties in Scotland to ensure that prices charged to tariff customers in any area specified in an Order made by the Secretary of State, comply with tariffs, which do not distinguish between different parts of that area. This constraint, however, must not disadvantage electricity suppliers in Scotland from competing with others authorised to supply electricity or supply under any exemption.

Northern Ireland

In Northern Ireland, the DG of Electricity Supply for Northern Ireland and the Department of Economic Development are each given specific functions under the 1992 Electricity Order. Under Article 4(2)(a), general duties are placed upon both the Department and the DG to ensure that all reasonable demands for electricity are satisfied. Article 4(2)(b) imposes a duty upon the DG and the Department to ensure that licence holders must be able to finance their licensed activities, in addition to the promoting competition in the generation and supply of electricity.[160] Most of the remaining general considerations applicable to the carrying out of the regulatory duties mirror those of the Secretary of State and the Director under section 3(3) of the Electricity Act.

9.13.5 The Licensing Regime

The regulatory regime is based on the powers of the Secretary of State and the Director, the system of control as laid down in the Electricity Act, and the conditions contained in the licences granted under the Act. Further, the DGs of Fair Trading and the Competition Commission have roles to play with regard to the wider context of the economic regulation of industry in general. As with other sectorally regulated industries, the electricity supply industry is concerned with two important questions of balance. The first regards the balance between regulatory intervention and market competition, while the second is the balance between the interest of the consumer and providing the regulated sector with sufficient return on capital to attract investment.

9.13.5.1 The Grant of Licences

Under section 6(1) of the Electricity Act, the Secretary of State, or the Director with the Minister's consent, may grant a licence to any person for the purpose of generating, transmitting, or supplying electricity. It is unlawful for any person to undertake any such activities unless he is authorised to do so by a licence or by virtue of an exemption under section 4(1) of the Act.

The initial licences were granted by the Secretary of State but since then this power has been delegated to the Director. The general authority sets out the discretion that the Director enjoys in granting licences, in addition to the specific conditions or limitations that they must contain.

9.13.5.2 The Licence

Licensing is intended to control the activities of participants in the electricity supply industry, in addition to ensuring that there is a common standard for quality of electricity and ensuring that the product is homogeneous. Under the licensing system, there is no possibility of a company competing on the ground that its electricity is in some way superior to other electricity. All electricity is identical and the only difference can be in respect of price. The government controls, by regulation, the quality and safety, but under the privatisation of the industry, the government is not involved in establishing the price.

Each licence granted is specific to generation, transmission, and supply. Licensees may hold more than one form of licence and it is common for licensed generators and regional electricity companies to hold, either themselves or in a company in the same group, a second-tier licence.

9.13.5.3 The Generation Licence

The Exemption Order exempts certain classes of generator from the requirement to have a licence. Otherwise, all generators of electricity require a licence. The generation licences granted to the privatised power generators in England, Wales, and Scotland, apart from the operation of nuclear power stations, are essentially the same. They contain conditions that require:

1. Compliance with the Grid Code, Distribution Code, and the Fuel Security Code
2. Adherence to the Pooling and Settlement Agreement
3. Prohibition on cross-subsidies between businesses and separate accounting for the generation and second-tier supply businesses
4. The offer of terms of connection to the system of electric lines operated by the generator
5. Compliance with health and safety regulations

The licence also contains authority for the licensee to carry out street works in addition to the power to acquire land or rights by compulsory purchase for the purposes connected with their licensed activities.

The prices charged by the generators are not subject to control. However, with some exceptions, all generation capacity must be sold to the pool, which, for each half-hour, will determine the system marginal price.

Independent generators are not required to account separately for their businesses, nor are there any prohibitions on cross-subsidy or discrimination. In addition, the licence may also contain special powers relating to acquisition of land and rights in land and street works.

9.13.5.4 The Transmission Licence

The transmission licences are held by the NGC, Scottish Power, and Scottish Hydro Electric. As transmission systems are natural monopolies, the transmission licences contain a system of price control as discussed above, in addition to provisions for separate accounting and prohibition of cross-subsidy. The licences also require the licensee to offer non-discriminatory terms for the use of the transmission system. There are specific conditions imposed on the NGC to maintain and comply with the Grid Code and the Distribution Codes. Paragraphs 25–76 and 25–77 provide for the transmission licences to contain special powers for licensees to carry out street works and to acquire land or land rights by compulsory purchase, if connected to the licensed activities. However, under section 9 of the Electricity Act, all holders of transmission licences are under a duty to:

(i) develop and maintain an efficient, coordinated, and economical system of electricity transmission; and
(ii) facilitate competition in the supply and generation of electricity.

9.13.5.5 Public Electricity Supply Licences

Public electricity supply licences are held by each of the regional electricity companies in England and Wales, Scottish Power, and Scottish Hydro Electric. These licences authorise the companies to supply electricity in their franchise area. The form of these licences reflects the fact that the licensees have *de facto* monopolies in their areas of electricity distribution (with the exception of some private distribution systems that may exist). Consequent upon the planned introduction of the competitive supply market to all electricity consumers by June 1999, these public electricity supply licences were revised to protect small consumers from the effect that the market dominance of PESs might have upon the supply to small consumers. In addition, the licence was revised to introduce detailed arrangements in which the second-tier suppliers must also operate under conditions contained in their own licences. Further, the new form of public electricity supply licence contains conditions for the protection of consumers. The principal features of the new electricity supply licence are as follows:

1. A duty to keep separate accounts for the separate regulated businesses and a prohibition on cross-subsidies
2. The regulation of distribution charges
3. Inclusion of conditions that apply to supply charges
4. An obligation regarding economic purchasing
5. Restrictions on PESs regarding their own generation capacity
6. Provisions regarding non-discrimination
7. Conditions regarding security of supply

8. A duty to comply with both the Distribution Code and the Grid Code
9. A duty to join the pool and comply with the Pooling and Settlement Agreement
10. A duty to be a party to the new arrangements required to facilitate the competitive supply market

Some of these obligations are contained in the Electricity Act itself under subsections 16 to 23 (paras. 25–45). However, the PESs' special powers and duties, as set out in their licences and in the Electricity Act, only apply in relation to their authorised areas. Any supplies made by a PES outside its areas are authorised by a second-tier supply licence.

9.13.5.6 Second-Tier Supply Licences

A second-tier supply licence is required by any person who supplies electricity within the meaning of section 4 of the Electricity Act, unless either the person is a PES supplying within its authorised area, or the person is exempt under section 5 of the Act. Under certain circumstances, as provided in paragraphs 25–59, a second-tier supply licence holder must supply domestic and other small consumers on request. In addition, the second-tier supplier must, if authorised in England and Wales, join the pool and become a party to the Pooling and Settlement Agreement in addition to complying with the Distribution Codes and the Grid Code. If the licensee operates its own distribution system, it is required to grant access to it to other suppliers.

This opening up of the second-tier supply market to all consumers has created a need for the Director to include major modifications to the terms of the second-tier supply licences. The new licence imposes conditions specifically designed to protect small consumers and it introduces detailed arrangements for the implementation of the widened competitive supply market, including provisions for new metering and settlement requirements under paragraphs 25–59.

9.13.5.7 Exemptions from Licensing

The current Exemptions Order came into effect on March 31, 1998 and is concerned with exemptions from the requirement for generators and suppliers of electricity to be licensed. Although the exemptions are complex, the principles are generally as follows:

1. An exemption from a generation licence includes two classes—small generators and offshore generators. The small generators do not require a licence if they do not supply from any one generating station of more than 10 MW or 50 MW in the case of a generating station of a declared net capacity of less than 100 MW. It is important to note that there can be disregarded power supplied to certain classes of consumer including single consumers or a group of consumers with the required corporate connection with each other. However, the electricity must be used by the consumer on the same site as the generating station. Offshore generation is exempt where the supply is to offshore installations, as specified in the Order.
2. In the case of electricity supply, there are four categories that are eligible for exemption. If a supply does not exceed 500 KW, this is regarded as a small supply and is exempt under the circumstances specified in the Order.

Any onward supply of electricity, that has been supplied to that supplier by a licensed supplier or under certain circumstances as an 'on-site' or 'private wires' supply, is also exempt. In addition, any electricity supplied for the purposes of meeting a temporary interruption where it is supplied by a supplier who generates the electricity itself or who receives it from an exempt supplier is also exempt. Such exemptions are all subject to the detailed conditions of the Order. Any on-site or private wires supply attracts an exemption to suppliers who supply several consumers on the same site or who supply consumers off-site through private wires, subject to a maximum total of 100 MW. They need not be corporately connected in any way. The provisions in the Order concerning on-site or private wire supply exemptions are very detailed as supply through private wires necessarily involves 'distribution' that may become licensable in order to facilitate any third-party access. Offshore supply exemption reflects the same principle as that available in relation to the generation of electricity by covering electricity generated offshore, which is supplied only to premises in an offshore installation.

9.13.5.8 Modification and Enforcement

Under the procedures contained in sections 11–14 of the Electricity Act, the Director enjoys a powerful lever in the context of his obligations to promote competition. The Director enjoys the power to initiate a review of the licence terms, under which, in the absence of agreement, the Director may refer the matter to the Competition Commission for investigation.

In addition, the Secretary of State enjoys an express power to modify a licence by using powers available to him under the Fair Trading Act 1973. First these apply where the Competition Commission concludes that, under section 56(1) of the Fair Trading Act, a monopoly situation exists, which is to the detriment of the public interest. If the Secretary of State makes an order under the Act, he can include in the Order a requirement to modify the licence conditions. Secondly, if, under section 73(1), the Competition Commission reports that any merger involving a licensee may operate against the public interest, then the Secretary of State again may make an Order in relation to that finding, which can include a modification to the licence.

Enforcement proceedings against a licence holder who is in breach of its licence may only be brought by the Director, normally after informal negotiations with the licensee. There are two alternative enforcement procedures. Under section 25(1) of the Electricity Act, the Director is under a duty to 'make such provision as is requisite' if he considers that the licensee is either contravening, or is likely to contravene, a relevant condition or requirement. This provision is, therefore proactive. Even if the facts have not been established sufficiently to justify a final order under section 25(1), it is nevertheless important for the Director to take immediate action; a provisional order may be made, which takes immediate effect, but cannot subsist for more than 3 months. Neither a final order nor a provisional order may be made if it would contravene any of the duties of the Director under section 3 of the Electricity Act, or if he is satisfied that the licensee will take appropriate measures to comply with the order and is doing so.

If a final order is made or a provisional order is confirmed, then the licensee has 42 days from the date of service of the order to apply to the High Court to question its validity. The three principal consequences of non-compliance with an enforcement order are as follows:

1. Proceedings may be brought by the Director under section 27(7) for an injunction or other appropriate measures.
2. The licensee may be liable in third-party actions. However, there may be a defence if the licensee can show that he took all reasonable steps and exercised due diligence to avoid contravening the order.
3. Ultimately, the licence may be revoked.

9.13.5.9 Revocation of a Licence

Licence revocation may take place on failure to comply with a confirmed provisional order or final order within 3 months. In addition, the licence may be revoked in the event of the appointment of a receiver or the making of an administration order or a winding up order. The licensee may also request the revocation of its licence.

In Northern Ireland, the same terms Articles 14 to 18 of the Electricity Order 1992 provide for the modification of licences as apply to licensees in England, Wales, and Scotland. Enforcement is similarly dealt with under Articles 28–31.

9.13.6 The Powers and Duties of PESs

All PESs are under a duty to make electricity supplies available to consumers. The physical means by which the supply is delivered are also the responsibility of the operators. The framework of duties and powers that are applicable to them in addition to those contained in their public electricity licences are set out in the Electricity Act under sections 9 and 16–23. A summary would indicate that there is a general duty to develop and maintain an efficient, coordinated, and economical system of supply under section 9 of the Act. This duty is enforceable by means of a final or provisional order and is a 'relevant requirement' for the purposes of sections 25–27 of the Electricity Act.

Subject to the provisions of Part I of the Electricity Act, there is a specific duty to supply on request by an owner or occupier of any premises in its franchise area, and to give a supply of electricity to those premises and as far as is necessary for the purpose, to supply electrical lines and plant under section 16(1). Detailed provisions are set out in section 16(2)–(4), which apply to the requisitioning of a supply by a customer. The PES may impose conditions upon the making of the supply and these are set out in section 16(3) and (4).

Under section 22(1) of the Act, a person may enter into a special agreement with the supplier if the maximum power to be made available exceeds 10 MW or if it is reasonable for a supply to be made under an agreement rather than under the published tariff, which is regulated under the conditions of the licence. The Director has the power to determine what is reasonable for a special agreement and the terms that the supplier may wish to impose.

Exceptions to the duty to supply include, inter alia, where a consumer is provided with a supply by a second-tier supplier, or an exempt supplier. The duty to supply, under those circumstances, is not enforceable. Nor is it, in addition, if the supply is prevented by *force majeure* or where to do so would place the supplier in breach of any statute or regulations that are concerned with the safety of supply.

Certain powers are granted to the PESs under the Electricity Act, which are as follows:

1. The PES is empowered under section 19 of the Act to recover expenditure incurred in providing electricity lines or plant for the purpose of connecting a consumer.
2. Under section 20, the PES may also require security as a condition of providing a supply.
3. It is the Public Electricity Supply Code which governs the relationship between the supplier and the consumer. The Code covers such matters as the recovery of charges, disconnection, restoration of supply, and damage to or interference with electrical plant.

The New Electricity Trading Arrangements (NETA) clearly bring together both buyers and sellers. This contract market will also permit large users to make direct purchases from generators. NETA replaces the pool with electricity being traded like any other product, in addition to bringing the spot market for electricity and gas closer in line.

Key features of the system are the replacement of the Pooling and Settlement Agreement with a BSC. The Code sets out the precise obligations of the licensed generators and suppliers, all of which are required to subscribe to the Code in place of the obligation to subscribe to the Pooling and Settlement Agreement. Generators now no longer have to go through the pool bidding system to sell power. Parties are now able to enter into bilateral contracts in a series of linked power markets.

Each generator effectively determines when to operate rather than being directed to do so by the SO. Generators are responsible under the Code for ensuring that their net contracted position is matched by the output, the suppliers are responsible for ensuring that their net contracted positions meet the demand of their consumers. Should the amount of electricity a generator produces, or a supplier actually supplies or uses within the trading period, not match up with their contracted positions, then a balancing charge will have to be paid as part of the compulsory settlement process. This settlement balancing charge is intended to provide a strong incentive to both generators and suppliers to remain in balance. Should there be an imbalance, the system security is maintained by the grid operator who has the power to back charge the original party creating the imbalance.

There is now no single price for power, but rather one that is negotiated between the parties within each individual market. There will now be several prices for electricity. The cost is dependent on the ability to balance and avoid a balancing charge.

The BSC is administered by the NGC which ran the pooling system; so it is in the best position to develop the bilateral market. The new bilateral market is also subject to regulation by the Financial Services Authority to ensure transparent and competitive practices.

There are three separate bilateral contract markets. It is expected that the majority of power will be traded under long-term contracts in excess of 3 years. Such long-term contracts will suit the needs of the generators inasmuch as they guarantee cash flow in addition to suiting the suppliers' needs to meet baseload demand. Nevertheless, not all power can be traded under long-term contracts as the balancing of supply and demand requires more flexibility. Accordingly, a separate bilateral market will be established for medium-term contracts, that is, for contracts between 6 months and 3 years. These will be used by suppliers as demand patterns change. Under the competitive retail market, the suppliers no longer have the security of a monopoly customer base. Should a supplier lose any retail market share, there would be a strong incentive not to buy power from generators on a long-term contract.

Contracts for under 3 months are generally considered to be short-term contracts. These allow for even greater flexibility to meet a fluctuating seasonal demand, in addition to helping suppliers take account of gaining or losing market share. However, such short-term contracts provide less security for the generators. This could have an effect on financing new plant. However, in a maturing market and increasing confidence in new technology, the financing of merchant plant encourages the financing of such plant.

As mentioned above, electricity cannot be stored, and so it is essential that there is a balancing mechanism for the market. Accordingly, there is a balancing market selling power for delivery within a few days, or within the day itself, which permits continuous balancing. The balancing market is administered by the SO who aims to maintain the system integrity.

There is also a separate market for ancillary services, which allows a generator to seek two separate revenue streams. However, Stephen Dow[162] considers that this introduces a risk to the integrity of the system 'by allowing generators to choose whether to trade in the ancillary services market rather than having the system operator call on generators to provide ancillary services'.

The intention of the market is to have a separate price for ancillary services. However, by allowing the SO to make demands on generators under the emergency provisions in the BSC, and given that the value of the ancillary services is closely linked to plant location, there is a risk that some generators could effectively set their own price. One of the main problems with the new scheme is that there is no other example of a system that uses bilateral trading to the same extent. One of the criticisms of the pool system was that there was not enough pressure on the generators to reduce the wholesale price of power under the pool system. It is hoped that the bilateral trading system will bring greater pressure on the generators to reduce their costs, but only time will tell.

9.13.7 Possible Distortions of the Market

Historically, the coal contracts compelling the generators to purchase feedstock at higher prices than the market value brought a distortion into the generation market.

The higher costs were to support the coal industry and provide a diversity of fuel sources under government policy. However, in the new liberalised market, such contracts are difficult to reconcile with an open market. A further distortion under the Electricity Act[163] is the Non-Fossil Fuel Levy. Under the levy, suppliers are compelled to take a certain quantity of electricity generated from renewable sources, which includes nuclear power. Under the Electricity Act,[164] the minister is empowered to apportion the generation market and guarantee a share for renewables, a large proportion of which will come from nuclear sources. The cost of this support is permitted to fall on the consumer.

9.13.7.1 The Sale of Electricity to Consumers

The granting of access to the competitive supply market for electricity was complete by June 1999, and with it brought substantial organisational and accounting problems. Many of these difficulties centred upon the fact that the supplies made by each second-tier supplier to each customer were required to be recorded and transmitted to the settlements system of the pool, translated into units of half-hourly consumption, and billed to the relevant second-tier supplier. It was recognised that small consumers, both domestic and small businesses, needed some level of protection. The result was the introduction of major changes to the terms of the licences of PESs and of second-tier suppliers. These changes were intended to make the enlarged competitive supply market work and protect consumer interests.

9.13.7.2 The Protection of Consumers

There have always been conditions on the licences of PESs that prohibit discrimination in the levying of connection and use of system charges for the use of the distribution systems and in electricity supply to 'franchise' customers. However, a new condition has now been included, which provides that PESs will not show discrimination 'in supplying or offering terms for the supply of electricity to customers in any market in which it is dominant'. In addition, terms must not be 'unduly onerous'. Nevertheless, terms that could be considered as discriminatory may be introduced so long as they are reasonably necessary to meet competition. The decision lies with the Director to determine whether the licensee is dominant in any market, whether there is established competition, and whether any terms offered are predatory. Should the supplier be deemed to be 'dominant' in any area or in any classes of customers, the licensee must give the Director prior notice of any change in terms of either a tariff or a designated supply contract. However, the Director may disapply this condition in respect of any area or class of customer if he considers that the circumstances no longer exist.

Special protection exists under the revised public electricity supply licences to protect electricity consumers at 'designated premises'. Such designated premises include domestic premises or, subject to some exceptions, premises where the annual electricity consumption is less than 12,000 kW h. This additional protection is provided under a designated supply contract. The contract is of standard form, and must set out all the terms of supply, including various different payment options and the rights to any termination. The terms and conditions in such contracts can be reviewed at any time by the Director, as provided in the public electricity supply and second-tier supply licences, which apply to designated supply contracts.

9.13.8 The Electricity Market Reform White Paper

The Coalition Government provided details of the first of its four Electricity Market Reform proposals in the Budget 2011, confirming that a floor price for carbon in the electricity-generating sector will be introduced from April 1, 2013. The Government believes that greater low-carbon investment in electricity generation is required to meet the United Kingdom's carbon emissions reduction targets and to ensure security of supply. A perceived barrier to investment has been the low and uncertain 'carbon price'. In other words, the true environmental costs of fossil fuels have not been properly reflected in the actual cost of using these fuels. As a result, fossil fuels may be cheaper than the alternatives, particularly when the high upfront costs associated with alternative fuel sources are taken into account. The Government, therefore, propose to impose a new layer of taxation on the supply of fossil fuels to electricity generators, with the level of taxation being linked to the carbon content of the fossil fuel in question. The additional tax burden is intended to create incentives for generators to use non-fossil fuels. The rate of tax will increase over time. The proposals are in line with the Government's promise to use tax policies to make the United Kingdom 'greener' and to encourage investment in new technologies, but they are inconsistent with the promise to simplify the tax system and reduce the tax compliance burden for companies.

9.13.9 Implementing the European Union's Third Energy Package

The UK government have introduced new regulations that fundamentally change key aspects of the regulatory regime for large and small market players and include some licence modification powers and enforcement powers over unlicensed network operators and suppliers, as well as codifying the Third Energy Package unbundling requirements for gas and electricity transmission operators, including the controversial third country provisions, the unbundling requirements for gas and liquefied natural gas (LNG) storage facilities, and the extension of the third-party access regime to offshore facilities.

For the first time, the national regulator is explicitly required to comply with decisions of the new pan-European regulatory agency, Agency for the Cooperation of Energy Regulators (ACER), and the European Commission is decreasing the scope for matters to remain purely a matter for domestic regulation.

The Electricity and Gas (Internal Markets) Regulations 2011 provide for five distinct areas:

1. Consumer protection
2. The role of OFGEM as the National Regulatory Authority (NRA) for Great Britain
3. Transmission and distribution networks
4. Gas infrastructure
5. Licence-exempt undertakings and private networks

We shall take these in turn to consider the effects of the EU Third Energy Package.

9.13.10 Consumer Protection

The Third Energy Package contains measures that are intended to protect consumers and reinforce energy retail market competition, which has been slow to take off in a number of jurisdictions. While the United Kingdom has had open competition down to the domestic level for over 10 years, the new requirements will necessitate changes to existing practices so that consumers can switch energy suppliers within 3 weeks of agreeing to a contract with a new supplier. In Great Britain, there is currently no statutory minimum time frame for switching supplier, although there are other consumer protection measures such as cooling-off periods to allow consumers to reconsider decisions to change supplier. To fully implement the requirements of the Third Energy Package, the government has, for licensed suppliers:

- Introduced a licence requirement to include a three week switching right in supply contracts.
- Introduced a licence requirement for suppliers to take reasonable steps to improve their systems and processes.
- Obliged incumbent suppliers to co operate with prospective suppliers to enable customers to switch suppliers.

9.13.11 OFGEM's Role as the NRA for Great Britain

The government has designated OFGEM as the NRA for Great Britain. The Northern Ireland Authority for Utility Regulation (NIAUR) is designated as the NRA for Northern Ireland. OFGEM is required to work closely with NIAUR for the purposes of representing the United Kingdom's interests in dealings with the ACER.

In it's role as NRA, OFGEM will be given formal duties to monitor:

- The investment plans of TSOs
- Contractual arrangements between suppliers and large non domestic customers to ensure they do not restrict competition
- The speed with which TSOs and distribution system operators (DSOs) make connections and repairs
- The implementation of rules relating to the roles and responsibilities of TSOs, DSOs, suppliers, and customers and other market participants.

The government will also amend dispute resolution procedures to extend the scope of complaints that can be made against transmission system operators (TSOs), DSOs, owners of LNG import, and gas storage facilities.

9.13.12 Transmission and Distribution Networks

The Third Energy Package included unbundling requirements for the separation of transmission undertakings from supply, electricity generation, or gas production undertakings.

Legislative changes introduced by the draft Regulations will prohibit those who control a TSO from exercising rights over licensed supply, electricity generation, and gas production undertakings and vice versa. The existence of such rights will not prevent a TSO from being granted certification by OFGEM, but the exercise of these rights may allow a court to void decisions that could result in discrimination.

9.13.13 Gas Infrastructure

The provisions of the Third Gas Directive seek to improve the operation of the gas storage market and LNG facilities. The requirements of the Directive include the legal and operational unbundling of gas storage system operators to ensure their independence. The Directive also strengthens the requirement to grant third-party access to storage facilities that are technically and/or economically necessary for providing efficient access to the system.

9.13.14 Licence-Exempt Undertakings and Access to Private Networks

The Third Electricity and Gas Directives require Member States to implement a system of third-party access to electricity and gas transmission and distribution networks 'based on published tariffs, applicable to all eligible customers and applied objectively and without discrimination'. Essentially these measures, which give customers a choice of supplier, are intended to encourage competition and drive down costs.

Following the decision of the European Court of Justice (ECJ) in the *Citiworks case* in May 2008, the UK Government is obliged to make provision for third-party access to private unlicensed electricity and gas transmission and distribution networks rather than just licensed ones.

CASE STUDY

Cityworks concerned a complaint brought by an electricity supplier seeking access to a private electricity network at Leipzig airport to allow it to compete with a monopoly supplier. Access had been denied on the basis that the network was a small, private, unlicensed network serving only certain customers and had been exempted under German law from the obligation to provide third-party access. The ECJ ruled that this exemption was incompatible with the Second Electricity Directive (now replaced by the Third Electricity Directive). As the third-party access regime is currently only applied to licence holders in Great Britain. The UK Government is making changes to apply the third-party access requirements to unlicensed electricity and gas networks.

9.13.15 Next Steps

The vast majority of these new regulations were approved and take effect from March 3, 2012.

ENDNOTES

1. Approximately 15% of total greenhouse gas emissions, measured in terms of total global warming impact (DTI, Cm. 4071, October 1998).
2. DECC (2009), 'The UK Renewable Energy Strategy', p. 30.

3. For stations between 20 and 50 MW capacity.
4. Under the Electricity Act 1989, Section 36.
5. Government Response to Fourth and Fifth Reports of the Trade and Industry Committee, Cm. 4071, October 1998.
6. Since 1979, over 900 reports have been produced by the select committees.
7. Following the publication of the report from the Energy Select Committee on the coal industry, the response of the government was to subsume the Department of Energy into the DTI, which had the effect of disbanding the Select Committee itself.
8. Section 7(3).
9. Section 12.
10. Section 3(1)(a).
11. Section 3(1)(b).
12. Section 3(4).
13. Section 2(1)(a).
14. Section 2(1)(b) and (c).
15. Section 43.
16. (1982) 1 *WLR* 887; (1982) 3 All *ER* 579.
17. The 1998 Act consolidated earlier legislation in this respect.
18. As designated under the Continental Shelf Act 1964, Section 1(7).
19. Criminal Jurisdiction (Offshore Activities) Order, 1987, SI 1987 no. 2198.
20. Section 11(5) of the Petroleum Act 1998.
21. *Hegarty v E.E. Caledonia* (1997) Lloyd's Rep. 257, CA.
22. (2000) Env LR 221, QBD.
23. Council Directive 92/43 (1992) O.J. L206/7.
24. SI 1976, no. 766 (as amended).
25. SI 1988, no. 1213.
26. SI 1995, no. 1436.
27. Under Model Clause 21.
28. Under Model Clause 23.
29. *Mid-Essex Gravel Pits v Secretary of State for the Environment and Essex County Council* (1993) JPL 229.
30. Circular no. 2/85: *Planning Control over Oil and Gas Operations*.
31. *The Countryside and Rural Economy* (1992), PPG 7.
32. As defined by DOE Circular 9/84.
33. Planning (Hazardous Substances) Act 1990 and DOE Circular 11/92.
34. This also applies to outside territorial waters by virtue of Section 4 of the Continental Shelf Act 1964.
35. SI 1998, no. 1224.
36. SI 1987, no. 1331.
37. SI 1998, no. 698.
38. *The Times*, 25 May 1995.
39. Although the United Kingdom had not declared an EEZ.
40. Greenpeace decided to apply to an English court because of the stricter interpretation of the rules of standing in the Scottish courts (Mark Poustie [1995], 'Sparring at oil rigs: Greenpeace, Brent Spar and challenges to the legality of disposing of disused oil rigs at sea', JR, 542).
41. OSPAR Decision 98/3.
42. This decision does not cover pipelines and there are no other international guidelines on the decommissioning of disused pipelines.
43. Article 1 of the EC Framework Directive on Waste (Council Directive 75/442/EEC as amended by Council Directive 91/156/EEC).

44. General guidance on the definition of waste and its interpretation is provided in DOE Circular 11/94. Welsh Office Circular 26/94 and Scottish Office Environment Department Circular 10/94.
45. The definition of special waste is set out in the Special Waste Regulations 1996.
46. This convention was implemented into UK law by the Control of Pollution (Landed Ships Waste) Regulations 1987 as amended.
47. In Scotland, similarly a offence is created by the COPA 1974 as amended by the Water Act 1989.
48. As detailed in the Water Resources Act 1991.
49. *Energy Paper 67*. HMSO, April 1999.
50. Terence Daintith (1984), *United Kingdom Oil and Gas Law*, Loose leaf with updates. Sweet and Maxwell: UK.
51. Section 1(4), now 1998 Act, Section 1(b), para 3-809.
52. Later the British Coal Corporation (BCC).
53. This declares that no rights of entry on land are implicitly conferred by the Act. There is a procedure for obtaining such rights under Section 3 (now Section 7 of the 1998 Act).
54. The Secretary of State has issued a draft CBM extraction agreement in this respect.
55. Malcolm Grant, *Encyclopaedia of Planning Law*. Sweet & Maxwell: UK, Loose leaf and updated; Victor Moore (2000), *A Practical Approach to Planning Law*. Blackstone: UK, etc.
56. Local Government Act 1972, Schedule 16.
57. Now the GDO 1995.
58. See Section 9.12.2.2.
59. Victor Moore (2000), *A Practical Approach to Planning Law*. Oxford University Press: UK.
60. *Grampian Regional Council v City of Aberdeen District Council* (1984) 47 P&CR 633.
61. Now known as *Grampian* conditions.
62. *R v Plymouth City Council (ex parte Plymouth and South Devon Co-operative Society* (1993) JPL 1099. *Good v Epping Forest District Council* (1994) JPL 372. *Tesco Stores Ltd v Secretary of State for the Environment* (1994) JPL 919.
63. *Tesco Stores Ltd v Secretary of State for the Environment* (1995) 1 WLR 759.
64. *Environmental Law*, Chapter 7.
65. EC Directive 90/313. O.J. L 158/56 of 23 June 1990.
66. Article 3(3).
67. Article 3(2)2, 1st indent.
68. Article 3(3).
69. Article 3(2), 4th indent.
70. Gerd Winter (1996), *European Environmental Law: A Comparative Perspective*. Dartmouth: UK.
71. SI 1992 no. 3240.
72. Stephen Tromans (1999), 'Environmental Protest and the law', *O.G.L.T.R.*, 342.
73. Tromans, ibid, p. 346.
74. Article 10.
75. Article 11.
76. *Steele and others v U.K.*, Case no. 67/1997/851/1058. *The Times*, 1 October 1999.
77. *Coastal Planning and Management: A Review*. HMSO: London, 1993.
78. NPACA 1949, Section 15.
79. The three County Conservation Councils.
80. Section 20.
81. Section 35.
82. Countryside Act 1968, Section 15.
83. Under Section 29 of the 1981 Act.
84. See infra.

85. Department of Trade and Industry – Oil and Gas Directorate (New Licensing), January 2000.
86. E&P Forum, 'Guidelines for the Development and Application of Health, Safety and Environmental management Systems', Rep. no. 6, 36/210, July 1994.
87. In conversation with an inspector from the Environment Agency, he stated: 'With the chronic lack of resources the Agency concentrates on those sites which we fear may be in breach of their licence conditions. Those sites which are registered under one of the EMSs we know the emissions have been audited and their management systems have been independently verified'.
88. E&P Forum are in discussions with national regulatory authorities regarding the drafting of achievable, industry-specific environmental rules and regulations.
89. The 1986 Act has since been amended by the Gas Act 1995.
90. See Roggenkamp, Editor, *Energy Law in Europe*, Chapter 13, by Stephen Dow, pp. 901–971.
91. No financial institution would lend money to consortia without the security of a gas contract.
92. In addition, the government had the right to approve development plans under the model clauses.
93. These duties are performed by the Competition Commission, formerly the Monopolies and Mergers Commission; the Office of Fair Trading.
94. Although these are shared with the Minister, in practice, the authority is wholly delegated to the DG.
95. The Gas Act 1986, Section 4 (as amended), sets out the duties of the DG; *see also* the Gas Act 1995 and the Utilities and Competition Act 1992, in addition to the Utilities Act 2000.
96. Gas Act 1986, Section 19 (as amended), and Gas Act 1995.
97. Under the Petroleum and Submarine Pipelines Act 1975, Section 23, and the Pipelines Act 1962, Section 10, in addition to the Petroleum Act 1998, provide for access to private pipelines by negotiation.
98. Statutory third-party access is a qualified right.
99. Each user will have a range of nominated capacity, and if the independent pipeline company was to sign up as many transportation agreements as possible to the pipe's total physical capacity, this could lead to problems when nominations increased with demand.
100. Providing the tariff reflects the cost of service and a return on capital.
101. Roggenkamp, *Energy Law in Europe*, Chapter 13, pp. 901–971.
102. Given the dependence on the natural monopoly, this is an essential element.
103. The licence requirements are minimum conditions for participation in the industry.
104. The conditions in the licence also reduce the risk that insufficiently expert companies will enter the market to the detriment of the security of supply.
105. Although the statute does not identify British Gas in particular, in practice, the company was the only public gas supplier permitted to provide gas on the mainland of Britain.
106. *See* Roggenkamp, ibid, p. 941.
107. This would create a problem as far as the requirement for expertise is concerned. However, they could enter the gas market as a joint venture with a recognised expert in the field.
108. Such large consumers were well placed to look after their own interests.
109. Created under the Gas Act 1995.
110. Essentially, the Code is a contract, the terms of which have been approved by the DG, between the suppliers/shipper and the transmission company.
111. This issue fundamentally affects the revenue of suppliers and, as such, is critical to the success, or otherwise, of the operation of the liberalised market.
112. Currently the privatised transportation arm of British Gas, TransCo.

113. TransCo simply receives payment for the operation of moving the product.
114. However, this also had largely unforeseen effects.
115. Where make-up rights can be invoked.
116. George (Rock) Pring, James Otto, Koh Naito (1999), 'Trends in international environmental law affecting the mineral industry', *Journal of Energy and Natural Resources Law*, 17, 39.
117. Coal Mines Regulation Act 1872.
118. *R v Earl of Northumberland* (1567) Plaid 310, p. 336.
119. Directive 88/609/EEC.
120. Convention on Long-Range Transboundary Air Pollution (Geneva), UKTS 57 (1983), Cmd. 9034; ILM (1979), 1442. In force from 16 March 1983. Protocols of 1984, 1985, 1988, and 1991.
121. *The Prospects for Coal*, Cm. 2235 (1993).
122. *The Review of Energy Sources for Power Generation*, Cm. 4071 (1998).
123. *Sustainable Development: The UK Strategy*, HMSO, 1994.
124. MPG 6, *Guidance for the Mining of Aggregates.*
125. *Mid Essex Gravel Pits v Secretary of State for the Environment and Essex CC*, 3 (1993), JPL, 229.
126. Sections 1, 7, and 12 of the 1994 Act.
127. Revised explanatory note on p. 27.
128. Derived from Section 2 of the 1975 Act.
129. The Coal Industry Act 1994, Section 43.
130. Section 43(3)–(7).
131. Section 2(3).
132. *Opportunities for Change*, February 1998.
133. MPG 3.
134. PPG16.
135. PPG15 as amended by Appendix E of the DOE Circular 14/97.
136. 'Coal Mining Award Scheme', MPG3, May 1999.
137. Now incorporated into the Town and Country Planning Act 1990.
138. *Abandoned Mines and the Water Environment*, National Rivers Authority Water Quality Series no. 14, 1994.
139. *R v British Coal Corporation* (1993), (1994) *Water Law 48.*
140. Anne Jones (1996), 'Regulation, crime and pollution from abandoned coal mines', *Journal of Environmental Law* 8(1), 43.
141. Attorney General's Reference no. 1, (1995) 1 WLR 599 at 615.
142. For example, as museums or stores.
143. The Mines (Notice of Abandonment) Regulations 1998, DETR April 1998, SI 1998, no. 892.
144. Section 78B(1).
145. There is much case law in the area of 'causing' or 'knowingly permitting' starting with the landmark case of *Alphacell v. Woodward* (1972) AC 824. The latest is the *Empress Garage case* (1998).
146. *R (on the application of National Grid Gas plc, formerly Transco plc) v Environment Agency* (2008) Env LR 4.
147. This caused great concern to the lenders, who feared that they might be found liable for the costs of remediation of their clients' contaminated sites.
148. Electricity (Class Exemptions from the Requirement for a Licence) Order 1997.
149. These are commonly for a duration of 15 years.
150. Electricity Association (2001), Electricity Industry Review 5.
151. Centre for the Study of Regulated Industries, Industry Brief. *Regulation of the UK Electricity Industry.*

152. The BSC is the responsibility of Elexon, a wholly owned subsidiary of NGC.
153. Electricity Act 1989, Section 96(4).
154. Section 96(2).
155. Section 5, paras 25–35 and 25–36.
156. Section (1).
157. Sections 36 and 37.
158. Energy Act 1976, Section 14.
159. Electricity Act, Schedule 3, para 1.
160. Article 4(2)(c).
161. A merchant plant is a plant that sells into the market with no long-term offtake contracts in place.
162. Martha Roggenkamp, Editor (2001), *Energy Law in Europe*. Sweet and Maxwell: UK.
163. Electricity Act 1989, Sections 32 and 33, and regulations made thereunder.
164. Electricity Act 1989, Sections 32 and 33.

10 Energy Law in Norway

10.1 INTRODUCTION

Norway is a large hydropower producer, and a significant exporter of oil and gas. This chapter describes the development of the energy sector in Norway. As a member of the European Economic Area, it implements most of the EU commercial legislation through the European Economic Agreement (EEA).

The energy resource base in Norway is dominated by two energy resources, petroleum and hydroelectricity. Petroleum resources are mainly found on the Norwegian continental shelf. These offshore resources have increasingly been developed since the 1970s when it became both technologically feasible and economically viable. However, with regard to hydroelectricity, the situation is exactly the opposite. As Norway is a mountainous country, it has made good use of its waterfalls for centuries but industrial as well as large-scale development of hydropower has taken place only since the beginning of the last century.

10.2 GOVERNMENT IN NORWAY

There are three levels of government in Norway situated at central, regional, and local levels. The highest levels of legislative and financial powers rest centrally with the unicameral Parliament. The highest judicial power rests with the Supreme Court under the Constitution of May 17, 1814. Under the Constitution, the highest executive power formally rests with the Monarch; however, according to constitutional practice, the real executive power rests with the Cabinet. The Cabinet oversees the Ministries and the directorates, who, in turn, may be instructed by a superior organ. However, as the Ministries and directorates are established to execute the more detailed state powers within defined sectors, they would normally only receive policy guidelines from any superior organs.

The legislative and executive powers to regulate the Norwegian energy sector lie within the state. The electricity sector is heavily influenced by state regulation and control, as are Norwegian petroleum activities, which also refer to offshore petroleum deposits. However, the development of the energy infrastructure on land involves the municipalities who have executive powers under the Planning and Building Act.[1]

10.3 THE INDUSTRY STRUCTURE OF ELECTRICITY SUPPLY IN NORWAY

Hydropower accounts for 99% of the electricity produced in Norway; in fact, Norway is the sixth largest hydropower producer in the world and the largest in Europe. The total installed capacity in hydropower plants is about 28,000 MW. In addition, 260 MW of thermal power and 280 MW by wind power plants had been installed by the end of 2005. Norway's domestic electricity production is based almost entirely on hydropower, accounting for approximately 96% of the energy mix. Under the Norwegian Industry Concession Act, only public undertakings may acquire concessions for the ownership of hydropower resources other than small-scale resources. Public undertakings are defined as undertakings where Norwegian public entities directly or indirectly control at least two-thirds of the shares and the capital, and are organised in such a manner that genuine public ownership exists.[2] Therefore, Norwegian large- and medium-scale hydropower is subject to Norwegian public ownership, which has been the case from the beginning of the development of the sector 100 years ago. The same ownership restrictions do not apply to other energy sources such as wind power.

The largest Norwegian electricity producer is Statkraft SF, which is 100% state owned. Statkraft owns approximately 36% of the Norwegian electricity-generating capacity with the other 52% owned by a large number of local municipalities and county authorities, mostly exercised through ownership interests in public undertakings. Private companies still own roughly 12% of the generating capacity on the basis of previous concession legislation.[3]

Unlike most other European countries, Norway has three grid levels rather than two: the central, the regional, and the local distribution grids. The central grid, which for most practical purposes is the Norwegian transmissions grid, is operated by the wholly state-owned enterprise Statnett SF. Statnett is designated as transmission system operator (TSO) with particular system responsibility under the Norwegian Energy Act.[4] It owns approximately 87% of the central grid and operates the remaining parts of the central grid on the basis of rental agreements.

While the ownership interest in Statnett is held by the Ministry of Petroleum and Energy (MPE), the ownership interest in Statkraft is held by the Ministry of Trade and Energy. Consequently, the clarification in Article 9(6) of the New Electricity Directive that two separate public bodies are not deemed to be the same person entails that the organisation of public ownership interests in Statnett and Statkraft, at the outset, complies with the full ownership, unbundling the requirements of the New Electricity Directive.

Regional grids and distribution grids are owned by a large number of companies, mostly owned, in turn, by county and municipal authorities. Companies with private ownership are more common within the electricity-trading sector but private companies do exist within most parts of the resource chain.

Regulatory responsibility and supervision is mostly delegated from the MPE to the national regulatory authority, the Norwegian Water Resources and Energy Directorate (NVE). This includes activities such as granting concessions for the building and operation of electricity production plants and grid infrastructure, and the setting of overall grid tariff levels. However, NVE does not have regulatory responsibility for the Norwegian upstream petroleum sector.

The Energy Act represents the main statutory instrument of Norwegian electricity market regulation, comprising all parts of the resource chain from electricity production to consumption.[5] When the Energy Act came into force in January 1991, full competition was, in principle, introduced into the Norwegian electricity sector.

The regime has been developed into a Nordic competitive electricity market that can be characterised as a well-functioning and advanced regulatory market model.

The Act itself is a framework act that sets out the general concession requirements for local area licences applicable to monopoly grid operators as well as electricity producers and traders, marketplace licences, and import and export licences. The key group of licences are the trading licences, which provide the power to govern grid company operations and tariffs as set out in the Regulations under the Energy Act.[6]

The Norwegian electricity market regime generally complies with most of the provisions of the New Electricity Directive, although some amendments may be necessary, in particular relating to the New Electricity Directive's comprehensive new set of provisions governing national regulatory authorities.

10.3.1 Electricity Trading

Electricity can be traded bilaterally or on the Nordic power exchange Noord Pool Spot.[7] This offers both Elspot (day ahead) and Elbas (intraday) markets. Elspot trades must be in by 12 midday. In order to participate in Noord Pool Spot's physical markets, participants must sign a Participant Agreement under which they accept Nord Pool Spot's Trading Rules. Market participants are also required to achieve a balance between commitments and rights for each hour in each Elspot area.[8] The Elbas market opens up for hourly trade and consequently also provides a balancing market. The Norwegian TSO Statnett also administers a regulating power market where participants bid for regulating power, and a regulating power options market for balancing purposes. In the latter options market, which is operated on a weekly basis, participants may commit to future bids in the regulating power market for a fee paid by Statnett.

Financial trading is organised by the power derivate exchange National Association of Securities Dealers Automated Quotations (NASDAQ) OMX Oslo ASA.[9] Through the brand name NASDAQ OMX Commodities, the financial power-trading exchange provides trading and clearing services for power derivatives and carbon contracts. This provides price-hedging possibilities for participants also engaged in physical trade, as well as trading opportunities for participants solely engaged in financial transactions.[10]

10.3.2 The Third-Party Access Regime

The regional grid and distribution companies operate the third grid under a trading licence.[11] According to section 4-1 of the Energy Act, the objective of the licensing regime is to facilitate an efficient market and an efficient development of the grid. The licence contains conditions in respect of non-discriminatory market access, impartial behaviour, and calculation of tariffs.[12] In other words, licensees must ensure market access for all customers requiring grid services on a non-discriminatory and objective point tariffs and terms.[13] More information on how to achieve this result is given in the Control Regulation.

Grid tariffs are set by the grid operators on the basis of yearly income frames that are determined by the regulator, NVE, the overall principles of which are that the grid revenues will cover the costs of operation and depreciation of the grid over a period of time, and also give a reasonable rate of return on invested capital on the assumption of an efficient operation, utilisation, and development of the grid. Tariffs refer to the customer's connection point to the grid and those that are independent of power purchase and sales agreement must be non-discriminatory and objective.[14] Again further information is given in the Control Regulation Part 5.

Distribution grid companies with local area licences are required to ensure that customers within their grid area are supplied with electricity from the grid.[15] In addition, grid companies are required to invest in new grids in order to connect to new production or supply facilities.[16] However, the grid companies may be exempt from the investment obligation to the benefit of electricity production facilities if the grid investment is not considered to be socio-economically efficient, but this will only be granted in exceptional circumstances. On the production side, the investment obligation will be of great importance for new small-scale renewable production facilities when project financing may be difficult to obtain.

10.3.3 Market Entry for Supply and Generation

The Energy Act 3-1 provides for the licence requirements for the construction, ownership, or operation of electrical installations such as electricity generation and transmission facilities. In practice, electricity distribution grids are owned by a separate area licence requirement pursuant to section 3-2 of the Energy Act. In addition, the Qualifications Regulation sets out the qualification requirements for the staff of companies with licences under sections 3-1 and 3-2 of the Energy Act.[17] The Planning and Building Act sets out the construction requirements for these projects including an impact assessment.[18]

Ownership of hydropower resources over a certain capacity[19] is subject to specific restrictions under the Industry Concession Act, and consequently foreign and private market players cannot be granted hydropower licences; they may only participate as minority shareholders in Norwegian public companies. Trading licences are to be obtained by companies trading electricity in their own name under a highly standardised system under which companies sign a standardised Participant Agreement in order to trade electricity on Nord Pool Spot.

10.3.4 Public Service Obligations and Smart Metering

With respect to grid companies, Statnett is subject to a number of obligations in its role as TSO with system responsibility. Designated senior officers (DSOs) are also subject to a number of obligations and these obligations, to a large extent, correspond to the tasks appointed to these grid operators under the EU Electricity Directive. A regulation concerning smart metering has also recently been adopted by Norway, which sets out as a main rule that grid companies must install advanced metering systems at all measuring points within their licensing area by January 1, 2017.[20] The new regulations require that the grid companies will gradually submit plans for procurement and installation of metering systems by 2012 and report periodically on progress to NVE. They will need to have installed metering systems in 80% of their measuring points by January 1, 2016.

10.3.5 Cross-Border Interconnectors

Norway is a major energy exporter to the rest of Europe and the cross-border interconnector capacity from Southern Norway in 2012 amounts to a total of 3700 MW, including 2050 MW capacity to Sweden, 950 MW to Denmark, and 700 MW to the Netherlands (the NorNed cable). Interconnector capacity from Mid and Northern Norway includes a further 1400–1700 MW to Sweden, 120 MW to Finland, and 50 MW to Russia.[21] New projects planned include cables to Denmark (Skagerrak 4) for which a licence was granted in June 2010, Germany (NORD.LINK/NorGer), the Netherlands (NorNed 2), the United Kingdom (NSN), and Sweden (SydVest linken).

10.4 THE DEVELOPMENT OF THE PETROLEUM SECTOR

Exploration and production of oil and gas is governed by the Petroleum Act with the intention to regulate all important aspects of the petroleum activities on the Norwegian continental shelf. The regulatory emphasis of the Petroleum Act is on the relationship between the authorities and the licensees, in addition to the relationship between the licensees and third parties in the form of liability regimes adapted to the petroleum sector. The regulation of the relationship between licensees and suppliers of goods and services is another important factor. However, this relationship, as in most developed countries, is regulated by freedom of contract. Despite that, there is a significant element of standard form contracts and 'agreed documentation'.

Norway proclaimed sovereignty over the Norwegian continental shelf with regard to exploration and development of subsea natural resources by Royal Decree in May 1963, and agreements to divide the continental shelf according to the median line principle under the 1958 Convention on the Continental Shelf were concluded between Norway and the United Kingdom in March 1965, and between Norway and Denmark in December 1965. The Agreement with Russia regarding the Barents Sea was not concluded until much later, in 2010.

Domestic legislation was put in place with the Petroleum Act 1963.[22] This remains the main statute with regard to the oil and gas activities and now regulates scientific research of the seabed and exploration of subsea natural resources other than petroleum products. The starting point of the licence system was the 1965 Decree, passed under section 3 of the 1963 Act, which authorised the first round of licences.[23] Large discoveries of oil and gas were made in 1969 in the Ekofisk field, which gave the Norwegian government the incentive to strengthen the licence terms under the 1972 Decree.[24] A number of large fields were discovered during the 1970s and 1980s under this Decree including the Statfjord, Gullfaks, Oseberg, and Troll fields. In 1985, a further Petroleum Act was passed, which continued the licensing system but included a codification of government practice under the system.[25] However, the main purpose of the 1985 Act was to regulate other aspects of petroleum activities and introduced, inter alia, a chapter on the licensee's liability for petroleum pollution damage. A further new chapter was added in 1989 on compensation rules with regard to any losses that Norwegian fishermen may suffer as a result of petroleum activities.[26]

The new Petroleum Act replaced the 1985 Act in 1996.[27] The 1996 Act introduced more flexible rules on, inter alia, the duration and geographical scope of production licences. This more flexible approach was a response to the expansion of petroleum activities out into the Norwegian Sea where the depths were much greater and the distances to infrastructure was much more. In such cases, the 1996 Act provided for more benevolent licence terms,[28] but introduced more detailed regulation in respect of decommissioning.

This wide range of government policy has led to a high degree of intervention in petroleum activities, which rest upon two main pillars. Firstly, all phases of the activities are subject to direct government control through the licence system, which means that no major activity can go ahead without government approval. Secondly, the state regulates indirectly by organising the activities through state-owned licence groups.

10.4.1 The Licence System

As in most jurisdictions, the licence system consists of, firstly, the exploration licence, the production licence, and the pipeline licence, with each licence covering different phases of the activities. Secondly, the development plan must be approved with all its connected terms in respect of the licensee's plans for the development of the petroleum deposit.

The exploration licence is regulated under Chapter 2 of the Petroleum Act and is granted for a limited period in a limited designated area. The licensee has no right to drill for petroleum, nor any right to obtain future production licences. The only right the licensee has is to conduct various types of surveys of the seabed in order to identify the potential for possible deposits.

The production licence is regulated under Chapter 3 of the Petroleum Act and gives the licensee an exclusive right to exploration, exploration drilling, and production of petroleum within the limited area for a limited period. Although the licensee is given certain rights within the production licence, subsequent activities are subject to government regulation and control including the development plan, which must be approved by the Ministry under section 4-2 of the Act. The contents of the plan are subject to extensive requirements, which make it possible for the government to control all important aspects of the development, including the efficient recovery of the petroleum in place, the tie-in of the development to existing or future infrastructure, and safety and environmental aspects. It is at this point that all the considerations under resource management must be taken into consideration as a substantial degree of regulatory flexibility will be lost once the licensee has made his investments.[29] The production licence does not give the licensee the right to build or operate a pipeline as that requires a different licence under the Petroleum Act section 4-3. Pipelines are normally constructed to transport oil from several fields due to the high investment costs, and therefore pipeline licences are awarded to all the field owners transporting petroleum through the pipeline. This means that the ownership structure of the pipeline corresponds to the ownership structure of the co-mingled stream of petroleum. This is to encourage the pipeline owners/shippers to charge low transportation tariffs.

Finally, the abandonment plan for decommissioning is the last element in the licensing system. This is regulated by Chapter 5 of the Petroleum Act and requires the operator to submit such a plan to the Ministry for their decision prior to the expiration of a production licence or the end of use of a fixed installation operated under licence. The Ministry may decide among the options permitted under international law as discussed earlier.

All agreements that licensees enter into with each other are subject to government approval, in particular, those involving unitisation agreements under section 4-7 of the Act, the transportation and processing agreements under section 4-8 of the Act, and also the gas sales agreements. As Norway has developed into a major gas producer with connected infrastructure to European markets, the Norwegian legal framework has developed certain characteristics that distinguish it from other jurisdictions as follows:

- Licence groups have been organised by the government since the third licence round in 1974. The government has also appointed an operator for each licence group, who is responsible for the practical management of licence activities on behalf of the group.
  Formally, a production licence is granted to each individual company of the group with one of the standard terms being that the company enters into a Joint Operating Agreement (JOA) with the other members of the group. Since 1974, these agreements have been made by the Ministry, with the terms being updated with each round.
- From the outset, the state has been closely connected with comprehensive participation in the groups. The Norwegian State Oil Company, Statoil, was established in 1972[30] with the purpose of increasing state revenues, increasing state influence in the activities, and increasing the 'know-how' compared to what could have been achieved through the normal licence and tax systems. However, in 1985, Statoil was reorganised with a financial arrangement being established between the state and Statoil whereby Statoil's participating interest was split into a Statoil economic share and a state economic share, the State Direct Financial Interest (SDFI). Since this was only a financial arrangement between Statoil and the state, Statoil managed the SDFI in the licence groups on behalf of the state. However, these changes since 1985 mean that state participation has less significance than before as an instrument for state influence on the activities of the groups but it is still important in respect of the 'total government take' together with taxes, royalties, and fees.
- In 1987, a gas negotiating committee (GFU) was established. This committee consisted of three Norwegian companies, and under the leadership of Statoil, the GFU negotiates supply contracts with foreign buyers. The contracts do not relate to individual fields but to the whole of the Norwegian continental shelf and are signed by the GFU companies as sellers, but under the Ministry's approval. The Ministry then allocates contract volumes among producing fields on the advice of a special supply committee (FU). The FU was established in 1993 and consists of all the major gas producers including foreign oil companies.

10.4.2 Decommissioning

The Norwegian continental shelf is characterised by deepwater and severe weather conditions, which make the removal of offshore installations technically difficult and very costly. These conditions are very different from those in the Gulf of Mexico and most other offshore petroleum fields other than the UK continental shelf. These conditions also raise issues of technical feasibility and safety of personnel and other users of the marine environment.

Norway has acceded to the International Monetary Organisation (IMO) Guidelines and has ratified the Oslo/Paris (OSPAR) Convention and is therefore bound by this international framework. The Petroleum Act Chapter 5 transposes the international requirements into Norwegian law. Chapter 5 of the Petroleum Act establishes a decision-making process based on the principle of a case-by-case evaluation, starting with the abandonment plan prepared by the licensee. On the basis of the plan, it is the MPE who decides on both the removal and disposal issues. This provides for a more flexible and case-specific control by the government of the decommissioning phase. It is an important characteristic of the Norwegian licensing system that the government has a major degree of control over the whole system. A further characteristic is that of direct state participation administered by Statoil. This is supplemented by the allocation of costs of removal of installations under the Removal Cost Act, under which the state pays a direct share of the total costs of removal, in addition to the share incurred as a result of the SDFI involvement. However, the removal costs incurred by the other licensees are not deductible for tax purposes, therefore excluding removal costs from the ordinary tax system. This also has the effect that every abandonment case must be presented to Parliament as a budgetary matter.

The abandonment plan is provided for under section 5-1 of the Petroleum Act under which both production installations and pipelines are subject to the plan. Therefore, this means that under the geographical scope of the Petroleum Act section 1-4, installations on land that are functionally connected to the offshore activities are also subject to the plan. However, the Ministry of Petroleum has limited the options in relation to these installations. Under section 5-3, paragraph six, the Ministry of Petroleum can only decide on 'continued use in petroleum activities, including State takeover for such purposes. All alternatives are covered by other authorities depending on the relevant legislation'.

The licensee must present the plan to the Ministry of Petroleum no earlier than 5 years, but no later than 2 years prior to either of two trigger events. These are (1) the permanent disuse of the installation or (2) the expiry of the licence, whichever event comes first. Even though any number of alternative disposals may be presented with their evaluation, the licensee must recommend the preferred option supported by evidence.

If the decision is that the installation should be left in place, a special alternative is that the state takes over the installation as provided for under the Petroleum Act section 5-6. This option only applies to installations permanently placed on the seabed and not to mobile ones. This distinction is based on a functional criterion, whereby, if the installation is designed to serve only one petroleum field, the installation is permanently placed; however, if it is designed to serve several petroleum fields, it is mobile. If the installation is placed on the continental shelf, then it is for the King[31] to decide to what extent compensation shall be given for the takeover. If the installation is placed on land, or on that part of the seabed subject to private property rights, then the licensee can claim full compensation under the law.

10.4.3 Liability

There are four liability regimes provided for under the Norwegian Petroleum Act that deal exclusively with petroleum activities.

1. Section 10-9 imposes vicarious liability on the licensee.
2. Chapter 8 provides for special compensation rules with regard to losses that Norwegian fishermen suffer as a result of petroleum activities.
3. Chapter 7 regulates liability for pollution from petroleum activities.
4. Section 5-4 regulates liability for the abandonment phase.

In addition, these specific liability rules are supplemented by the normal tort law rules and the liability regulation under the Maritime Act and the more general Pollution Act that would apply in certain areas of petroleum activities.

The Maritime Act regulates tort issues related to ships and vessels, which include those operating in the petroleum sector, including mobile drilling platforms.[32] The Pollution Act regulates tort issues relating to all pollution incidents including those arising from petroleum activities. However, it is the Petroleum Act that regulates most of the liability issues that arise from petroleum activities; the Act contains certain specific characteristics, which may deviate from Norwegian tort law.

Although the liability regimes differ in structure, the one constant factor is that it is the licensee who remains the main responsible party because licences are granted to a group of licensees and one of those licensees is appointed as the operator who performs most of the petroleum activities on behalf of the group. However, under the Petroleum Act section 5-4, the licensee is not the only main party responsible as this provision has certain characteristic features that separate it from the other liability regimes under the Act.

As with a number of other jurisdictions, liability in the petroleum sector complies with the principle of 'channelling'. This is where the practical work is carried out by a number of subcontractors other than the licence group, which may in effect dilute the liability of those entities who have the greatest economic interest in the enterprise, other than the state. As a result, the Petroleum Act prescribes that each licensee must ensure that anyone working for the licence group complies with the Act, including the regulatory decisions taken under the Petroleum Act.[33] The licence group are financially strong entities and are able to obtain adequate insurance cover for their activities on the open market and so will be better placed to pay any compensation awarded to a damaged third party.

Vicarious liability is covered under the Petroleum Act section 10-9 and covers the liability for damage caused by a person, either legal or natural, who performs work for the licensee. The scope of the liability is both wide and several:

- The general liability for damage caused to third parties as a result of petroleum activities. 'Petroleum activities' are described under section 1-6 of the Petroleum Act to cover all activities connected to the development of petroleum fields on the Norwegian continental shelf. These cover exploration, exploration drilling, production, and abandonment. The term not only covers the main activities in each phase but also the related land-based activities such as planning and the building of installations. However, section 10-9, although generally formulated, does not apply to losses resulting from petroleum pollution and to losses that Norwegian fishermen suffer as a result of petroleum activities.
- The licensee is liable for anyone working or performing work or services for him. This covers all contractors and subcontractors and their employees. A direct contractual relationship with the licensee is not necessary as long as they are part of the hierarchy of entities performing work related to the relevant licence. In this respect, vicarious liability is much wider than under the normal vicarious liability under Norwegian law.
- Although the contractor's liability may be either strict or based on negligence, the licensee's liability is independent of this.

Liability for pollution damage caused by petroleum activities is regulated by Chapter 7 of the Petroleum Act. The scope is regulated through a functional definition of the term *pollution damage* under section 7-1. Accordingly, the provisions only apply when damage or loss is caused by pollution resulting from leakage or discharge of petroleum from an installation or well. This includes the costs of reasonable measures taken to prevent or remedy such damage or loss. The definition does not define the damage or losses that may result from such leakage or discharge of petroleum, or the injured parties, but such damage or loss does include lost fishing opportunities for fishermen.

In addition, the geographical scope is subject to special regulation under section 7-2. This section relates to where the damage occurs rather than where the damage originates. Therefore, the installation that caused the damage may be situated anywhere within Norwegian jurisdiction in addition to an installation situated outside the Norwegian continental shelf. If the installation were situated outside the Norwegian continental shelf, such cases, subject to the liability regime, would mainly seek to protect Norwegian interests.[34] However, this does not apply to the territory of a state that has acceded to the Nordic Convention on Environmental Protection of February 19, 1974.

Section 7-3 of the Petroleum Act imposes strict liability on the licensee for petroleum pollution damage, which includes the channelling provisions. Only if the operator fails to satisfy a claim may it be directed towards the other licensees. The liability of the licensee is, in principle, unlimited, subject to the discretionary power of the court to reduce the liability, partly or in total, if *force majeure*, or other similar event contributed to the petroleum pollution damage.

Liability during the abandonment phase is regulated under section 5-4 of the Petroleum Act which, although overlaps with the liability regime as described earlier, also involves other principles.

As stated earlier, it is the MPE who makes the decision as to whether an installation should remain in place or be completely removed. Should the Ministry decide that the installation should stay in place and be used for petroleum activities by the licensee, and if the Ministry then decides to take over the installation under section 5-6 of the Petroleum Act and continue petroleum activities, then the liability regime as described earlier prevails.

The alternatives to the points mentioned earlier are partial or total removal. It is the licensee, the owner, and the user who are the designated parties responsible for the implementation of the Ministry's decision under section 5-3 of the Petroleum Act. Section 5-4 simply states that it is the responsible party under section 5-3 that is liable for damage caused during the implementation of the decision, provided that such damage resulted from negligence or willful misconduct by a responsible party. Inasmuch as several entities may be responsible for implementing the Ministry's decision, several entities may be liable under section 5-4, which makes them jointly and severally liable. The scope of the liability is unlimited.

Under section 10-7, the Ministry requires security from the licensees for their fulfillment of economic obligations resulting from the petroleum activities, including economic obligations to third parties. This duty is particularly important during the decommissioning stage when licensees are winding up their operations.

10.4.4 Suppliers of Goods and Services to the Norwegian Offshore Sector

Offshore development projects in Norway are, like other jurisdictions, never built sequentially, starting 'at one end' and completing the various tasks in a sequential order. The installation is normally split into packages, planned, and constructed in parallel. By this method, time is saved. Inasmuch as the net present value of the final year of production is drastically lower than that of the first, time is of essence. Separate contracts governing each of these phases covering the functions relating to all or several[35] of the packages are signed. Most suppliers under most of these contracts use subcontractors for various aspects of the contract. Therefore, there is a hierarchy of contracts within a development project. As this creates a complex and closely linked bundle of contracts, Norway has developed standard form contracts such as the Norwegian Fabrication Contract 1992 (NF92), which is the most widely used set of conditions for offshore fabrication and is also used 'back to back' for the subcontractors.

Should any dispute be raised under the contract, the regulation ensures that any problems with disputes are not permitted to interfere with the smooth running of the contract. Once the potential disagreement has been identified, that is put on one side to be addressed separately while the work and relationship under the contract continues.

Within the contract, there are standard clauses on risk, liability, and insurance, all of which are closely interlinked. The contract deviates from ordinary tort law considerations of causation as a basis for liability for compensation and adequacy, inasmuch as the system is built on the principle of 'knock for knock'.[36] In this way, the loss is more predictable and open to a more rational insurance system.

Risks of damage are allocated to one of three 'interest zones', plus a fourth special zone related to the object of the contract.[37]

1. Under the Norwegian Fabrication Contract 1992 (NF92) Article 20.1, 'the contractor zone' regulates the situation where the contractor, his employees, or subcontractors (referred to collectively as the *contractor group*) suffer personal injury or physical damage related to the contract work. Simply put, the contractor bears the loss, irrespective of whether the loss is caused or contributed to in any way by the company or any of its employees, contractors, subcontractors, or those involved in the project (collectively referred to as the *company group*).
2. NF92 Article 30.2 (the company zone) provides for regulations that mirror those found on the contractor zone. This zone covers the entire contract hierarchy except for the parties linked to the company via the contract in question.
3. A third party suffering loss because of damage related to contract work can claim under tort law, and the contract cannot modify this. However, once the contract has been brought into the contractual relationship, the third party may be included in the company or the main contractor zone. Any remaining genuine third parties form a separate risk zone and claims arising out of loss or damage suffered by anyone other than the contractor group and the company group in connection with the work shall be allocated to a contractor at the outset.

10.5 INDUSTRY STRUCTURE

In respect of the petroleum activities on the Norwegian continental shelf, the ultimate regulatory authority is exercised by the Norwegian Parliament (Stortinget). The overall responsibility for ensuring that the petroleum activities are carried out under the regulatory framework rests with the MPE. Below the MPE, there are two further authorities, the Norwegian Petroleum Directorate (NPD) and the Petroleum Safety Authority (PSA). The main functions of the NPD relate to resource management whereas the responsibilities of the PSA obviously relate to health, safety, and the environment. Policy and legislation concerning the taxation of petroleum is within the remit of the Ministry of Finance (MoF), but the annual tax assessments are carried out by the Oil Taxation Office.

The legal basis for government regulation of the oil and gas sector is provided in the Petroleum Act. This also includes the licensing system.[38] The legal basis for taxation of offshore oil and gas activities is found in the 1975 Act Relating to Taxation of Subsea Petroleum Deposits.[39] The level of state participation in Norwegian oil and gas industry is very high, with the Norwegian state being the largest player on the Norwegian continental shelf, by way of its shareholdings in Statoil ASA, and also through the state's Direct Financial Interest (the SDFI), whereby the state participates directly in various production licences.[40]

The Norwegian offshore licensing system comprises various licences, approvals, agreements, and other mechanisms.

The production licence is the core document in the licensing system, and it grants the licensee an exclusive right to explore for, develop, and produce petroleum in any blocks covered by the licence. These licences are normally awarded through annual bidding rounds during which companies can apply for licence grants individually or in groups.

If the production licence is granted to a group of companies, then one company is appointed as the 'operator' and becomes responsible for the execution of the day-to-day management of the petroleum activities on behalf of the consortium. The licence is awarded for an initial period of up to 10 years. Within that period, a specific work programme must be fulfilled. The licence period then would normally be extended. After the initial period, an area fee would be payable based on the size of the licence acreage.

One of the conditions of the award of a production licence is that the licensee should enter into a JOA in a standard form prepared by the MPE. The JOA will govern the relationship between the companies in the group and will form the basis for the day-to-day running and management of the activities, allocation of costs, and decision-making processes. All the petroleum produced is allocated to the licensee in accordance with their participating interests.

Once petroleum has been discovered in commercial quantities, in order to develop the discovery, the licence partners must submit a plan for development and operation (the PDO) to the authorities. The PDO will set out, inter alia, the development plan of operation, estimated development costs, and a production profile for the whole field deposit. The PDO must then be approved by the MPE, and must also be presented to Stortinget if the estimated investment is more than 10 billion Norwegian Kroner.

On the basis of the PDO, the NPD will then issue annual production permits that allow the licensees to produce defined volumes of petroleum. The licensees are also required to submit a plan for decommissioning and cessation of the petroleum activities to the MPE. On the basis of this plan, the MPE will decide on how the facilities are to be disposed at the end of the term.

Construction and operation of transportation and processing facilities are subject to a plan for installation and operation (PIO), which again must be approved by the MPE. However, a PIO is not required if the facilities are already covered by PDO.

Gassled is the Norwegian offshore transportation infrastructure system and is organised as a joint venture. Gassled was historically owned by the main shippers of natural gas from the Norwegian continental shelf, but recently, ExxonMobil, Total, Norske Shell, and Statoil have announced the sale of their ownership interests to financial investors, with Statoil retaining a 5% ownership interest.[41] Standard provisions apply to access to the Gassled facilities and the system is operated by Gassco, which is a 100% state-owned company that is not a shareholder in Gassled. Gassco exercises its operator duties as an independent system operator (ISO) under the provisions of the Petroleum Act and in accordance with an operator agreement with Gassled.

The Norwegian domestic downstream gas market is not yet mature and is considered to be an emergent market under the Second Gas Directive. The unbundling of the downstream gas market has therefore not been considered to be a pressing issue in Norway.

10.5.1 Gas Trading in Norway

All gas produced on the Norwegian continental shelf is the property of the licensees of any particular block. They are obliged to market, transport, and sell their gas, with gas sales being carried out on a bilateral basis. There is little financial trading in natural gas in Norway and any trading is usually carried out on a physical basis.

Although gas volumes are increasingly being sold on short-term contracts, most Norwegian natural gas is sold on long-term take or pay contracts mainly to the United Kingdom and other European countries. Typically, such long-term contracts are for a period of 20 years or more. Traditionally, most of these long-term contracts are based on price formulae linked to other energy sources, mainly oil prices. However, recently there has been a move towards the use of more gas indices in long-term contracts. Short-term contracts are typically concluded on the basis of market standards such as the National Balancing Point[42] and the European Federation of Energy Traders contracts. Licensees are required to submit to the authorities quarterly information regarding their gas delivery obligations, including information on volume profiles and the main contractual terms.

As there is no gas hub trading in Norway, there is no downstream balancing regime as such. However, in the upstream sector, Gassco is responsible for balancing the inlet and outlet of natural gas and for maintaining the pressure necessary in the transportation system. If they wish to use the upstream system, then shippers need to reserve capacity with Gassco. They will need to provide information on volume as required under the access regime (see later), but there are no particular notification requirements with regard to the contractual volumes.

10.5.2 Third-Party Access Regime to Gas Transportation Networks

The Petroleum Act Regulation June 27, 1997, no. 653 regulates access to the Norwegian upstream pipeline system. This follows the principles of third-party access laid down in the upstream pipeline provision in the Second Gas Directive. The New Gas Directive is yet to be implemented but the MPE does not anticipate that any changes to the current upstream regime will be necessary.

'Any natural gas undertaking and eligible customers who have a substantiated reasonable need' for access to the system, such as shippers, are entitled to access to the system on non-discriminatory and objective conditions.

The access regime is operated by Gassco, as the system operator. The access regime consists of a primary and a secondary market, in which the shippers reserve capacity through Gassco in the primary market. If the spare capacity is lower than the requested reservations, then the right to use the spare capacity is allocated according to an allocation formula. In the secondary market, natural gas undertakings and eligible customers can sell their reserved capacity to other 'natural gas undertakings and eligible customers with duly substantiated reasonable need' for transportation.

This can take place either through the Gassco-organised market or by bilateral agreement.

Gassco, on behalf of Gassled, enters into a standard agreement with the shipper for the transportation and/or processing of the natural gas. There is a separate Tariff Regulation[43] that establishes the tariff and is operated on a ship-or-pay basis. There are two elements to the tariff: the capital element, which is stipulated by the MPE, and will give the investors a 7% interest on the invested capital before tax; and the operating element, which is the cost base, and is laid down by the operator. The transportation system is divided into different zones, with different tariffs applying to each zone.

Onshore, the Natural Gas Act regulates the Norwegian distribution of natural gas. This consists of a small distribution network that is linked to the Karstø gas-processing facility on the west coast of Norway. The downstream Norwegian gas market is relatively immature and therefore there is no specific legislative framework addressing transportation tariffs.

10.5.2.1 Liquefied Natural Gas and Gas Storage

There is no onshore gas storage facility in Norway. Domestic gas consumption is limited in Norway as there are only a few small-scale facilities for receiving liquefied natural gas (LNG) on the west coast. In addition, there is the Snøhvit LNG facility at Hammerfest in Northern Norway. This is a full-scale facility for exportation, and processing of gas from the Snøhvit field in the Barents Sea.

Upstream LNG facilities are provided for within the Petroleum Act under the same regulatory regime as other petroleum activities. Downstream LNG facilities are provided for under the Natural Gas Act June 28, 2002, no. 61. As part of an emerging market under Article 28(2) of the Second European Gas Directive, LNG is subject to less detailed regulations and there are no general obligations.

10.6 ENVIRONMENTAL ISSUES

In 2008, Parliament entered into a cross-party agreement that Norway would be carbon neutral by 2030. They agreed to reduce Norwegian greenhouse gas emissions by 15–17 million tCO_2e by 2020. The main policy measures agreed on were to increase the use of renewable energy sources and carbon capture and storage (CCS). As Norway is not a member of the European Union, the EU Climate Change Package must be implemented into the EEA prior to adoption by the Norwegian Parliament into domestic law.

10.6.1 Emissions Trading

Emissions trading in Norway is regulated by the Emissions Trading Scheme (ETS) Act of December 17, 2004, no. 99, and supplemented by the Regulation of December 23, 2004 relating to emission allowance trading and the duty to surrender emission allowances.[44]

Norway did not implement the EU Directive 2003/87/EC, as the Directive was seen by Norway together with Iceland and Liechtenstein, to fall outside the scope of the EEA. This means that the Norwegian ETS was a separate market for the trading of emission allowances until January 2008 when the Norwegian ETS was merged with the EU scheme, thus allowing Norwegian businesses to trade allowances within the European Union. However, banked allowances were only valid for the period 2005–2007 and could not be saved for the 2008–2012 commitment period.

The new European Union Emissions Trading Scheme (EU ETS) Directive will be fully implemented in Norway and the allowance award rules will be the same as those provided for in the Directive, with one exception.

The offshore petroleum enterprises will not be allocated free allowances. However, Norwegian businesses will be able to trade and surrender both EU Allowances and Certified Emission Reductions in addition to allowances issued directly by the Norwegian government under the Act.

Norwegian energy companies generally welcome this development so that involvement in the EU ETS becomes possible. One major energy company is exceptionally supportive, notwithstanding some perceived problems in the past, and is working hard towards establishing an acceptable carbon price.

10.6.2 Renewable Energy

The EU Renewable Energy Directive is relevant within the EEA and is to be incorporated into the EEA Agreement and, therefore, will be applicable in Norway. Even though practically all electricity generation in Norway is from renewable sources, the extended scope of the Renewable Energy Directive compared to the previous Directive 2001/77/EC to include other categories of energy generation may cause problems for Norway in meeting their targets. Norway will aim to increase its total share of gross domestic consumption generated from renewable sources from 60% to 67% by 2020. They hope to achieve this through energy efficiency measures, possibly by the use of electric and hybrid cars within the transport sector. The decrease in use of fossil fuels for heating and total decrease in energy consumption as a whole is a further aim.

Norway has huge potential for further development of electricity production from renewables, in particular, with wind power and the upgrading of existing hydropower facilities. Further investment in new and small-scale hydropower is anticipated.

A new Act came into force on July 1, 2010, which provided for the production of renewable energy offshore.[45] This Act provides for the development and operation of renewable energy production offshore with a regulatory framework for a concession regime. In practice, the first of these areas will most likely be for wind farms, and this is currently being considered by the government.

In order to facilitate the planned certificates market under the Norwegian Green Certificates Act of June 24, 2011, no. 39, proposed regulations were set up at the end of 2011. This was to implement the agreement between Norway and Sweden on June 29, 2011 for a common market in green certificates. The certificates apply to investments in electricity production from renewable energy sources as defined in the Renewable Energy Directive and are based on the principle of technology neutrality. The agreement came into force on January 1, 2012 and will provide for the implementation of the Renewable Energy Directive to be applicable in Norway through the EEA Agreement.

These targets for development of new electricity production from renewable sources necessitate a major increase in the cable capacity between Norway and the EU Member States.[46] Statnett has stated that, in addition to the need for the development of new cross-border interconnectors, this will cost 40 billion Norwegian Kroner over the next 10 years.[47]

10.6.2.1 Biofuels

The Regulation no. 922, section 3-16 of June 1, 2004, regarding limitation in use of noxious and environmentally dangerous chemicals and other products, states that a seller of fuel must ensure that at least 3.5% of the total volume sold in a year for use in road traffic consists of biofuel. This will increase to 54% when sustainable criteria for traded biofuel are set. This is based on the EU Biofuel Directive, which is considered to be relevant to the EEA Agreement and so implemented into Norwegian law.

10.7 ENERGY EFFICIENCY

Owing to the high costs of heating Norwegian houses and buildings in the cold climate, energy efficiency has always been part of the government's policy to reduce greenhouse gas emissions. The government has introduced new requirements for construction of new houses and buildings in order to reduce energy consumption for heating. It has also introduced an energy label system for all houses and buildings when they are sold. To offset this, the government has established a number of support schemes for energy efficiency that are administered by the state enterprise Enova SF. Energy efficiency is also a priority for major energy companies in Norway who are developing new technology to make the production of energy more efficient.

10.8 CARBON CAPTURE AND STORAGE (CCS)

The possible storage locations for carbon dioxide in Norway have been identified almost entirely offshore in empty or hydrocarbon producing reservoirs or aquifers. The process of implementing the European CCS Directive into Norwegian law started in 2009 and is yet to be completed, and the operation is currently regulated by either the Petroleum Act or the Act Concerning Subsea Natural Resources. In a 2011 report to the Parliament,[48] the government set out its plan for the development and deployment of CCS over the following 5 years. The 2011 report explains the government support to the Mongstad refinery project[49] with the aim of mitigating Norway's emissions of CO_2 and a clear intention to develop CCS with a view to contribute to deploy its techniques outside Norway. The Mongstad project has two stages, the first being a technology centre being constructed with start-up in 2012, where the aim is to test, verify, and demonstrate technology suitable for deployment at large-scale carbon dioxide facilities. The owners of the CO_2 Technology Centre, Mongstad DA, are currently the Norwegian State (75.12%), Statoil (20%), Norske Shell (2.44%), and Saol (2.44%). In parallel to the Mongstad project is the ongoing project of Statoil and the Norwegian state (represented by the fully state-owned entity Gassover SF) for establishing a full-scale CCS facility for the capture of carbon dioxide from the combined heat and power plant at Mongstad. It is planned that the captured carbon dioxide will be transported by pipelines offshore and stored on the Norwegian continental shelf.

There are already existing carbon dioxide storage projects in Norway at the Sleipner field and the Snøhvit field, where carbon dioxide is removed from the gas in order to meet quality specifications and is then injected into the reservoir. However, in these cases, the carbon dioxide is treated as an integrated part of the petroleum activities and is regulated by the petroleum legislation.

Major oil companies in Norway see CCS as a useful way of reducing their costs with regard to the CO_2 emissions tax. The main source of CO_2 being from gas, capturing it reduces not only the amount of emissions but also increases the quality of the gas produced. These two factors together make the operation commercially worthwhile. As captured CO_2 does not qualify as an emission, the opportunity to earn ETS credits also creates a commercial incentive to engage in CCS activity. Government support for the development of further technology in this field is both sought and given.

A CCS project from a land-based source, for example, a gas-fired power station, falls outside the petroleum legislation and therefore separate legislation is required. The focus of CCS is to develop offshore storage of CO_2 from gas-fired power stations, which are the biggest single-point emitters of CO_2 in Norway. There is little public opposition to CCS in Norway as the CO_2 will be captured from gas-fired power stations located close to the coast, transported, and stored offshore.[50] The fact that the offshore oil industry has been operating on the Norwegian CCS since the 1970s, with few serious accidents, has also contributed to the general acceptance by both politicians and the public.[51]

Although the EU CCS Directive qualifies as EES relevant, it has yet to be incorporated into the EEA Agreement. As the Directive only establishes a minimum legal framework for Member States if they wish to engage in storage activities, and does not require them to do so, the implementation would be relatively easy. Norway and Iceland have carried out pioneering work with demonstration projects[52] suggesting that the legal framework provided by the Directive is welcomed. Although it is not envisaged that the implementation of the Directive into Norwegian law will be problematic, one area which is of concern is the issue of long-term liability. The Directive prescribes that the operator of a storage site is liable for the site until responsibility has been transferred to the 'competent authority'. This can happen after a minimum period of 20 years. It is for the 'competent authority' to decide when the stored CO_2 'will be completely and permanently contained' and may agree that transfer of liability may take place before 20 years. Under the Directive, post-closure long-term liability provides for a duty to carry out monitoring, reporting, potential corrective measures, the surrender of allowances if leakage occurs, and preventative and remedial actions under the EU Environmental Liability Directive.[53] Ergo, both a comprehensive duty of action and a degree of economic liability, is imposed on the operator. However, the economic liability under the Directive only imposes an obligation on the operator to give up ETS allowances in case of leakages, which would be below the current legal standard of economic liability with regard to third parties under Norwegian law.

Under the Norwegian Pollution and Waste Act, there is a strict liability requirement to take reasonable measures to prevent, abate, and clean-up pollution, including accidental emissions and discharges. There is also strict liability for the owner or operator of a property, installation, or activity for damage and loss caused by pollution from such source. The Petroleum Activities Act imposes strict liability for any discharge of petroleum from a petroleum installation and also for any type of damage to third parties including the fishing industry. It is highly likely that these statutes will be extended to cover the implementation of the CCS Directive.

As mentioned earlier, the CCS Directive prescribes under Article 18 that the operator's post-closure liability can be transferred to the competent authority after a minimum of 20 years, subject to the fulfillment of certain conditions. However, Article 18 is contrary to the 'polluter pays' principle. In addition, there is no provision in Norwegian law where the public purse is obliged to accept the transfer of environmental liability from private commercial entities.[54] In some cases, the liability has been transferred but only when the private entity has ceased to exist or the liability is time barred. However, considering that the state owns the offshore pore space and regulates, and supervises all CCS activities, it is expected that implementation of this part of the Directive will not create an insurmountable problem.

To mitigate the effects of transfer of liability to the public purse, Article 19 of the Directive imposes requirement of financial security on potential operators in order to ensure that all obligations under the Directive can be met. However, the Directive does not define 'financial security'. In 2004, Norway adopted the Financial Security Act,[55] which does define financial security as an agreement where the ownership of an asset is transferred for the purpose of securing the fulfillment of financial obligations. The Act was the result of the implementation of the Directive 2002/47/EC and applies to agreements between public authorities, central banks, other financial institutions, and legal persons. This statute provides a large degree of autonomy to all the parties involved to make their own arrangements as to the kind of financial security to be set up and rights to use that security. This is also likely to be the situation with regard to Article 19 of the CCS Directive. However, given the potentially large sums involved, the government is likely to require the transfer of ownership of private property or bank guarantees rather than security in shares.

Norway has been using the capture, transport, and injection of CO_2 in depleted oilfields for the last 30 years in the form of enhanced oil recovery (EOR), and on the Norwegian continental shelf for the last 15 years.[56] Although the techniques for CCS are similar to those for EOR, the CCS Directive only mentions EOR in Section 20 of the preamble as enhanced hydrocarbon recovery (EHR). Also the Directive does not cover EOR processes where CO_2 is not permanently stored but there is no prohibition of this activity. Therefore, Norway will probably maintain the provisions of the Petroleum Activities Act with regard to EOR and impose the use of CCS technology where they can.[57]

ENDNOTES

1. Act 77 of 14 June 1985.
2. Act 16 of 14 December 1916, §§1 and 2.
3. Norwegian Ministry of Petroleum and Energy, *Facts 2008—Energy and Water Resources in Norway*, p. 78, www.regieringen.no/en/dep/oed/Documents-and-publications/reports/2008/fact-2008. Accessed September 2010.

4. Act 50 of 29 June 1990, §6-1. Also appurtenant Regulation no. 448 of 7 May 2002, concerning system responsibility in the power system.
5. Energy Act, §1-1.
6. The Energy Regulation no. 959 of 7 December 1990 is a Royal Decree, and subordinate regulations such as the Control Regulation no. 302 of 11 March 1999 and the Electricity Supply and Grid Services Regulation no. 301 of 11 March 1999. Other important regulatory instruments relevant to the electricity market include the Industry Concession Act 16 of 14 December 1917 and the Watercourse Regulation Act 17 of 14 December 1917, the Planning and Building Act 71 of 27 June 2008, and the Competition Act 12 of 5 March 2004.
7. Owned by the Nordic TSOs.
8. Section 8 of the System Responsibility Regulation.
9. This was formally known as *Noord Pool ASA*.
10. www.nasdaqomxcommodities.com. Accessed November 2011.
11. Granted under Section 4-1 of the Energy Act.
12. Section 4-1 of the Energy Act and Chapter 4 of the Energy Regulation.
13. Section 4-4(b) of the Energy Regulation.
14. Section 4-1(2)(2) of the Energy Act and Section 4-4(d) of the Energy Regulation.
15. Section 3-3 of the Energy Act.
16. Section 3-4 of the Energy Act.
17. Regulation no.263 of 10 March 2011.
18. Act 71 of 27 June 2008; Energy Act, Section 2-1.
19. In practice, large- and medium-scale hydropower production.
20. Regulation no. 726 of 24 June 2011. This introduces a new Chapter 4 concerning advanced measuring and control systems in the Electricity Supply and Grid Services Regulation.
21. Statnett (2010), *Netutviklingsplan*, p. 70. www.statnett.no/Documents/kraftsysternet/Nettutviklingsplaner/Statnetts%20nettutviklingsplan%2020.pdf. Accessed October 2011.
22. Act 12 of 21 June 1963 relating to exploration for and exploitation of submarine natural resources.
23. Royal Decree of 9 April 1965 relating to exploration for and exploitation of petroleum.
24. Royal Decree of 8 December 1972 relating to exploration for and exploitation of petroleum.
25. Act 11 of 22 March 1985 relating to petroleum activities.
26. Amendment Act of 9 June 1989.
27. Act 72 of 29 November 1996.
28. Arnesen, Hammer, et al. (2001) 'Energy law in Norway', Roggenkamp et al., Editors, *Energy Law in Europe*. Oxford University Press: Oxford.
29. 'Investment on the Norwegian continental shelf tends to be high due to the great water depths and tough weather conditions. There are also additional costs connected to the development of an infrastructure of pipelines' (Arnesson, pp. 729–830).
30. Decision of Norwegian Parliament of 2 June 1972.
31. Here, the Norwegian Cabinet exercises this decision on behalf of the 'King'.
32. However, the Maritime Act excludes leakage or discharge of petroleum during drilling operations under Section 507.
33. Petroleum Act, Section 10-6.
34. Chapter 7 applies if damage is sustained by Norwegian vessels, fishing gear, or Norwegian installations.
35. For example, engineering and transportation.
36. Arnesen, Hammer, 782–830.
37. Arnesen, Hammer, 782–830.
38. Act 72 of 29 November 1996.

39. Act 35 of 13 June 1975.
40. The SDFI is managed by the State-owned company Petoro AS.
41. Only the ExxonMobil transaction is approved by MPE in December 2011, and the Total, Norske Shell, and Statoil transactions are still pending an MPE approval.
42. Normally used in the United Kingdom.
43. Regulation no. 1724 of 20 December 2002.
44. The ETS Regulation.
45. Act 21 of 4 June 2010.
46. Statement by Statnett on the Renewable Energy Directive given to the Ministry of Petroleum and Energy by letter on 24 March 2009.
47. (2010) http://www.statnett.no/Documents/Kraftsysternet/Nettutvikingsplaner/Statnett %20nettutviklingsplan%202010.pdf. Accessed January 2012.
48. (2011) 'Full scale CCS' St meld nr 9 (2010–2011), www.reggjeringen.no/nb/dep/oed/dok/regpubl/stmeld/2010-2011/meld-st-9-20102011.html?id=635116. Accessed January 2012.
49. There is a combined heat and power plant at the Mongstad refinery.
50. Bugge & Ueland (November 2011), Case studies on the implementation of Directive 2009/31/EC on the geological storage of carbon dioxide. www.ucl.ac.uk/cclp. Accessed January 2012.
51. Ibid, pp. 729–830.
52. The CarbFix project in Mongstad, Norway, and Iceland.
53. Directive 2004/35/EC.
54. Bugge & Ueland (2011), Case studies on the implementation of Directive 2009/31/EC on the geological storage of carbon dioxide. www.ucl.ac.uk/cclp. Accessed December 2011.
55. Lov 26 March 2004, nr 17 om finansiell sikkerhetsstillelse.
56. Statoil uses EOR in both the Sleipner field and the Snohvit field.
57. Bugge & Ueland, pp. 729–830.

11 Energy Law in Australia

In the Southern Hemisphere, Australia is the foremost state to consider with regard to the energy sector and the environment.

11.1 INTRODUCTION

Australia is self-sufficient in energy and also enjoys a healthy export market. In 2009–2010, net exports of energy accounted for 68% of the domestic energy production, and domestic consumption accounted for the remaining 32%. In fact, Australia is the world's ninth largest energy producer, accounting for about 2.5% of world energy production and 5% of world energy exports. Given its large energy resource base, Australia will have little problem in remaining self-sufficient in energy while continuing to enjoy a healthy export market.

11.1.1 Energy Resources

Australia has a diverse energy base including both renewable and non-renewable resources. While enjoying approximately 33% of the world's uranium, 10% of the world's black coal, and almost 2% of the world's conventional gas resources, Australia has proportionately a small amount of the world's crude oil.[1] In addition, Australia has large amounts of widely distributed wind, solar, geothermal, hydroelectricity, wave power, and bioenergy resources. However, apart from hydroelectricity and wind energy, these renewable sources of energy are largely underdeveloped.

11.1.2 Energy Production

Between 2009 and 2010, Australia's primary energy production was from coal, which accounted for 61% of the total energy production; this was followed by uranium, which accounted for 19%, with gas at 12%. Although Australia produces large amounts of uranium, none of it is for domestic use—all of the uranium production is for export. Crude oil and liquefied petroleum gas (LPG) account for only 6% of the total energy production in Australia, with renewable energy sources following at a mere 2%.[2]

Energy consumption in Australia is dominated by black and brown coal, which accounts for 37% of total primary energy supply in 2009–2010, followed by oil at 35%, gas at 23%, and renewable energy sources at 5%. Although Australia's energy consumption continues to increase, the rate of growth has been slowing because of energy efficiency policies and an increase in worldwide energy needs.

11.1.3 Energy Policy

As we have seen in other jurisdictions, government policies play an important role in shaping the energy market, and these can affect both the pace of the growth in energy demand and the type of energy used. Australian policies on energy efficiency have slowed the pace of growth in domestic energy demand. Likewise, policies designed to enhance energy security may encourage the diversity of types of fuel used in an economy or the sources of the energy. We have been looking at policies in other chapters, which address environmental issues such as climate change, and seen that these target a greater uptake of renewable energy technologies.

In December 2011, the Australian government issued a Draft Energy White Paper, *Strengthening the Foundations for Australia's Energy Future*, which identified four main policy priorities for the government:

1. Enhancing energy policy through regular evaluations,
2. Furthering competitiveness and efficiency in the energy market through reforms,
3. Furthering the development of energy resources with an emphasis on gas,
4. Promoting the transition towards clean energy technologies.

The final paper was published on March 30, 2012.

Also published in December 2011 was the National Energy Security Assessment (NESA), which found that Australia's overall energy situation would be shaped by the strength of new investment in the future and the price of energy. Both of these will be influenced materially by global trends.

In addition, the Strategic Framework for Alternative Transport Fuels was also published in December 2011. This document sets out a long-term strategic framework to support the market-led development of alternative transport fuels in the context of maintaining liquid fuel security but with a move towards a low-emission economy. The Renewable Energy Target mandates that 45,000 GW h of Australia's electricity supply will come from renewable energy sources by 2020, with a carbon price being introduced from July 1, 2012.[3] This renders large emitters of carbon financially liable for their emissions.

In addition to federal policies, there are a number of state-based policies that affect energy markets, including some aimed at promoting the transition to cleaner energy technologies. The state governments can also influence the development of energy sources through the issue of exploration and exploitation and operating permits.

11.2 INDUSTRY STRUCTURE OF THE ELECTRICITY MARKET

The current structure of the electricity market in southern and eastern Australia was shaped by the industry reforms in the early 1990s, with a key element being the establishment of the National Electricity Market (NEM), which began operating in 1998. The NEM provides for market-determined power flows across the Australian Capital Territory, New South Wales, Queensland, South Australia, Victoria, and Tasmania. Neither Western Australia nor the Northern Territory is connected to the NEM because of their geographical distance from the east coast.

The NEM operates as a wholesale spot market in which generators and retailers trade electricity through a gross pool managed by the Australian Energy Market Operator (AEMO).

11.2.1 The Regulators

The AEMO was established in 2009 as an independent organisation with the remit to work in the long-term interests of the Australian energy consumers by developing markets that offer affordable, safe, and reliable energy supplies. AEMO operates with the Australian Energy Regulator (AER), which oversees the economic regulation and compliance with the national laws and rules, reports on generator-bidding behaviour in the NEM, and regulates electricity transmission and distribution networks in the NEM.

The Standing Council for Energy and Resources (SCER), as part of the Standing Committee of Officials, is responsible for developing policies related to the gas and electricity markets.

The Australian Energy Market Commission (AEMC) is responsible to the Council of Australian Governments (COAG) through SCER, while the AER is accountable to the Commonwealth Government as a constituent entity of the Australian Competition and Consumer Commission (ACCC).

A Memorandum of Understanding among the ACCC, the AER, and the AEMC guides the interactions among these three bodies and their functions within the Australian energy sector.[4]

The *Australian Energy Regulator (AER)* is Australia's national energy market regulator and an independent state authority. The AER is funded by the Commonwealth, with staff, resources, and facilities, provided by the ACCC.

The AER operates under the Competition and Consumer Act 2010 and its functions are set out in the national energy market legislation and rules. These mostly relate to energy markets in eastern and southern Australia. The functions include the following:

1. Setting the prices charged for using energy networks electricity poles and wires and gas pipelines to transport energy to customers
2. Monitoring wholesale electricity and gas markets to ensure suppliers comply with the legislation and rules, and taking enforcement action when necessary
3. Publishing information on energy markets, including the annual state-of-the-energy-market report and a more detailed market-compliance report, to assist participants and the wider community
4. Assisting the ACCC with energy-related issues arising under the Competition and Consumer Act, including enforcement, mergers, and authorisations

The AER assumed new responsibilities in July 2012 for regulating retail energy markets. The responsibilities are wide ranging and include monitoring and enforcing compliance with the National Energy Retail Law (NERL); authorising retailers to sell energy; approving retailer's policies for dealing with customers in hardship;

administering a national retailer of last resort scheme, to protect customers and the market if a retail business fails; and reporting on retailer performance and market activity.

The *Australian Energy Market Commission* (AEMC) is the rule maker and developer for Australian energy markets. As an independent, national body, the AEMC makes and amends the detailed rules for the NEM and elements of the natural gas markets. To further support the development of these markets, AEMC also provides strategic and operational advice to the Australian Government's Ministerial Council on Energy. Since its establishment in July 2005, the remit of the AEMC has expanded to include wholesale markets and transmission regulation and access to natural gas pipeline services, including gas retail functions in some jurisdictions.

In 2010, Western Australia became a participating jurisdiction in the National Gas Law. During 2012, it is expected that the AEMC will assume responsibility for the National Energy Retail Rules under Ministerial Council on Energy (MCE)'s National Energy Customer Framework package of reforms.

11.2.2 Commonwealth and State Relations

Australia is a federation of six states and two territories. Including the Commonwealth, there are nine governments. In general, it may be said that responsibility for land use decision making and hence, historically, environmental protection, has lain with state governments. The Commonwealth has no direct legislative powers in relation to the environment because in 1900, when the Commonwealth Constitution Act was passed, environmental protection was not an issue that occupied the minds of the legislators. 'The drafters of the Constitution would have emphasised the immunity of the continent, the difficulties in "overcoming" it, rather than the fragility of many of its ecosystems or the problems in managing it once it had been overcome.'[5] Since then, proposals to insert an 'environmental' head of power into the Constitution via referendum have not been pursued. Alternatively, some of the powers that the Commonwealth does possess may validly be exercised for environmental purposes.

The Commonwealth does have exclusive legislative jurisdiction over Australian External territories, and land owned by the Commonwealth within the states. In the self-governing territories, the Northern Territory, Australian Capital Territory, and Norfolk Island, ministers have been granted executive authority in respect of environmental protection and conservation, and the Territory Legislative Assemblies have power to enact their own laws with respect to those matters over which they have been granted executive authority.

Generally, it could be argued that the states can legislate on all matters not specifically reserved to the Commonwealth by the Constitution. In the event that Commonwealth legislation under one of the 'heads of power' conflicts with existing or future state legislation, the Commonwealth legislation will prevail.[6]

11.2.2.1 The Tasmanian Dam Case

The Tasmanian Parliament authorised work to begin on a dam project by enacting the Gordon River Hydro-Electric Power Development Act 1982 (Tas). The site chosen for the project lay in an area of national and world heritage significance. To stop the scheme going ahead, the federal Labor Government enacted the World Heritage Properties Conservation Act 1983 (Cth) and passed regulations thereunder that effectively prohibited any construction work in the area from proceeding without the consent of the appropriate federal Minister. The High Court of Australia upheld the validity of the Commonwealth Act under section 109 of the Constitution, and therefore, the Tasmanian Act ceased to have any effect. This case determined that the Commonwealth may validly enact domestic legislation in relation to 'external affairs' if the subject matter of the legislation is of 'international concern', or if, in an appropriate manner, it implements the purposes of any international treaty or agreement. Where legislation relies on the implementation of a treaty, then no independent requirement of 'international concern' is necessary. In this case, the validity of Commonwealth legislation designed to prevent the Tasmanian Government from constructing a dam on the Gordon River in the South West Tasmanian wilderness was upheld in so far as prohibitions contained in the legislation were appropriate to the Commonwealth carrying out its international obligations under the Convention for the Protection of the World Cultural and Natural Heritage (The World Heritage Convention).

It was previously thought that where the Commonwealth claimed to act on the basis of an international treaty, that treaty or agreement must impose obligations on the Commonwealth to take action. However, the High Court denied that there was such a necessity, even though the Convention does oblige parties to the Convention to take action to protect property of 'outstanding universal value' situated on its territory.

The Tasmanian case also clarified certain aspects of the power of the Commonwealth under section 51 (xx) of the Constitution to legislate with respect to 'foreign corporations, and trading or financial corporations'. Firstly, a corporation is a 'trading' corporation if a substantial part of its activities are trading activities, and the Tasmanian Hydro-Electric Commission was found to be a trading corporation because one of the principal functions of the Commission was the sale of electricity. Secondly, it is clear from the case that the power under section 51 (xx) is not confined to regulating the trading activities of trading corporations as non-trading activities may also be regulated so long as they are being carried out for the purpose of engaging in trading activities. This interpretation by the High Court gave the Commonwealth power to prohibit the construction of the dam and associated works, which, although not trading activities themselves, were to be undertaken pursuant to engaging in a trading activity, namely, the sale of electricity. The scope of this power is very significant as almost all major developments are carried out by trading corporations, either on their own account or on behalf of natural persons or governments.[7]

The legal response regarding the energy sector has been to concentrate on the adverse effects of energy use, such as pollution, land degradation, and ecological impacts, rather than seek to regulate the sources of energy chosen. Mining for coal, petroleum, natural gas, and uranium, as well as hydroelectric and thermal power station developments are all part of the overall scheme for energy production and are governed by their own specific legislation. The purpose of this legislation is primarily to encourage and regulate the development of such resources rather than display some wider environmental awareness.

Nevertheless, while the federal government consistently refuses to ratify the Kyoto Protocol unless the United States and developing countries do so,[8] it will be difficult to construct an energy policy that is consistent with the principles of environmentally sustainable development. In addition, the general public in Australia has been quite vociferous in its opposition to the federal government's position on Kyoto and a number of reports have been produced recommending ratification.[9] All of the major non-governmental organisations (NGOs) have rejected the Australian government's refusal to ratify, and even the Australian Catholic bishops have called on government to do so.[10] In their annual social justice statement, *A New Earth—The Environmental Challenge*, the bishops stated that Australians owe a duty to their Pacific Island neighbours to curb their excessive life style as the Pacific Islanders, survival is at risk from rising sea levels caused by greenhouse gas emissions leading to climate change.

The federal government claims that it is addressing climate change in a number of different ways, one of the most significant of which is the setting up of the Australian Greenhouse Office (AGO). This Office was established in 1998 as a separate agency dedicated to cutting greenhouse gas emissions. The Agency administers the Commonwealth government's climate change package, 'Safeguarding the Future: Australia's Response to Climate Change'.[11]

The package includes:

- The Greenhouse Challenge, which is a voluntary industry programme to reduce greenhouse gas emissions, drive continuous improvement, and enhance knowledge and understanding of the best ways of managing greenhouse gas emissions ($27.1 million)
- The Renewable Energy Equity Fund, which is an investment programme to encourage the commercialisation of research and development in renewable technologies ($ 19.5 million)
- The Renewable Energy Commercialisation Programme (RECP), which is a grant programme to support innovative renewable energy equipment and technologies ($ 29.6 million)[12]

11.3 ENERGY SECTORS

Apart from the obvious and traditional sources of energy such as coal, gas, and hydroelectric power, there is a growing involvement with 'renewable energy' sources. The term itself is defined in the Renewable Energy Authority Victoria Act 1990 (Vic) as including 'energy which comes from sources such as the sun, wind, wave, tides, the hydrological cycle, biomass, and geothermal sources'. In addition to promoting the research and development of such sources, the objectives and functions of the Renewable Energy Authority extend importantly to energy conservation.

11.3.1 Uranium Mining

Control over the mining of prescribed substances[13] associated with nuclear production is governed by the provisions of the Atomic Energy Act 1953 (Cth) and the Nuclear Non-Proliferation (Safeguards) Act 1987 (Cth), and covers all states and territories.

The Australian Nuclear Science and Technology Act 1987 (Cth) establishes an Australian Nuclear Science and Technology Organisation, the functions of which are, inter alia, to carry out research and supervise the activities of those persons who are mining, treating, or selling uranium.[14] Persons who discover prescribed substances must notify the minister,[15] and the minister can require information to be supplied about prescribed substances.[16]

Applications for mining in the Ranger Project Area in the Northern Territory are subject to special procedures to take account of the provisions of the Aboriginal Land Rights (Northern Territory) Act 1976 (Cth); section 41.[17] The proposal to mine uranium in the Alligator Rivers Region of the Northern Territory, which is where the Ranger Project Area and Kakadu National Park are situated, was the subject of a full-scale enquiry under the Environment Protection (Impact of Proposals) Act 1974 (Cth), which ultimately recommended against using the terms of the Atomic Energy Act 1953 (Cth) to control operations because of the need to introduce strong environmental conditions into the development. Although the project was ultimately authorised, it gave rise to a number of other statutes to protect the environment.

In addition, state legislation may operate concurrently with the Atomic Energy Act in the control of prescribed substances.

11.3.1.1 New South Wales

The Uranium Mining and Nuclear Facilities (Prohibition) Act 1986 (NSW) is the basis of nuclear legislation in New South Wales and it prohibits prospecting and mining for uranium and the construction or operation of nuclear reactors and other facilities in the nuclear fuel cycle.

11.3.1.2 Northern Territory

The Northern Territory legislated in 1979 to control the environmental effects of uranium mining generally in the Alligator Rivers Region (Uranium Mining (Environmental Control) Act 1979 [NT]). The Act controls mining operations in respect of all 'prescribed substances' as defined in the Atomic Energy Act 1953 (Cth) (section 4).

Authority to mine for prescribed substances may only be granted or refused by the minister if satisfied that the grant or refusal will assist in protecting the environment from harmful effects of mining (section 13 of the Act).

11.3.1.3 South Australia

In South Australia, the Roxby Downs mine is subject to special legislation, the Roxby Downs (Indenture Ratification) Act 1982 (SA).

11.3.1.4 Victoria

The Victorian Government introduced ultimate control over uranium mining in the Nuclear Activities (Prohibitions) Act 1983 (Vic) by banning entirely the exploration, mining, or quarrying of uranium or thorium, unless mined in prescribed amounts in the course of mining or quarrying for some other material. The Act also prohibits the construction or operation of facilities, including nuclear reactors, waste disposal, conversion, enrichment, fuel fabrication, and reprocessing facilities under section 8.

Public statutory undertakers are prohibited from developing nuclear energy sources, and no financial advancements may be made out of any public funds for any nuclear-related activities. Under section 3 of the Act, the objective of all these prohibitions is to assist the Commonwealth of Australia in meeting its international nuclear non-proliferation objectives.

11.3.1.5 Western Australia

The preferred approach of the government of Western Australia is to include a uranium mining franchise agreement as a schedule to a special Act of Parliament: the Uranium (Yeelirrie) Agreement Act 1978 (WA). This requires that all mining activities conform to relevant codes of practice under clause 13, and the mining concern must submit an environmental management programme in respect of the corporation's operations, which must include what measures will be taken to protect Aboriginal and historic sites.

Australia has access to abundant clean energy sources that are used for heating, electricity generation, and transportation. Building on the relative success of the Green Power programme, the federal government enacted the Renewable Energy (Electricity) Act 2000 (Cth). The object of the Act was to encourage an additional 2% of electricity generation from renewable sources by 2010. Owing to historic hydroelectric schemes in Tasmania and New South Wales, 10% of the electricity supply in Australia already comes from renewable sources even before the introduction of the Act.

Under the Act, accredited power stations are given a 1997 eligible renewable power baseline. These power stations may create renewable energy certificates (RECs) when they generate power using renewable energy sources, as defined under section 16 of the Act, that exceed the 1997 baseline.[18] Certificates may also be created by installations of solar hot water heaters installed after January 2001 and which replace non-renewable source heaters.

A generator must be registered and accredited before a certificate can be issued in relation to the power created by it. Once the installation is approved, it is allocated a unique identification code under Part 2 of the Act. Also Part 2, Division 4 of the Act provides for anybody accredited by the Australian Securities and Investment Commission or similar body may act as a broker in renewable energy certificates among renewable energy generators, liable parties, and third parties.

Under the Act, it is 'liable entities'[19] who make an acquisition from the National Electricity Market Management Commission (NEMMCO) that are required to achieve individual renewable energy targets based on their projected market share consumption.

It is expected that a market in RECs might develop where one liable entity is in possession of excess RECs and another has too few RECs to meet its obligation under the Act. An early assessment of the legislation has indicated that the Act will not deliver the hoped for 2% increase in renewable energy and currently the increase stands at 0.9%, which is not impressive.[20]

11.4 CARBON CAPTURE AND STORAGE

It is the Australian government that has jurisdiction over Commonwealth waters and the states and territories have jurisdiction over onshore areas and their coastal waters up to 3 nautical miles. The development of legislative and regulatory systems in each jurisdiction is a matter for the jurisdiction concerned.

The Commonwealth is in the final stages of developing the Offshore Petroleum and Greenhouse Gas Storage (Injection and Storage) Regulations, which will be the principal regulations for regulating offshore injection and storage operations. After the consultation period, the majority were in favour of the draft regulations that are currently being amended.

11.5 FUTURE DEVELOPMENTS

Once the Regulations are finalised, it is expected that they will cover six linked elements:

- Significant risk of a significant adverse impact test
- Declaration of a storage formation
- The site plan for greenhouse gas injection and storage
- Incident reporting
- Decommissioning
- Discharge of securities

11.5.1 Long-Term Liability Issues

Under the Offshore Petroleum and Greenhouse Gas Storage Act 2006, the Commonwealth will take over common law liabilities no less than 20 years after injection ceases, subject to the responsible Commonwealth Minister being satisfied as to any risks posed.

Once injection ceases, the title holder applies for a closing certificate. The Minister must make a decision within 5 years on whether to grant this certificate, and will only grant a certificate if the post-injection monitoring shows that the stored substance does not pose a significant risk to human health or the environment. The closing certificate will also require a prepayment by the operator of monies to fund a long-term monitoring programme. Once the closing certificate is issued, the title holder's statutory obligations cease but any common law liabilities will continue. After the closing certificate has been issued for at least a period of 15 years, and subject to the behaviour of the stored substance being predicted, the Commonwealth will take over the common law liabilities.

There is a difference in the treatment of long-term liability among all states and territories and the Commonwealth, which is currently being investigated by the CCS working group. This will have implications for any cross-boundary migration of the stored substance.

ENDNOTES

1. Australian Government Department of Resources Energy and Tourism, Bureau of Resources and Energy Economics (February 2012), 'Energy in Australia 2012', www.bre.gov.au. Accessed April 2012.

2. ABARES (2011), *Australian Energy Statistics*, Department of Agriculture Fisheries and Forests, Australia.
3. The intention is to have a fixed price for at least 3 years, and possibly up to 5 with annual increases.
4. A very good figure outlining the Australian Energy Market Governance Structure can be seen at http://www.aemo.com.au/en/About-AEMO/About-AEMO/Who-is-AEMO. Accessed April 2012.
5. Crawford (1992), 'The constitution', Bonyhady, Editor, *Environmental Protection and Legal Change*. The Federation Press: Sydney.
6. *Commonwealth v Tasmania* (1983) 158 *CLR* 1; 46 *ALR* 625; *Botony MC v Federal Airports Corporation* (1992) 109 *ALR* 321.
7. Bates (1997), *Environmental Law in Australia*, 4th edition. Butterworths: UK.
8. The government has claimed that without such ratification the Protocol will not be effective and will damage Australia's economy (Rosemary Lyster and Adrian Bradbrook (2011), *Energy Law and the Environment*. Cambridge).
9. *Incremental Electricity Supply Costs from Additional Renewable and Gas Fired Generation in Australia* (23 August 2003), http://www.originenergy.com.au/about/files/MMAReport.pdf. Accessed April 2012; Environmental Business Australia (July 2002), *The Business Case for Ratification of the Kyoto Protocol: Consultation Draft*.
10. Media release, Bishop's Committee for Justice, Development, Ecology and Peace, September 2002.
11. http://www.greenhouse.gov.au/ago/safeguarding.html. Accessed April 2012.
12. Lyster & Bradbrook, ibid, p. 86.
13. Mainly uranium.
14. Section 4.
15. Section 36.
16. Section 37.
17. Part IV of this Act enables traditional owners to prevent exploration and mining on their lands or to negotiate the terms on which it will occur.
18. Renewable Energy (Electricity) Act 2000 (Cth), Section 18.
19. A 'liable entity' is one that makes a 'relevant acquisition' of electricity that is either a wholesale acquisition or a notional wholesale acquisition.
20. Lyster & Bradbrook (2011), *Energy Law and the Environment*. Cambridge University Press: Melbourne, Australia.

12 Energy Law in India

India is considered to be one of the emerging economies. India, through her diverse religious background has, culturally, always had a healthy respect for the environment. With the ever-increasing demand for energy to support the burgeoning economy, there has been a growing consciousness about pollution and eco-imbalances. This developing consciousness has evolved into a policy to consider the five Es:

- Ecology
- Economics
- Energy
- Employment
- Equity.

To this extent, in order to comply with their international obligations under the Stockholm Conference in 1972, the Indian Government has amended its Constitution to accord constitutional sanctity to the ecosystem.[1] Article 48A of the Amendment declares, 'The State shall endeavour to protect and improve the environment and to safeguard the forests and wildlife of the country'. Article 51A(g) of the Amendment stipulates that 'It shall be the duty of every citizen of India to protect and improve the natural environment including forests, lakes, rivers, and wildlife and to have compassion for living creatures'. Therefore, the amended Constitution of India casts a fundamental duty on the state as well as every Indian citizen to preserve, protect, and maintain the purity of the environment in the country. These new constitutional provisions, together with the provisions under Part IV of the Constitution, lay down a foundation of sustainable development by outlining a blueprint of social and economic improvement. To this end, the higher judiciary in the form of the Apex Court of India has given the fundamental right to life, guaranteed by Article 21 of the Constitution, an innovative and purposeful interpretation.

Because of the lack of observance of these Articles of the Constitution by developers, the lack of political will on the part of politicians, and a show of indifference on the part of the executive to implement and enforce these Articles, the higher judiciary have assumed the role of 'Ombudsman'.[2]

12.1 LEGISLATIVE FRAMEWORK

The Indian Constitution provides for a federal polity and structure within the overall framework of a parliamentary form of government. Although the states have some degree of autonomy, the ultimate authority lies with the central government. It is Part XI of the Constitution that deals with the division of legislative powers between the Union and the states.

Parliament has the power to legislate for the whole country while state legislatures are empowered to make laws for their respective states.[3] The Constitution contains three lists and it depends on which list the subject matter is on that determines where the legislative authority lies. Under this system, there are certain subjects with respect to which Parliament has exclusive power to make laws such as, inter alia, major industries,[4] major ports,[5] oilfields and minerals, oil resources, petroleum and petroleum products, dangerously inflammable liquids and substances,[6] mines and mineral development,[7] interstate rivers and river valleys,[8] fishing and fisheries beyond territorial waters,[9] maritime shipping,[10] and atomic energy.[11] In respect of the development of energy projects, the states have legislative powers for planning,[12] gas and gas works,[13] and the regulation of mines and mineral development subject to the provisions of the Union List with respect to regulation and development under the control of the Union.[14] Electricity is a concurrent subject that finds place under entry 38 in the seventh schedule of the Constitution. Hence, at central level it is the Indian Parliament, and at state level, the state legislatures, who have the authority to legislate on the subject of electricity.

12.2 REGULATION OF ELECTRICITY

The Electricity Act 2003 was enacted by the Indian Parliament to bring about a qualitative transformation of the electricity sector in India. The Act consolidates all laws relating to generation, transmission, distribution, trading, and use of electricity. The Electricity Act and the Conservation Act 2001 primarily govern the electricity sector along with the regulations framed there under them. The Electricity Act provides for the Central Electricity Authority (CEA) at the federal level and the State Electricity Regulatory Commissions (SERCs) at the state level. There is also the Appellate Tribunal for Electricity (APTEL).

Under the Electricity Act 2003, the central and the state governments have the power to make rules in order to carry out the provisions of the Act. These comprise some elements of the Indian Electricity Rules 1956 and the Electricity Rules 2005.

The CEA, the Central Electricity Regulatory Commission (CERC), and the SERCs have been delegated power under the Act to make regulations on certain matters stipulated in the Electricity Act consistent with the Electricity Act and the Electricity Rules.

The Atomic Energy Act 1962, inter alia, regulates the generation of electricity from atomic energy and the provisions of this Act have an overriding effect over provisions of the Electricity Act in cases of inconsistency.

The Coal Mines (Nationalisation) Act 1973 governs the production of electricity generated by coal-fired thermal power plants, and therefore is important legislation in respect of the energy sector. Only those companies that are engaged in the generation of electricity, iron and steel, cement, and syngas obtained through coal gasification[15] can carry out coal mining in India for captive consumption. The Coal Mines (Nationalisation) Act 1973 permits coal mining for captive consumption for generation of electricity, while the New Coal Distribution Policy 2007 facilitates the supply of assured quantities of coal to various categories of consumers at predetermined prices. Under this policy, while the different sectors and consumers have been treated on merit, keeping in view the regulatory provisions, provision is made for the allocation of coal blocks for the generation of electricity.

12.3 THE POLICY FRAMEWORK

The central government has issued the following main policy framework:

- The National Electricity Policy 2005
- The Rural Electricity Policy 2006
- The Tariff Policy 2006
- The Hydro Policy 2008
- The Revised Mega Power Project Policy
- The Ultra Mega Power Policy
- The Jawaharlal Nehru National Solar Mission (JNNSM)
- The Three Stage Nuclear Power Generation Programme

Certain states have also published policies for electricity generation from renewable sources of energy.

With respect to foreign investment, 100% is permitted under an automatic route without the approval of the Government of India. This includes investment for the purposes of generation, transmission, distribution, and electricity trading. However, direct foreign investment in generation of electricity from nuclear energy is prohibited.

12.4 MARKET ORGANISATION

The Electricity Act unbundled the various elements of electricity production under the following heads:

- Generation—production
- Transmission—the bulk transport of electricity over the national grid's high-voltage cables
- Distribution—the transport of electricity from the national grid through local medium- and low-voltage networks for sale of electricity to the final consumer
- Trading—the purchase of electricity for resale

In India, the transmission and distribution system is a three-tier structure comprising regional grids, state grids, and distribution networks. It is the Power Grid Corporation of India Limited (POWERGRID) that discharges the functions of the Central Transmission Utility (CTU). POWERGRID is a central government–owned public sector company, which is also a transmission licensee at interstate level. The state grids are primarily owned and operated by state-owned transmission utilities (STUs) and the state distribution networks are substantially owned by state-owned distribution licensees (DISCOMS). These state grids are also interconnected with other state grids in order to facilitate transmission of power across borders. These regional grids are currently being integrated to form a national grid.

Grid management in India is geographically divided into five regions and the central government has established the load despatch centres at national as well as regional (interstate) levels. POWERGRIDs subsidiary, Power System Operation Corporation Ltd (POSOCO) discharges the functions of the national load despatch centre (NLDC) and the five regional load despatch centres (RLDCs). At intrastate level, the state governments are required to establish load despatch centres (state load despatch centres—SLDCs).

12.4.1 Generation

As of September 30, 2011, the installed capacity of electricity generation from all sources in India is 182,345 MW.

The generation of electricity in India is completely delicensed[16] and the private sector is permitted to set up coal-, gas-, or liquid-based thermal projects, hydel projects, and wind, solar, and other renewable energy-based projects of any size. But, as mentioned earlier, the use of nuclear energy for electricity production by the private sector is prohibited. Currently, only a government company is permitted to build and operate atomic energy stations.

12.4.2 Transmission

Under the Electricity Act, a licence is required to carry out the business of transmission of electricity except in rural areas designated by the state government. The Act prohibits the licensees from engaging in trading and distribution of electricity apart from the CTU and STUs. Private companies are allowed to engage in transmission of electricity but an application for a licence must be made to the SERCs for intrastate transmission.

As transmission is a regulated activity, it involves the intervention of certain players. The central government is empowered to make and modify, on a regional basis, demarcation of the country for the efficient, economical, and integrated transmission and supply of electricity. Such demarcation is made to facilitate voluntary interconnections and coordination of facilities for the interstate, regional, and interregional generation and transmission of electricity.

12.4.3 Distribution

Distribution of electricity also needs a licence under the Electricity Act. The SERCs have framed regulations for the grant of distribution licences, and applications must be made in accordance with these regulations. However, again no licence is required for the distribution of electricity in rural areas. A distribution license does not require a separate trading license and the licensee may appoint a franchisee to distribute power without the need for the franchisee to apply for a distribution license.

12.4.4 Sale of Electricity

Under the Act, the sale of electricity is permitted by any generating company subject to the cross-subsidy surcharge. In the event of a sale by a generating company to a licensee, the Electricity Act provides that such a sale can either be at a tariff to be agreed between the parties under a power purchase agreement or on the basis of competitive bidding. Both of these would need to be approved by the CERC in the case of interstate and the SERCs in the case of an intrastate sale. Sale of electricity also takes place through trading platforms such as power exchanges. However, such sales also require prior approval from the CERC in the case of interstate sales, and the SERCs in the case of intrastate sales.

12.5 RENEWABLE ENERGY POLICY

There is no comprehensive law on renewable energy in India; however, there are various statutes, policies, and directives that are relevant to renewable energy projects.

Under the Energy Act, the central government is empowered to specify energy consumption standards for notified equipment and appliances, establish and prescribe energy consumption norms and standards for designated consumers, and prescribe energy conservation building codes for the efficient use of energy.

State governments may amend the energy conservation building codes prepared by the central government to suit regional and local climate conditions.

12.6 THE NATIONAL ACTION PLAN ON CLIMATE CHANGE (NAPCC)

The National Action Plan on Climate Change (NAPCC) consists of several targets on climate change issues and addresses urgent and critical concerns through a directional shift in the development pathway and have laid out eight missions.

1. *The National Solar Mission*: The ultimate objective of this Mission is to make solar energy competitive with fossil-based energy options, and by increasing the share of solar energy in the total energy mix, it aims to empower people at the grass-roots level. A further aspect of this Mission is to launch a research and development programme facilitating international cooperation to enable the creation of affordable, more convenient solar energy systems and to promote innovations for sustained, long-term storage and use of solar energy.
2. *The National Mission for Enhanced Energy Efficiency*: The energy Conservation Act of 2001 provides a legal mandate for the implementation of energy efficiency measures through the mechanisms of the Bureau of Energy Efficiency (BEE) in the designated agencies in the country.
3. *The National Mission on Sustainable Habitats*: This Mission aims at making habitats sustainable through improvements in energy efficiency in buildings, management of solid waste, and a modal shift to public transport. It aims to promote energy efficiency as an integral component of urban planning and urban renewal.
4. *The National Water Mission*: By 2050, India is likely to be short of water. This Mission aims at conserving water, minimising wastage, and ensuring more equitable distribution and management of water resources. It also

aims to optimise water use efficiency by 20% by developing a framework of regulatory mechanisms. It calls for strategic accommodation of fluctuations in rainfall and river flows by enhancing water storage methods, rainwater harvesting, and more efficient irrigation systems such as drip irrigation.
5. The National Mission for Sustaining the Himalayan Ecosystem is to preserve the ecological security of India.
6. The National Mission for a Green India aims at enhancing ecosystem services such as carbon sinks, and builds on the campaign for afforestation and increasing land area under forest cover. This Mission is to be implemented through the Joint Forest Management Committees under the respective State Departments of Forests. It also implements the Protected Area System under the National Biodiversity Conservation Act 2001.
7. The National Mission for Sustainable Agriculture aims to make Indian agriculture more resilient to climate change by identifying new varieties of crops and alternative cropping patterns. Although there is a great deal of traditional knowledge in this area, it will also be supported by information technology and biotechnology.
8. The National Mission on Strategic Knowledge on Climate Change aims to work with the global community in research and technology development by collaboration through different mechanisms. To support this, there is a Climate Research Fund, which encourages initiatives from the private sector for developing innovative technologies for mitigation and adaptation.[17]

A further Mission is to be published, which will offer a policy and regulatory environment to facilitate large-scale capital investments in biomass-fired power stations.

Given the growing concerns for climate change and energy security, these missions will substantially increase the share of renewable energy in the country's energy mix.

According to the CERC Regulations 2010 (terms and conditions for recognition and issuance of renewable energy certificates [RECs] for renewable energy generation), a central-level agency would register renewable energy generators participating in the Scheme. These certificates would be for the equivalent of 1 MW h of electricity passed into the grid from renewable energy sources. In August 2011, the CERC set the floor price and the forbearance price for solar and non-solar RECs as applicable from April 1, 2012. These prices will remain valid until the financial year 2016–2017.

To facilitate the operation of these and other initiatives in the Missions as set out in the NAPCC, the states will play a major role. In 2009, states were required to initiate the preparation of state action plans including the creation of an institutional and operational framework for implementing the missions and aligning them with other developmental priorities of the state. To achieve this aim, the Ministry prepared guidelines for the states including the following:

- Implementing an inclusive and sustainable development strategy, protecting poor and vulnerable sections of society from adverse effects of climate change

- Undertaking actions that deliver benefits for growth and development while mitigating climate change
- Ensuring and improving ecological sustainability
- Building climate scenarios and investing in knowledge and research to reduce uncertainty and to improve knowledge about appropriate responses
- Assessing the impact of climate change on existing vulnerabilities, and identifying and enhancing risk management tools for addressing climate change
- Setting out mitigation and adaptation options and evaluating them in accordance with cost effectiveness, cost benefit, and feasibility
- Implementing state-planned and voluntary community-based adaptation while building broader stakeholder engagement
- Addressing state-specific priority issues, while also creating an enabling environment for implementation of NAPCC at state level
- Establishing appropriate institutional arrangements and building capacities, keeping in view co-ordination, interdepartmental consultations, stakeholder involvement, and integration with regular planning and budgetary processes
- Linking up with national policies and programmes for consistency and to identify financial and policy support that may be available

Currently the state of Karnataka has submitted their draft report to the government for approval.

12.7 WIND ENERGY

India is among the five leading nations in wind power generation.[18] The government has made available a package of fiscal and financial incentives for wind energy projects depending on the scale and capacity of the project. These incentives include concessions such as 80% accelerated depreciation, concessional custom duty on specified terms, excise duty exemption, sales tax exemption and income tax exemption for 10 years. In addition, the government is encouraging foreign investors to set up RE-based power generation projects on a build–own–operate basis.

12.8 SOLAR ENERGY

Most parts of India receive 4–7 kW h of solar radiation per square metre per day, with 250–300 sunny days per year.[19] It is therefore, not surprising that India's Prime Minister recently approved a US$19 billion plan to make India a global leader in solar energy over the next 30 years. This project will see a huge expansion of installed solar capacity while seeking to reduce the price of solar-generated electricity. The first plant was commissioned in August 2009 in West Bengal and the latest plant was commissioned in March 2011 in Andhra Pradesh.

In January 2008 and March 2008, respectively, the Ministry of New and Renewable Energy (MNRE) announced that grid-interactive solar projects could benefit from a green building initiative up to a maximum of INR 12 per kW h for solar photovoltaic (SPV) projects commissioned by December 31, 2009 and INR 11.4 for SPV projects commissioned after that date. To simplify the process, a single agency has been set up to deal with all solar power developments country wide. Any private developer will now only have to deal with one single agency for all solar projects from registration and allocation to the signing of a power purchase agreement wherever the project is situated, right across India.[20]

A National Solar Mission has been approved, and the JNNSM aims to increase the use of solar energy and make it affordable, with the objectives to be achieved via a three-stage process over the five-year plans. The immediate aim of the JNNSM is to focus on setting up an environment conducive to solar penetration in the country at all levels.

Phase I (2010–2012) focuses on promoting off-grid systems to serve consumers without access to commercial energy, in addition to grid-based systems.

Phase II (2013–2017) entails increasing capacity to create conditions for increased and competitive solar energy penetration in India.

To achieve these aims targets have been set as follows:

- To increase the capacity of grid connected solar power generation to 100 MW by 2013
- To create favourable conditions for solar power manufacturing capabilities, in particular, solar thermal power for indigenous production
- To achieve a solar thermal area of 15 million m^2 by 2017 and 20 million m^2 by 2022

The MNRE is supporting the installation of solar lights throughout all state agencies, banks, and system integrators. MNRE is also implementing a scheme to promote off-grid applications of solar energy such as solar lanterns, home lights, other small-capacity photovoltaic systems, and solar water-heating systems with a 30% capital subsidy, a loan at 5%, or both through the National Bank for Agriculture and Rural Development (NABARD).

Incentives by the states include the state of Rajasthan's policy for promoting solar energy production.

12.9 HYDROPOWER, GEOTHERMAL, AND WAVE ENERGIES

About 25% of the total installed power capacity in India is accounted for by hydropower.[21] Hydro projects up to 25 MW station capacity are categorised as small hydropower projects, which are now governed by the policies and tariffs as decided by the SERCs. There are direct subsidies granted to specific grid-interactive power programmes, while the Minister of Power is responsible for large hydro projects.

Under a 2009–2010 Indian Government programme for the research and development of new technology, money was made available for chemical sources of energy, hydrogen energy, geothermal energy, and ocean energy. These projects include the tidal power project at Durgaduani in West Bengal, and India's first geothermal power plant, with an initial capacity of 25 MW in Andhra Pradesh's Khammam district by 2012.

12.10 WASTE TO ENERGY

Under the Municipal Solid Waste (Management & Handling) Rules 2000 (the MSW Rules 2000), every municipal authority is responsible for the implementation of various provisions of the rules within its territorial jurisdiction and also to develop an effective infrastructure for collection, storage, segregation, transportation, processing, and disposal of MSW. The Rules deal with specifications for landfill sites, water quality monitoring, and ambient quality monitoring.

In August 2011, the MNRE announced that it was implementing a programme for setting up five new projects on energy recovery from municipal solid wastes.[22] However, in 2005, the Supreme Court of India, in the Almitra Patel case,[23] had prohibited the government from sanctioning any further subsidies to such plants as the waste to energy plant had not lived up to its promise. The Court ordered the government to constitute a committee comprising experts, including those from the non-government sector, to inspect the plant's record. Since then, any subsidies have been kept on hold. Nevertheless, in April 2011, the MNRE granted an administrative approval to the Programme on 'Biomass Co Generation (non-bagasse) in Industry' for implementation during 2011–2012. The objective of this programme is:

- To encourage the deployment of biomass cogeneration systems in industry for meeting their captive thermal and electrical energy requirements by supplying surplus power to the grid
- To conserve the use of fossil fuels for captive requirements in industry
- To bring about reduction in greenhouse gas (GHG) emissions in industry
- To create awareness about the potential and benefits of alternative modes of energy generation

The Minister of Petroleum and Gas announced that GAIL (India) Ltd has initiated a collaborative experimental research project with the Indian Institute of Petroleum (IIP) at Dehradun on 'Scale Up Studies for the Conversion of Waste Plastic and Low Polymer Wax to Value Added Hydrocarbons'. The Minister also announced that the objective of the project is to develop a feasible system for conversion of waste plastics and low-polymer wax/polyethylene (PE) wax to value added hydrocarbons, fuels, and petrochemicals.[24]

12.11 BIOFUELS

A new policy on biofuels was announced in 2009 that sets an indicative target of 20% blending of biofuels, both for biodiesel and bioethanol by 2017. Some subsidies may be forthcoming in the future. No subsidy will be available for projects based on municipal solid waste until the Supreme Court of India decides otherwise (see Section 12.10).

12.12 CARBON CAPTURE AND STORAGE

India ratified the Kyoto Protocol in August 2002 but only committed to be legally bound to GHG emission cuts post 2012. India does accept that it will have to adopt carbon restrictions in the future as part of their obligations to a multilateral agreement or under international duress, and so is moving towards creating a legislative mandate to facilitate this move.[25] However, so far, the government has not formulated any policies or legislation to regulate and implement carbon capture and storage (CCS). It is possible that a mandate for CCS would come within a broader legislative framework with the aim to limit India's GHG emissions. It is expected that any CCS activity would be undertaken by the government itself and with significant worldwide support. Analysts are of the view that linking CCS with clean development mechanism (CDM) is necessary before India allows the adoption of CCS under the CDM umbrella.[26]

12.13 POLICY AND THE CLIMATE CHANGE AGENDA

In response to the international obligations on climate change, the Government set up an Advisory Council on Climate Change, which is chaired by the Prime Minister. The Council has a broad-based representation from different stakeholders including, government, industry, and civil society. The Council sets out the policy directions for national actions and provides guidance on matters related to the coordination of the national action plan on the domestic agenda. The Council also reviews the implementation of NAPCC including its research and development agenda.

12.14 NUCLEAR ENERGY

Although the Government of India's policy goal regarding the development of new nuclear power plants is very positive, its current policy is to achieve the development of these new plants will be through the central government-owned public sector undertakings (PSUs). Currently, 20 nuclear reactors are in operation, which generate 4780 MW of electricity from atomic energy, and four are under construction, with two of them expected to be completed during 2011–2012. A further 20 nuclear power plants are planned by 2020.[27]

ENDNOTES

1. The Constitution (42nd) Amendment Act, 1976, added Articles 48A and 51A to the Constitution.
2. Venkat (2011), *Environmental Law and Policy*. PHI Publishing: New Delhi, India.
3. Constitution of India, Article 245(1).
4. List 1, 7th Schedule of the Indian Constitution, entries 7 and 52.
5. Ibid, entry 27.
6. Ibid, entry 53.
7. Ibid, entry 54.
8. Ibid, entry 56.
9. Ibid, entry 57.
10. Ibid, entry 25.
11. Ibid, entry 6.
12. Ibid, entry 20.
13. Ibid, entry 25.
14. Ibid, entry 23.

15. Both underground and surface.
16. Except for certain hydel projects.
17. www.indiaclimateportal.org/The-NAPCC. Accessed April 2012.
18. Bhattacharya & Dutta (2012), *Climate Regulation 2012 India.* HSA Advocates, http://login.westlaw.co.uk/maf/wluk/app/document?src=doc&linktype=ref&&context. Accessed April 2012.
19. Ibid.
20. Under the National Solar Mission.
21. Bhattacharya & Dutta, ibid.
22. Bhattacharya & Dutta (2011), HSA Advocates, India.
23. Writ Petition no. 888/1996 decided on 6 May 2005 by Honourable Supreme Court.
24. Bhattacharya & Dutta, HSA Advocates.
25. See Note 24.
26. See Note 24.
27. Sharma, Puranik, Arora. Khaitan & Co., India.

13 Energy Law in China

Another emerging economy with energy issues is China. The Chinese energy sector being based on coal, the related environmental problems are considered under Chinese government policy.

13.1 EVOLUTION OF CHINESE GOVERNMENT POLICY IN RESPECT OF THE ENERGY SECTOR

In 2007, China took three important steps towards defining its energy policy—the first being the publication of a draft new energy law on December 1, 2007; the second being the publication of its first ever energy White Paper on December 26, 2007 by the Information Office of the State Council; and the third being an announcement by an official of the State Administration of Taxation on January 10, 2008 that the Chinese government was preparing a radical overhaul of the resource taxation system.

13.2 FRAMEWORK OF ENERGY LAW IN CHINA

The new law created a much needed general framework for the energy sector in China with the aim of regulating the development, use, and administration of energy resources in China. The previous energy laws and regulations were inadequate to deal with the development of the energy sector because many of the provisions were out of date and contained numerous inconsistencies. The plethora of existing legislation included the Energy Conservation Law, the Renewable Energy Law, the Mineral Resources Law, the Coal Industry Law, and the Electric Power Law. All the current legislations needed to be updated to be consistent with the overall foundation set by the new energy law.

Two years before the publications in 2007, a task force, overseen by the National Energy Leading Group of the National Development Reform Commission (NDRC), which was made up of Chinese government ministries and commissions as well as experts, was tasked to develop the new law. The draft was published and sent out for public comment. The new law covers not only the traditional sources of energy such as coal, crude oil, and natural gas but also the less conventional sources such as wind, solar, geothermal, and biomass energy.

The intention was that the new energy law would be a comprehensive foundation for a wide range of issues affecting the energy sector. Therefore, the framework set out the law providing general guidelines but no detailed measures, which came later in the form of regulations or secondary legislation issued either by the central government, government agencies, or local legislatures.

Although China is the world's second biggest consumer and producer of energy, the per capita average of China's energy consumption is very low compared to that of other nations, and it has been a net importer of petroleum since 1993. As a result, the new law focuses on energy efficiency and security, both of which are viewed as being key to China's continued development. Consistent with China's eleventh 5-year programme, the development of the economy is also an important factor. The new twelfth 5-year programme[1] (which covers 2011–2015) established the goals of cutting China's carbon intensity by 17% by 2015.

Central to the new framework was the creation of a unified Department of Energy, which would directly report to and be under the supervision of the State Council. China divided the authority to regulate and manage the energy sector among various ministries and commissions, such as the NDRC, which is responsible for research, planning, and price setting, and the Ministry of Land and Resources, which licences the use of natural resources.

The formation of the unified Department of Energy was part of the drive to improve energy development, efficiency, supply, and security, with the new energy law providing a legislative basis for the creation of the unified entity.

Over the last decade, China has aggressively built up its strategic energy reserves, with the majority state-owned enterprises having also developed storage facilities. The new law went further by requiring oil enterprises operating in China to build their reserves (both energy resources and petroleum products) in accordance with figures already agreed with the Chinese government. In addition, energy storage used in the daily operation of the enterprise may not be counted towards its energy reserve obligation, and if an enterprise fails to meet its reserve obligation, its illegal income will be confiscated and it will be fined with an amount equal to one to five times its illegal income. Where such failure to meet the reserve is serious, the enterprise may be ordered to cease its operation or its business licence may be revoked. It is unclear whether this provision would be interpreted as being applicable solely to state-owned oil companies or also to privately owned enterprises and foreign contractors.

In respect of energy pricing, China has recently shown a willingness to move towards market pricing, but most prices remain regulated. The general principle of market access has been provided for as is the move to a free market pricing system whereby the level of government control will eventually decrease. Nevertheless, the overall pricing of energy transmission services and any energy that is important to public welfare continues to be closely regulated by the Chinese government.

13.3 RENEWABLE ENERGY

There are few specific provisions regarding non-conventional clean energies, beyond the call for a reduction in energy use to preserve the environment and recommendations that new energy resources replace the traditional energy resources; renewable energy replace fossil fuels; and low-carbon resources replace high-carbon resources. There is also provision for preferential pricing support to be given to assist the establishment of non-conventional clean energy and related technology.

However, the first 'Renewable Energy Law' in China was adopted in 2005 and came into force in 2006. This new legislation gave a huge boost to the development of renewable energy and in particular to the development of energy from wind power.

This Renewable Energy Law marked a major shift in energy policy towards more market supportive policies for renewable sources and stipulates, for the first time, that grid companies are obliged to purchase the full amount of the electricity produced from renewable sources.[2] The first implementation rules for the Renewable Energy Law came in 2007 and gave further impetus to the development of wind energy. Also released in 2007 was the 'Medium- and Long-Term Development Plan for Renewable Energy in China', which set out the government's long-term commitment to renewable energy up to 2020. It also put forward national renewable energy targets, priority sectors, and policies and measures for implementation.

For the first time, the Plan had actually set a target for a mandatory market share of electricity from renewable sources. By 2010 and 2020, electricity production from non-hydro sources should account for 1% and 3% of total electricity in the grid.[3] The Plan also requires larger power producers to source at least 3% of their electricity from non-hydro renewable resources.

In 2009, the Renewable Energy Law was amended not only by reiterating priority grid access for wind farms but also by including an obligation for grid operators to purchase a certain fixed amount of renewable energy, with penalties for non-compliance. The State Council Energy Department together with the State Power Regulatory Agency and the State Council Finance Department are responsible for determining the proportion of renewable energy in the overall generating capacity and they draw up detailed implementation plans for requiring grid companies to purchase the full amount of renewable energy produced. The 2009 amendment also requires grid companies to enhance the power grid's capability to absorb the full amount of renewable power produced. However, there is a 'Renewable Energy Fund' from which the grid companies can apply for a subsidy to cover the extra costs for integrating the renewable energy sources.

The Renewable Energy Law mandates that the difference in price between the costs of electricity from renewable energy and that from coal-fired powered plants is shared across the whole electricity supply sector. To achieve this objective, there is a €0.01 cents/kWh renewable energy premium added to the costs of each kW h of electricity sold. In 2008, the premium was raised to €0.04 cents to keep up with the renewable energy development in China.

To assist with the development of wind power, in 2009 the Chinese government introduced a feed in tariff that applies to the full operational period of 20 years. However, there are four categories of differing levels of support ranging from €5.4 to €6.5 cents. This feed in tariff replaces the old dual-track system of a concession-tendering process on the one hand and the government approval process on a project-by-project basis on the other.[4]

ENDNOTES

1. (2011), '12th five year plan on greenhouse emission control'. *Guofa no. 41.*
2. Mechoir (2010), *The Chinese Renewable Energy Law*, http://horizonmicrogrid.com/node/19. Accessed April 2012.
3. See Note 2.
4. See Note 2.

14 Summary

This chapter is a summary of environmental regulation and the energy sector. It reviews the analysis in the various chapters throughout the book and draws conclusions on the current state of environmental regulation of the energy sector in addition to identifying any possible future developments.

It is now over 100 years since the Swedish Nobel Prize winner Svante Arrhenius first predicted the effects of global warming due to emissions of pollutants into the atmosphere in his paper, 'On the influence of carbonic acid in the air upon the temperature of the ground' (*Phil. Mag. & J. Sci.* 237 [1896]).

14.1 INTRODUCTION

In the first section of this book, we considered the more theoretical arguments of the environmental regulation of the energy sector with the basic conclusion that although market mechanisms are a business-friendly form of regulation, there is still room for the more basic command-and-control system, which was then considered under the international conventions. In the second section, the international regulation of the energy sector was considered under the international conventions and treaties which covered mainly oil, gas, nuclear and the environment. The third section of this book looked at how some of the major energy producers and consumers of energy had interpreted those international conventions and treaties within their own national legislation.

Economists argue that the use of economic instruments is a straightforward, superior form of regulation and the superiority of these instruments is usually expressed in terms of their economic efficiency.[1] Lawyers, alternatively, are more accustomed to discussion of the theory of regulation and have difficulty in finding any merit in the use of such instruments within the established theories.[2] Nevertheless, the use of such instruments has found favour with politicians in the United States, the European Union, and the United Kingdom. Throughout the period of research for this book, there has been a continual restatement by politicians of the move away from the command-and-control system of regulation towards the use of economic instruments. The idea would be for countries where emissions can be cut more cheaply to sell emission rights to others where it would cost more to reduce emissions. The United States argues that as climate knows no boundaries, what actually matters is the overall emissions. It is claimed that such a system could cut the costs of CO_2 reductions in the developed world by up to 40%. Poorer countries would make gains, in return, in technology and efficiency. In November 1997, the United States signed what could be a model agreement of this kind with Argentina. Britain, Germany, and the United States have drafted trading rules and a few potential future polluters, such as India, have expressed cautious interest.

The basic conclusion of this book is that the use of market mechanisms is policy relevant: economic instruments do not afford straightforward or simple solutions but force both politicians and lawyers to think about environmental issues within a broader context. All individuals interested in effective environmental regulation need to address the nature of the regulation in the light of the anticipated real and substantial effects on the environment.

The connection between environmental policy and other fields of international policy create conflicts among various interests and interest groups, with these conflicts of interest inevitably hindering both problem solving and decision making.

The choice between multilateral action and unilateral action is highlighted by the debate on the concept of subsidiarity in the European Union with the added complication of the problem of extraterritorial decisions that are made by one body for application outside its borders. There is also the problem of enforcement in a multi-state structure and the need for stable structures, rather than incremental decision making in multinational environmental programmes.

It is clear that the European Union and the United States are not enough alike to allow direct exchange of legal models and theories. Nevertheless, in both the United States and the European Union, the institutional structures are in place and with regard to environmental problems, the structures are presently undergoing new testing.

14.2 THE SIMPLE ECONOMICS OF ENVIRONMENTAL POLICIES

Economists argue that an environmental problem arises from the existence of certain resources being held in common, that is, a common resource available for public use but without a structure in place to control the nature of the use. This is seen as a missing market, that is, the nonexistence of some mechanism for allocating a common resource such as air or water among competing users. From the economist's point of view, environmental problems exist wherever there are common resources that are inadequately regulated so that users fail to consider the full costs to others of their own use of the resource.

Timothy Swanson[3] suggests that the economist's objective for setting right environmental wrongs is simply to create mechanisms that replace the missing markets. The simplest of all the approaches is to privatise all common resources, the idea being that only unowned resources suffer from over-exploitation. However, this solution creates a number of problems: society may prefer that at least some resources remain in common ownership and then there is the issue of finding some initial basis for allocating the rights in the common resource between competing users, to name but two problems.[4]

The marketable permit approach also falls within the property rights concept. It is, however, more complicated inasmuch as the regulator plays a major part by creating a surrogate market in tradable permits. The regulator must then ensure that the permits themselves are complied with. This creates the need for sophisticated monitoring procedures to ensure the confidence of other players in the market. Additionally, the question of initial allocation must be addressed by the policymakers.

An alternative market mechanism to the property-based approach is the tax-based approach, which establishes a price on the use of the common resource, the theory being that as prices of formally unpriced resources are increased, more users will come to rely on other resources. This has been amply demonstrated by the differentiation in tax between leaded and unleaded petrol and the move to the use of natural gas as a feedstock for power generation as the sulphur content of fuels is taxed in order to internalise the cost of acid rain.[5] However, the tax-based approach again is not without its problems, inasmuch as the level of taxation is difficult to adjust so that it creates the desired effect and is also politically acceptable. The Landfill Tax came into force in the United Kingdom in November 1996 and has not been without its problems in the first years of implementation.

14.2.1 The US Experience

The history of air pollution control in the United States demonstrates that a government cannot achieve clean air by simply legislating it. After two decades of attempting to improve the air quality under the Clean Air Act 1968, at least one study indicated that illness and premature death resulting from breathing polluted air cost US $40–50 billion annually.[6] The Environmental Protection Agency reported that in 1989, 96 areas exceeded their ozone air quality standard and 41 areas exceeded their carbon monoxide standard. These reports provided an indication of the widespread failure of existing legislation and regulation. As we have seen, this led the United States to enact the Clean Air Act Amendments of 1990, under which the RECLAIM programme was developed.

During the political debate on the Amendments, it was the regional interests rather than party politics that dominated the acid rain debates. The high-sulphur coal burning areas of the Midwest and Appalachia opposed the rest of the country on the grounds of costs to reduce emissions and the possibility that some plant would decide to burn low-sulphur coal mined in other areas with a resulting loss of 22,000 mining and support jobs and indirect income. This led to a proposal for a nationwide tax on sulphur dioxide emissions, with the proceeds to be directed to subsidise the most polluting utilities to help offset the costs of the technology necessary to meet the new emission limits. This found little favour with states who did not have the problems of the Midwest and Appalachia, as they believed that a state should not benefit because it is a high polluter.

On the other hand, the West had its own agenda; besides opposing cost sharing, it wanted to ensure that the programme would accommodate its expected economic growth. The East also had its problems to consider, namely that of being the primary recipient of acid rain produced in the United States. Therefore, politicians from the Eastern states were the strongest supporters of the acid rain provisions and were not sympathetic to the cost-sharing proposals.

These competing interests are not exclusive to the United States who sought to construct a market for sulphur dioxide emission allowances that could accommodate disparate growth interests and clean-up costs within a framework of baseline emission reductions. It remains to be seen whether any unforeseen political and macroeconomic factors will impair the smooth running of the programme.

14.2.1.1 The RECLAIM Programme

In spite of the perceived problems of the offset programme,[7] the South California Air Quality Management District devised the most developed emission credit trading scheme to date, thus making the Los Angeles Air Basin a pilot scheme for the rest of the United States for the implementation and enforcement of such a programme. Supporters of such a scheme in Europe would do well to follow closely and perhaps adapt the scheme to European circumstances.

The Rules for this programme are now in place, providing for strict civil liability and a full range of enforcement options for the Regulators and more extensive and stringent monitoring, reporting, and record keeping (MRR) than under the old command-and-control system. In exchange for the freedom to select their own emission control methods, RECLAIM facilities must comply with the programme's rigorous MRR requirements. These requirements for SO*x* and NO*x* are highly complex and are central to the success of RECLAIM.[8] In general, the programme's MRR requirements depend on the size of the emitting source: the larger the source, the more demanding the MRR requirements.

Monitoring compliance may prove to be more difficult under a marketable permits system than under a command-and-control system.[9] Critics contend that a regulator under command and control need only ensure that the polluter has installed the control and that it is operating correctly, whereas to enforce a marketable permits programme the regulator must ensure that the facility complies with an emission cap, and to accomplish this, the regulator must be able to establish the facilities' emissions at any time. The regulators answer these criticisms by claiming that the majority of emissions will come from major sources and RECLAIM necessarily subjects these sources to the most stringent MRR requirements in order to ensure accurate monitoring. Regulators also claim that experience suggests that non-compliance under command and control is probably quite common.

14.2.2 European Union Policy

Since the early 1970s, the European Community and the Member States have largely relied on traditional command-and-control instruments to implement their environmental policies. The old style of economic incentives, based on a curative and reactive style of environmental policy, played a minor role. The other exception to command-and-control regulation is the use of subsidies.

The Fourth Environmental Action Programme began a reorientation of EC environmental policy, placing more emphasis on economic and fiscal instruments. The Community is now focusing on negative incentives in the form of charges and taxes, thereby setting a price on the utilisation of natural resources. The Community has as of now neglected transferable permits as an alternative to charges and taxes, mainly because of the nature of Community policy development, which is largely linked to policies already existing in one or more Member States.

Since the Fifth Environmental Action Programme, there has been a more positive shift towards the use of modern economic and fiscal instruments, which can be related to a number of factors: the Single European Act and the 'polluter pays' principle; and the introduction by some Member States of charges and taxes for fostering the implementation of environmental policy objectives, which provided the key for a Community-wide coordination of the introduction of economic instruments.

The Treaty as amended by both the Single European Act and the Treaty for European Union, together with the Environment Action Programmes, set out a framework for the development of a stronger environmental policy and required the integration of environmental considerations into other Community policies, especially in view of structural change stemming from the completion of the internal market. Also, the use of economic incentives and market mechanisms for the achievement of environmental policy objectives is consistent with the philosophy of the single market.

The development of any economic instruments of environmental policy must be addressed at Community level in order to cope with trans-boundary pollution, to prevent trade distortion, and to prevent Member States from postponing any action to avoid competitive disadvantages.

The integration of environmental considerations is of particular relevance in the field of harmonisation of indirect taxation, which impinges to a certain extent on the future development of economic instruments for environmental purposes.

The initial design of a tradable permit programme was devised to accommodate the competing interests of the various stakeholders concerned. However, the problem of transposition and enforcement was hard to solve by legal means alone. In this respect, the influence of the conflicts of interest as well as the different cultural mentalities of the Member States is paramount. The European Union faces immense implementation problems, and the idea behind the predominant use of directives for EU environmental measures is that they set a target to be reached by all Member States alike, but the Member States act through their own legal systems.

The transposition of EU environmental laws into national legislation displays a wide variation.[10] Even in a federal structure with a powerful central government, absolute compulsion is not possible. If all the EU directives were implemented correctly and promptly in all Member States, there could still be differences among the various Member States' approaches to sanctioning breaches of community norms. At present, the potential (legal) influence of the European Union ends outside the courthouse of the Member States, and all efforts to vest the European Environment Agency with the power to investigate environmental problems have failed so far.

These difficulties have resulted in differences in implementation among the Member States, and with the growing liberalisation of trade with the establishment of the internal market, this only serves to highlight the considerable problems of enforcement. Companies may well decide to transfer to those Member States that are less rigorous in applying EU environmental standards.

Economic instruments could provide an answer to the enforcement problem. Taxes and charges can be enforced relatively easily and stringent MRR requirements under a tradable permit programme would be in line with the European Union move to self-regulation.

14.2.3 Economic Instruments in United Kingdom

Consequent on the EU principle of mutual recognition and with the need to preserve the competiveness of domestic industries, the governments of the Member States have come under pressure to reduce the burden of national regulation. Among the

Member States of the Community, the United Kingdom has led the way in their twin-track approach of the practice of privatisation and the commitment to deregulation and self-regulation. However, this style of control would also appear to place greater weight on private enforcement of rights, thus increasing the role of the courts. As we have seen, Prof. Daintinth claims that it is at this point that deregulation assumes the problem of 'democratic deficit' inasmuch as the legitimacy of relying on the courts for judicial policing becomes suspect. Prof. Susan Rose-Ackerman also believes that reliance on private law suits is not sufficient to protect the environment.[11]

The challenge then is to so structure market mechanisms that they become self-policing. Given strict MRR requirements and public law sanctions for non-compliance, the burden of regulation remains with the facilities involved. To further such aims, the UK government is supporting self-regulatory efforts with subsidies for such programmes as the European Management and Audit Scheme (EMAS).

However, at first blush, the experience of the UK government with market mechanisms was not a great success except in the case of the tax differential for unleaded petrol.[12] The imposition of value-added tax (VAT) on energy was met with hostility from both sides of the House of Commons; and the initial proposals for a landfill tax were substantially revised because of industry lobbying following the white paper.[13] The tax came into force in November 1996 and although there have been some initial problems concerning 'leakage from the system', until the true extent of this 'leakage' is known, it appears to be operating well.[14]

As far as market mechanisms for the control of pollution in the aqueous environment is concerned, interest has largely centred on the suggested use of an effluent tax or water pollution charge. However, the Advisory Committee on Business and the Environment considered that such a system did not take sufficient account of local quality objectives and decided that a better approach would be the use of tradable permits. It is accepted that these could not be introduced without addressing some difficulties such as 'grandfather rights', and not only the interface with integrated pollution control (IPC) but also the interface with statutory controls on water quality.

Economic instruments raise as many questions as they do answers when considering their use for the regulation of environmental problems. The economic approach identifies the environmental problem as being one of restricting the exploitation of unregulated resources by means of the allocation of rights among competing users.

Given that the solution of environmental problems is often hindered by conflicts of interest among various interests and interest groups, policymakers need to address their concerns while considering the use of market mechanisms to implement environmental policy.

The development of any economic instrument must be addressed at Community level in order to cope with trans-boundary pollution, to prevent trade distortion and to prevent Member States from postponing any action to avoid competitive disadvantage.

Accepting this, the integration of environmental considerations is of particular relevance in the field of the harmonisation of indirect taxation, where necessary for the functioning of the internal market.

Market mechanisms are a declared policy objective of both the European Union and the United Kingdom as part of the move towards deregulation, whereby detailed prescriptive rules devised by government are replaced by general standards that may express objectives rather than the means by which those objectives may be achieved.

This is not a new concept to the United Kingdom, in particular in the area of Health and Safety on offshore installations following the Cullen Report. The general approach of the withdrawal of public authorities from detailed substantive regulation is also compatible with the European Union's need to eliminate trading barriers without damaging the protection of public interests.

Given this move towards deregulation and the recognised problems of enforcement, the *quid pro quo* should be stricter enforcement of the regulations that do remain, which is in direct contrast to the declared policy of the Chairman of the Environment Agency.[15] The concept of stricter enforcement in return for fewer regulations was recognised by Barry Mcbee, the then Chairman of the Texas Natural Resource and Conservation Committee,[16] in his talk at a breakfast meeting of the Environment Committee of the Greater Houston Partnership (Chamber of Commerce).[17]

14.3 A TECHNOLOGICAL ANSWER TO CARBON DIOXIDE EMISSIONS

The injection of carbon dioxide or sulphur dioxide to depleting oilfields is not new. This has been done since the 1980s for the purpose of enhanced oil recovery (EOR), but to use depleted oil and gas fields as a means of storing the captured gases is an activity that the law had not previously addressed.

Firstly, we consider the legalities of carbon capture and storage (CCS) under international and European legislation. We also analyse the proposal to use carbon sequestration as a 'carbon sink' under the Framework Convention on Climate Change and the Kyoto Protocol. Finally, we evaluate the current proposals in respect of inclusion of any credits earned in the European Union Emissions Trading Scheme (EU ETS).

Carbon dioxide would be captured at large, stationary point sources such as fossil fuel-fired power stations, liquefied, and transported, usually by pipeline, to a storage site (in some cases beneath the continental shelf). It would then be pumped down a well into an underground reservoir rock. Here, it would be held in place by natural geological seals that prevent it from moving out of the storage site.[18]

Given that power production is responsible for over 29% of global carbon dioxide > emissions—CO_2 being the main greenhouse gas (GHG)—and that in the United Kingdom, about 70% of electricity comes from fossil fuel generation,[19] it is a key challenge for energy policy, both in the United Kingdom and internationally, to mitigate the impact of power generation based on fossil fuels. One of the main reasons for the rapidly accelerating growth in global GHG emissions is the need for the developing world to generate ever more electricity to meet the growing energy demand associated with its population growth and economic development. In the largest developing countries of India and China, much of this extra power generated is based on coal. However vital it is to increase the share of power production from renewable sources, the reality is that conventional fossil fuels remain the most widely and cheaply available source of producing electricity.

According to the Stern Report,[20] 'extensive CCS would allow . . . continued use of fossil fuels without damage to the atmosphere . . .'.

Stern also states in his report that extensive CCS would not only allow this continued use of fossil fuels without damage to the atmosphere, but also guard against the danger of strong climate change policy being undermined at some stage by falls in fossil fuel prices.

The CCS process itself falls into three distinct stages: capture of the CO_2, which may occur through post-combustion flue gas separation, through oxygen-fuelled combustion, or through pre-combustion capture. The capture stage is followed by compression of the CO_2 into a liquid or supercritical form and then transportation to the storage site. The final stage is storing the gas, which usually involves injection into underground geological formations such as hydrocarbon reservoirs or aquifers, both onshore and offshore. Other storage mechanisms such as storage by direct injection into deepwater have also been investigated.

The process of capturing CO_2 raises relatively few legal issues. In essence, these are industrial chemical processes and so the development of plants to carry this out would likely be done within the existing framework of planning, health and safety, and environmental regulations in the relevant jurisdiction. The legal issues associated with CCS begin to appear once the CO_2 has been captured and stem from the classification of CO_2 as 'waste' and the existing restrictions on the handling and disposal of waste under local and international law. The storage of CO_2 raises further questions, principally of land ownership and of short-, medium-, and long-term liabilities for stored CO_2. The final barrier that CCS faces is one which runs through all of the stages, which is the paucity of stable financial incentives for the construction of what is inevitably a costly engineering project.[21]

14.3.1 International Law with Regard to CCS on the Continental Shelf

The United Nations Convention on the Law of the Sea 1982 (UNCLOS III) is the most significant international convention regarding the marine environment. Although the Convention came into force in 1994, it was not until 1997 that the United Kingdom ratified the Convention and the United States is yet to do so. The Convention aims to govern all aspects of the marine environment including delimitation, environmental control, marine scientific research, and economic and commercial activity, but the most important of these is delimitation. Under the Convention, different areas of jurisdiction are subject to different powers and duties of the coastal state. Historically, areas of delimitation for territorial waters and the continental shelf had been subject to separate treaties but the 1982 Convention codified and consolidated previous case law and legislation. Unique to the 1982 Convention was the introduction of the exclusive economic zone (EEZ), which extended the coastal state jurisdiction out from the territorial sea to 200 nautical miles from the coastal baseline.[22] Although most maritime lawyers would recognise that the EEZ includes both the territorial seas and the continental shelf, care should be taken as each area of delimitation enjoys subtly different rights and obligations. Under the Convention, the coastal state has the right to license certain activities including those of exploration and exploitation of natural resources both under the seabed and in the column of water above. However, when exercising these rights, the coastal state does have a duty to consider all users of the marine environment, the rights and duties of other states, and the duty not to cause pollution to the territory of other states or areas beyond its national jurisdiction.[23] This duty becomes particularly important while considering the long-term sequestration of CO_2 under the seabed, and the possibility of any leakage. Should the coastal state license the sequestration of CO_2 within their EEZ (if an EEZ had been claimed and registered by that state), in accordance with its national legislation, the state may also authorise the pipelines to transport the liquid CO_2 out to the marine site. As carbon sequestration for storage purposes is relatively new, as opposed to the use of gas for EOR, the coastal state may also authorise research and development within their jurisdiction, which would include trials for CCS. In narrower marine areas, the continental shelf and any 200-nautical-mile EEZ could well overlap, and the relevant coastal states would need to come to a bilateral, or more, agreement on where to authorise any sequestration of CO_2, and the degree of liability associated with the long-term storage and possible leakage.

As technology develops for deeper sea exploration and exploitation, the possibility of use of the subsea for CCS will become more relevant. The area beyond any EEZ remains part of the high seas with its associated freedoms for all to use, but the 1982 UNCLOS declared the high seas to be the 'common heritage of mankind' and the seabed was described as 'the Area' under Part XI of the Convention.[24] Part XI also introduces a new 'Authority',[25] which was set up by Resolution I of UNCLOS III for the establishment of the International Sea Bed Authority and for the International Tribunal for the Law of the Sea. However, Part XI of the Convention only refers to 'activities in the Area' to mean all activities of exploration for, and exploitation of, the *resources*[26] of the Area.[27]

Therefore, unless it could be argued that a particular geological formation of the seabed within the Area, which would permit carbon sequestration, is a 'resource', it would appear that carbon sequestration will not be governed by the Deep Sea Bed Authority.

The London Dumping Convention (LDC) is an international convention that mainly applies to the water column rather than any subsea activity, but the 1996 Protocol, which came into force March 24, 2006, represents a major change of approach to the question of how to regulate the use of the sea as a depository for waste materials, inasmuch as it introduces a general prohibition on dumping of waste materials except for materials on an approved list (Annex 1). Problematically, carbon dioxide was not on the original approved list. However, at the first meeting under the Protocol in November 2006, clarification and amendments to facilitate and/or regulate the sequestration of CO_2 were made. This followed an earlier meeting of the Legal Working Group held under the auspices of the International Maritime Organisation (IMO) during April 10–12, which discussed the compatibility of CO_2 capture and storage in subseabed structures after the Technical Working Party meeting during April 3–7.

The Protocol does not include pipeline discharges from land, operational discharges from vessels or offshore installations, or placement for a purpose other than disposal (usually accepted to mean EOR). However, carbon storage could be considered 'a placement for a purpose other than disposal'. If carbon dioxide is 'stored', then this would be acceptable under the original Protocol. However, the operator would need to show that he will extract the CO_2 at a later date for other uses such as EOR. The Protocol also contains a stricter precautionary approach than the one in the 1972 Convention, as it requires its Contracting Parties to apply the Precautionary Principle (Resolution LDC 44(14) 1991) instead of just being 'guided by' it. CO_2 is likely to fall within its scope because it applies to the introduction into the marine environment of 'wastes or other matter'. However, the test is 'whether it is more likely than not to cause damage to the marine environment'. In the experience of the Norwegians, they consider that underground sequestration in large amounts *may,* but is unlikely to (Norwegian Research Council Project No 151393/210), cause some damage 'locally' to the atmosphere, which is covered by the United Nations Convention on the Law of the Sea 1982. But if the sequestration is injected into a geological structure in the subsoil in such a manner that it is *unlikely* to escape, such an injection would pass the 'likely' test. Given that oil companies have been using this operation for the purposes of EOR for over a decade, it is expected that 'good oilfield practice' should suffice.

Owing to the political impetus for the use of CCS and the lack of a positive provision under even the Protocol, a group of interested parties[28] submitted a proposal to amend Annex 1 to permit the storage of CO_2 in subseabed geological formations. This resolution was adopted at a meeting of the Contracting Parties in November 2006 and the amendment came into force on February 10, 2007.

The amendment provided for an eighth category of waste and other matters to be inserted into Annex 1 consisting of 'Carbon dioxide streams from carbon dioxide capture processes for sequestration'.[29] The amendment provides a basis for the regulation of CO_2 sequestration in subseabed geological formations under the Protocol permitting requirements. However, Article 6 of the Protocol provides that 'Contracting Parties shall not allow the export of wastes or other matter to other countries for dumping or incineration at sea', and so the Legal and Technical Working Group on Trans-boundary CO_2 Sequestration Issues was tasked to look into the compatibility of Article 6 with CCS activities. In February 2008, the Group met and took the view that trans-boundary transport of CO_2 would not be permitted under Article 6 but proposed a further amendment to allow it.

The Oslo/Paris (OSPAR) Convention is a regional convention that covers the North-East Atlantic maritime area and replaces the Oslo Convention for the Prevention of Marine Pollution by Dumping from Ships and Aircraft 1972 and the Paris Convention for the Prevention of Marine Pollution from Land-Based Sources 1974. The original convention was not drafted with carbon storage in mind and in 2004, the Jurists and Linguists Group of OSPAR accepted an amendment to consider the subject. The placement of CO_2 arising from operations of offshore installations is not prohibited but is regulated as a placement for scientific research. The Convention does not distinguish between ocean storage and subsoil storage; therefore, if it does not cause 'pollution', there is no prohibition under Annexes I–III. However, the precautionary principle must be considered for any substance introduced 'directly or indirectly' into the marine environment. To this end, the Commission took decisive action at their meeting in Ostend, June 2007, by adopting amendments to the Annexes to the Convention to allow the storage of CO_2 in geological formations under the seabed. They also adopted a decision[30] to ensure environmentally safe storage of CO_2 streams in geological formations and OSPAR Guidelines for Risk Assessment and Management of the activity.[31] The Commission also adopted a Decision to legally rule out placement of CO_2 into the water column of the sea and on the seabed, because of any potential negative effects. This recognises CCS as a pragmatic approach covered by a bundle of measures to reduce the amount of CO_2 escaping into the atmosphere, with the Commission producing guidelines to manage the process. Following these Decisions, it is expected that further guidance on the selection of suitable depleted oil and gas fields and deep saline formations may be forthcoming. However, these amendments to the OSPAR Convention still require ratification by at least seven Contracting Parties, but this is not expected to happen until the new European Draft Directive comes into force.[32]

14.3.1.1 The Use of Sinks and Reservoirs

The Framework Convention on Climate Change[33] was signed in 1992 and was consequent to agreement at the Earth Summit in Rio de Janeiro. The provisions include obligations by Contracting Parties to formulate, implement, and publish national and regional programmes designed to mitigate climate change and climate change effects,[34] promote the transfer of technology,[35] and promote the development of sinks.[36] Article 4(2) has been interpreted by some to include specific commitments on sources and sinks.[37] However, Article 4(2) only applies to the developed countries as described in Annex 1 to the Convention and focuses on net emissions by source minus the removal by sinks.[38]

A particular feature of the Convention is the promotion of both 'sinks'[39] and 'reservoirs'[40] for the removal of GHGs from the atmosphere. Once such a gas has been emitted into the atmosphere from a source, it may be removed from the atmosphere and stored in a 'sink' or 'reservoir'. Under Article 4.1(d), such 'sinks' and 'reservoirs' include 'forests and oceans as well as other terrestrial, coastal and marine ecosystems'.

Again the Convention itself does not make any reference to CCS as such but under Article 4.1(c) States Parties are required to 'Promote and co-operate in the development, application and diffusion of technologies; reduce or prevent anthropogenic emissions of greenhouse gases in all relevant sectors, including the energy sectors' which could be interpreted to include CCS as a technology that would comply with this provision.

However, while considering CCS, the term *emission* could prove problematic as its definition under Article 1.4 is 'the release of greenhouse gases and/or their precursors into the atmosphere over a specified area and period of time'. However, CO_2 that is captured at source will not find its way into the atmosphere and so will not be considered an 'emission'. Nevertheless, *The Kyoto Protocol* provides for State Parties to undertake 'research on, and promotion, development and increased use of, new and renewable forms of energy, of carbon dioxide sequestration technologies and of advanced and innovative environmentally sound technologies',[41] which, under the Convention, include the energy sector.[42] However, although the Protocol itself does not define sequestration technologies, it is arguable that the meaning of the Latin word *sequestrare*, 'to commit for safe keeping', could equally refer to the seabed as to terrestrial sinks as in 'forests and oceans as well as other terrestrial, coastal and marine ecosystems'.

Owing to the concerns over any migrant leakages from a storage site under the seabed, the Protocol requires that the reductions in GHGs through the employment of sinks must be accounted for in a transparent and verifiable manner.[43] In the absence of a verifiable system, there would be no incentive to install a robust monitoring system once a credit had been granted.[44]

Under the Protocol, Annex 1 countries are permitted to implement projects that reduce their emissions at source and so generate emission reduction units (ERUs) or certified emission reductions (CERs). CO_2, which is captured at the power station before transportation to a recognised 'sink' or 'reservoir', would qualify as an emission reduction.

Under the clean development mechanism (CDM), as discussed in Chapters 2 and 3, Annex 1 countries are allowed to purchase project-based ERUs from the developing nation of the partnership. Even though the only technology proscribed under the CDM is nuclear, the CDM Executive Board at its twenty-second meeting considered that a workshop was required to discuss the issue of whether CCS should be included in the CDM programme. This workshop took place at the twenty-fourth session of Subsidiary Body for Scientific and Technological Advice (SBSTA) in Bonn in May 2006 and again in Nairobi in November 2006. Neither of these meetings was able to give a definitive decision as to whether CCS should be included in the CDM. A consultation was carried out and intergovernmental and non-governmental organisations were invited to express their views on wide issues regarding CCS and the CDM.

ERUs or CERs may be earned then by Annex 1 countries that put in place technology for reducing emissions at source as in CCS, but there is still some uncertainty as to whether Annex 1 countries may earn further credits under the CDM programme.

Economic and Monetary Unions (EMUs) and CERs may be converted into allowances under the *European Linking* Directive,[45] which came into force on November 13, 2004. The EU ETS[46] provides for a European market for GHG allowances to 'promote reductions of greenhouse gases emissions in a cost-effective and economically efficient manner'.[47] The trading scheme is divided into phases, the first one running from 2005 to 2007 and the second from 2008 to 2012. This will then coincide with the Kyoto Protocol's first commitment period.

During the first phase of the EU ETS, only Certified Reduction Credits earned through CDM projects were eligible to be traded because under the Kyoto Protocol, ERUs for Joint Implementation projects would not be issued until a crediting period starting after the beginning of 2008, these ERUs were not eligible until Phase II of the EU ETS. However, if the policies had been put in place to support carbon capture back in the 1980s, when the scheme was first technically established for EOR, and the international community had been more robust in their interpretation of international Conventions,[48] then CCS would have been eligible for the first phase of the EU ETS in addition to the technology being available to fast-developing countries such as China.

14.3.1.2 Developments at European Level

In November 2006, the European Technology Platform for Zero Emission Fossil Fuel Power Plants (ETP ZEP) published the 'Strategic Deployment Document',[49] which considers CCS technology as a key element to fulfil the objective of using the EU ETS as a powerful tool for reducing GHG emissions at lowest cost to society. The ETP ZEP also considered that a failure to include CCS in the EU ETS may deter investment from industry to support the technology, as perceived returns may well be considered insufficient.

Under Article 24 of the European Emission Trading Directive,[50] Member States may extend the ambit of the Directive to 'opt-in' (with Commission approval) any activities, gases, and installations not already covered previously and it is expected that a number of Member States may apply to 'opt-in' installations that will use CCS technology in the second phase. However, if they do so, not only the installation but also the transport and storage of the CO_2 must be opted-in together as one installation. To include CCS under Article 24 will also need new monitoring and reporting guidelines for CCS to be seen as a viable candidate under Article 24.

This encouraged the European Commission to commission a number of discussion papers including one from Norton Rose, at the Energy Research Centre of the Netherlands, Amsterdam.[51]

On the basis of the information gathered, on January 23, 2008, the Commission published its 'Integrated Energy and Climate Change' package, which included a proposal for a Directive on the geological storage of carbon dioxide and amending Council Directives 85/337/EEC, 96/61/EC, Directives 2000/60/EC, 2001/80/EC, 2004/35/EC, 2006/12/EC, and Regulation (EC) No 1013/2006.

CO_2 that has been captured as a by-product of power generation would currently be classified as 'waste' in many jurisdictions. Under European law, the issue is whether the producer of the CO_2 discards or intends to discard it. This also has wider implications as the Organisation for Economic Cooperation and Development (OECD) has adopted an almost identical definition. This means that the same issue will arise in other countries such as United States, Australia, and Japan. *Prima facie* transporting captured CO_2 for indefinite storage without any intention to recover it appears to demonstrate an intention to discard. This would make the activity of CCS in the European Union subject to the EU Waste Directive, the EU Waste Shipment Regulation, and the EU Landfill Directive, under which the injection of liquid waste into landfill sites is prohibited. The one exception is where the CO_2 has an immediate and certain further use in an ongoing process of production such as EOR.[52]

With regard to transportation, the principal issue relates to the transportation across jurisdictional borders. The Waste Shipment Regulation prohibits the export of waste for disposal outside the European Union. Within the European Union, any cross-border shipment is subject to the trans-frontier shipment control system, which is based on 'informed consent'. This would require notification to the competent authorities of dispatch, destination, and transit accompanied by a consignment note containing the prescribed information.

As mentioned, any onshore storage of waste will be subject to the EU landfill legislation. This issue has already been considered by the English courts when the Court of Appeal considered it in *Blackland Park Exploration Limited v Environment Agency*.[53]

The appellant, who owned and operated an onshore oilfield, used the associated water produced during extraction of the oil for re-injection into a separate injection well for the purposes of keeping the pressure at the correct level for EOR. The issue before the court was whether a separate stream of liquid waste brought onto the site from elsewhere and injected into the oil-bearing strata by means of the same injection well was, in fact, waste for the purposes of the Landfill Regulations. The judge at first instance considered the nature of the activity and whether the site was operating for the disposal of waste into or onto land and duly declared the site a 'landfill'. The Court of Appeal agreed that the bringing onto the site liquid waste which, once deposited, did not dissipate beyond the site brought the site within the Landfill Regulations. As Regulation 9(1) prohibits the acceptance onto the site of any liquid waste, the operation by Blackland Park Exploration Limited was unlawful. The Court did accept that the re-injection of water produced during the extraction of oil was not in breach of the Landfill Regulations.

The Framework Water Directive also prohibits storage of liquid waste in onshore aquifers. If enacted in its current form, the Draft CCS Directive would radically change the whole of this regulatory framework by removing CO_2 storage from all these restrictions.

Given the potential issues with onshore storage and the time it may take to adopt the main proposals of the Draft Directive, current CCS projects are looking to offshore storage in the near term, which will be subject to the international legal instruments outlined earlier. The UK Government is already addressing a new regulatory regime offshore by creating gas importation and storage zones under the Energy Bill, which would enable the Secretary of State to grant licences for companies to carry out carbon storage, including prospecting for and developing storage sites. The proposed framework would include enforcement and any financial security required much the same as the current licensing system for the exploration and exploitation of offshore oil and gas. However, offshore projects do face trans-boundary issues while crossing jurisdictional boundaries such as access and liability. The Draft EU CCS Directive does not seek to address such issues as the Commission considers this to be a subsidiarity point, but it does propose that the Commission should act as a coordinating body to ensure consistency of approach.

14.3.1.3 Liability Issues

Liability is often thought to be a major barrier to the development of CCS projects but industry and their insurers in the hydrocarbon sector of industry have taken on the risks of operating facilities with far greater potential to cause damage. The main issue with regard to liability is the potential leakage of CO_2 some time in the distant future, long after any potential benefits to the operator have passed and the company may have changed so that it may be impossible for any applicant to seek compensation from the person who benefited from the original storage of the CO_2. In the absence of any such respondent, the liability would pass to the public purse by default.

14.3.1.4 Possible Incentives and 'State Aid'

Both the United Kingdom and Norway have recognised the need for a commercial incentive for large-scale CCS deployment. Any effective protection of the climate against the effects of GHG emissions has only recently become accepted as a goal by society and politicians alike. This has not, as yet, been translated into practical restrictions on emissions on the scale needed,[54] and therefore there is no commercial logic for private enterprise to accept the extra costs involved. This then calls for some kind of incentive that is compatible with the European rules on state aid.

Coal-fired power stations will be the primary source of CCS, and under the Coal Regulation,[55] the European Union aims to make coal an integral part of the sustainable needs of energy security in the twenty-first century. Even though Article 1 of the Regulation emphasises that the rules for the grant of state aid to the coal industry have the aim of contributing to the restructuring process, aid is to be restricted to, inter alia, costs of coal for the generation of electricity. Also, to be compatible with the proper functioning of the common market, aid may be granted to cover exceptional costs.[56] To retrofit coal-fired power stations in order that they capture the CO_2 before the release into the atmosphere could be considered an exceptional cost. The Commission was required to report on the operation of the Coal Regulation by the end of 2006 with an evaluation of the restructuring of the coal industry and its effect on the internal market. The report must also assess how much coal the European Union needs as part of the strategy of sustainable development and energy security. It is expected that proposals may be included for the amendment of the Coal Regulation with respect to its application to state aid after January 2008 and this could well consider the need of set-up costs for the establishment of a workable CCS sector. However, although during 2003–2008 £52.8 million of Coal Investment Aid was paid to maintain access to viable reserves at 12 deep mines, the scheme was closed in 2008.

The primary control of state aid in the European Union is set out in Articles 87–89 of the European Treaty and it exemplifies the balancing of competing policies, which is carried out particularly in the energy sector. According to the European Commission,[57] the principal task of state aid control is 'to ensure that State intervention does not distort the competitive situation on the market through subsidies and tax exemptions' and there is a blanket prohibition under Article 87(1). However, there are many examples of large amounts of state aid being granted in the sector under EURATOM and the now expired European Coal and Steel Convention Treaties (ECSC). Currently, aid has been channelled to the sector through preferential tariffs and long-term contracts and the main situations in which this has been acceptable are when aid is deemed to be necessary to counterbalance adverse effects of liberalisation of the energy market, or alternatively, when aid is granted for the support of renewable and environmentally friendly energy.[58] Under Article 87(2), there are a number of automatic exemptions that may prove helpful to CCS projects, while Article 87(3) lists the forms of aid that may be compatible with the common market.

In respect of environmental aid, the Commission has issued guidelines[59] in an attempt to balance the promotion of environmental protection with possible negative effects on competition. At both international and European levels, the central principle is that the 'polluter pays', which requires that the costs of measures to tackle pollution are met by those responsible for creating the pollution and should, in effect, be internalised into their production costs. In the guidelines, it is the Commission that has the duty to classify any rules for proposed aid into either investment aid or operating aid.

Although the guidelines centre on promoting renewable energy, some of the case laws highlight the difficulty of deciding the relationship among renewable energy promotion, state aids, and the internal market programme, and some of the risks involved in balancing potential conflicting objectives in energy and environmental policy. In the case of *PreussenElecktra AG v Schleswag AG,*[60] which was considering the issue of state aid, the tension between the aim of locating preferential treatment for renewable energy within a competitive framework and the growing support for environmental measures became evident. The court ruled that a statutory obligation to purchase electricity generated from renewable energy sources in Northern Germany did not constitute state aid within the meaning of the Treaty merely because it was imposed by statute.

Under the guidelines, it has been established, therefore, that if state aid does exist, it may be permissible if it promotes renewable energy sources. These may include aid compensating high investment costs and aid in line with the rules applicable to energy savings. As far as CCS is concerned, the 'high investment costs' would certainly apply. There have also been a number of state aid decisions by the Commission in which Member States have been allowed both significant and long-term support for the generation of green electricity. In 2006, the Commission approved a number of schemes for 'feed in' tariffs for electricity from renewable sources. These decisions were based on the view that the measures in each different state did constitute state aid but under the Guidelines for State Aid for Environmental Protection,[61] aid was permitted for the generation of electricity up to the difference between the market price and the generation cost of this type of electricity. If the capture of carbon from coal-fired power stations were considered to be 'the generation of green electricity', then support for the 'high investment costs' would apply. However, the Commission has also assessed aid for the reduction of GHG emissions in a number of ways, the most relevant being the National Allocation Plans for the European Union Emissions Trading Scheme. In the United Kingdom, the Climate Change Levy (CCL) also attracted the attention of the Commission.[62] Under this scheme, which the Commission approved, energy-intensive sectors were offered a rebate of 80% for a period of 10 years to adapt to the new Kyoto targets, and improve energy efficiency and cut CO_2 emissions. In addition, the Commission approved aid for a Slovenian scheme that grants reductions in CO_2 taxation to operators of combined heat and power (CHP) installations and companies that enter into voluntary environmental agreements.[63]

All new technologies aimed at the reduction of GHGs are reliant on government policy. Mandatory regulations would create a 'level playing field' within the European Union, but support schemes would give a more positive incentive to industry who may otherwise decide to take their generation activities to a less restrictive jurisdiction.

14.4 INTERNATIONAL EFFORTS TO MITIGATE ENVIRONMENTAL DEGRADATION

The decision of the arbitration panel in the *Trail Smelter* case to use negotiated regulation of states' activities at the international level was for some an undermining of state sovereignty but the decision established an important principle that no state has the right to permit the use of its territory to cause injury to others. This led to a negotiated regulatory agreement. The Montreal Protocol took this a step further by regulating the production of, and trade in, specific substances, with their eventual phasing out under command and control. This was relatively successful, as the production of particular substances from known sources within a particular industrial sector could be monitored and enforced with the aim of eventual elimination. The problem came when the Framework Convention on Climate Change aimed to merely stabilise GHGs produced from a number of disparate sources that were hard to monitor and enforce. The international community then moved to a 'market-based system', provided for in the Kyoto Protocol, to encourage these disparate sources to consider that there was a 'business case' for reducing their emissions at least cost to themselves.

In the study of one particular flexible mechanism that uses 'sinks', the Stern Review argues that, to achieve the reduction in GHG emissions required to prevent global average temperature rising by more than 2°C, public policy measures that accelerate the deployment of low-carbon energy sources will be required. He states that the key to reducing emissions to meet the targeted levels of GHG concentrations is 'the development and deployment of a wide range of low-carbon technologies'.[64] He also says that 'in some sectors—particularly electricity generation, where new technologies can struggle to gain a foothold—policies to support the market for early-stage technologies will be critical'.[65]

If CCS is to form part of the future of our energy and climate change policy, then Governments needed to press for international and European agreements on a framework for the recognition and regulation of CCS and to bring forward early proposals for a long-term storage monitoring, safety, and liability regime.[66] No commercial organisation will commit to this activity without a positive Government policy statement and some form of incentive, which brings into question the issue of European state aid rules and the internal market.

The attempts of the Commission to balance the need for environmental protection and the protection of the internal market through the state aid rules have led to some interesting results inasmuch as it has shown considerable reluctance to take action under the state aid rules against any of the National Allocation Plans in spite of some evidence of incompatibility. However, the state aid controls in the energy sector have become clearer and more predictable over the last few years through the guidelines on state aid to environmental protection, especially, with regard to aid for renewable energy and eligible investment and costs.

In the package of measures published by the Commission on January 23, 2008, new guidelines were issued on the use of state aid for environmental protection. The objective of the new guidelines was to provide increased incentives for Member Sate and industry to invest in more environmentally friendly technologies. The new guidelines will replace the 2001 guidelines and come into force immediately on publication.

The over-allocation of allowances in the first phase of the EU ETS led to a marked drop in the price for credits. However, given a more restricted allocation under Phases II and III, the Deusche Bank predicts an eventual carbon price of €35.

In addition, as long as a developed scheme to auction allowances under the National Allocation Plans for the EU ETS Phase III are in compliance with the state aid rules, then the hypothecated revenue raised could provide incentives for commercial organisations to engage in research and development to realise an economically viable CCS regime.

14.4.1 CCS under EU Legislation

Under the EU framework, a number of Directives could possibly apply. However, the Netherlands legal taskforce reported on the subject in 2001 and concluded that although CO_2 falls under the Framework Directive on Waste, it did not apply as CO_2 was not a dangerous substance. They also concluded that injection of CO_2 in the deep underground does not fall under the jurisdiction of the Directive on Dumping of Waste Materials. However, following the recognition by the Commission in its Communication of January 12, 2007[67] that coal is a key contributor to the European Union's security of energy supply and will remain so, the Commission announced in its 'Energy and Climate Change Package' adopted on January 10, 2007, that they are preparing legislative proposals that aim to establish a regulatory framework for the capture of CO_2 and its geological storage. To this end, an internet consultation was conducted and the legislative proposals to regulate CCS were published in January 2008. The new European CCS Directive is described in detail in Section 8.22.4.

Even before CCS projects are approved, two EC Directives require that a Member State government assess the effects on the environment of certain plans and programmes. Firstly, the Strategic Environmental Assessment Directive[68] requires that such an assessment is carried out and subjected to public participation. This must then be taken into account by decision makers before any authorisation. Secondly, the Environmental Assessment Directive[69] requires that the environmental impact of projects is assessed and any methods of mitigating those effects must be evaluated before authorisation. Although CCS projects are not specifically mentioned in the list of categories where a mandatory assessment is to be made, the list does include all projects connected to the oil and gas sector.

The storage phase will largely depend on whether CO_2 falls within the definition of waste or a by-product, but the Netherlands Legal Taskforce stated CO_2 is not considered to be a dangerous substance as it occurs naturally within oil and gas fields, and the Commission has now removed CCS from the EU waste legislation. If the CO_2 is used for EOR, then it would likely be considered a by-product.[70] Further EC Directives that will impact on the storage phase of CCS projects are the Habitats Directive, the Water Framework Directive, the Environmental Liability Directive, and the EU Monitoring Guidelines. The current Marine Bill will also have an effect once it is in force. The capture and transport of CO_2 are currently covered by existing provisions and so the Commission saw no reason to bring in any new regulatory framework for these activities.

The new CCS Directive will establish the objectives and general requirements necessary to ensure an overall level of environmental integrity for CCS with the details of implementation left to the Member States. However, the requirement to provide third-party access will not be welcomed by industry given the extent of major capital investment needed. As for liability, any local leakage will be covered by the Environmental Liability Directive[71] and any long-term emissions resulting in climate change will be accounted for by the surrender of emission EU ETS allowances.

14.4.2 The Norwegian Experience

Being an energy-rich country with a strong environmental ethic, Norway has a long-standing policy on CCS. There are currently three major CCS projects within the Norwegian jurisdiction, two of which the Norwegian government is collaborating. The third is an agreement between Shell and Statoil Hydro to work towards developing the world's largest project using carbon dioxide for enhanced oil recovery. All these projects need government incentives that must be compatible with EU State Aid Guidelines.[72] In March 2007, the Norwegian government introduced a bill in Parliament for the establishment of a state-owned company that will safeguard the national interest in all projects concerning CO_2 capture, transportation, and storage as part of a consortium with industrial partners.

Norway also has extensive experience in using CO_2 for EOR and has been storing CO_2 in geological structures since 1996. There is a programme for monitoring all such projects, the data from which confirm that the CO_2 is firmly confined within the storage reservoir.[73] This is in accordance with the Petroleum Act and the Pollution Control Act 1981, which requires Statoil Hydro to monitor the CO_2 storage and report to the Norwegian Pollution Control Authority annually.[74]

14.4.3 Liability for Environmental Damage Caused

The conventional solution, as already in place under the decommissioning regulations for offshore oil and gas installations, is for the operator to lodge some kind of security. The problem is that the costs for decommissioning an existing facility will be known, whereas the costs for remedying the damage caused by leaks of CO_2 from storage sites are hard to quantify. Any assessment of such costs would necessarily be based on an educated guess, and it is well known that insurers do not like to be asked to insure unlimited and unquantifiable liability. Given these difficulties, it is suggested that the liability for any leakage be transferred to the public purse. Having said this, if the public purse is to bear the costs of any future liability, then governments would require the operator to create funds in order to offset any future liability. In fact, the draft CCS Directive adopts this approach and provides for the Member State, within whose jurisdiction the storage site is located, to take responsibility for the site at a given handover date or by default if the operator is unable to perform his obligations. The draft Directive also provides for the Member State to seek security from the operator for any future liabilities.

Despite this, even if leakage of CO_2 from underground deposits into the sea occurs, it is difficult to see how this can result in a type of damage to public or private interests that may result in a compensable loss. The environmental, health, and safety risks associated with injection of CO_2 into a geologic formation have been successfully managed for well over a decade in commercial oil and gas operations.

14.4.4 The UK Perspective

Unlike Norway, the United Kingdom has been slow to appreciate the potential benefits of CCS. Since the late 1980s, studies had been carried out to use capture of both carbon dioxide and sulphur dioxide from coal-fired power stations for the purposes of EOR.[75] This work was used for a report to the Department of Trade and Industry (DTI) in 1991.[76] Had the then government designed a policy to require the oil companies to factor in EOR to their exploitation programmes, both the regulatory structure and possibly the pipelines would now be in place for the sequestration of CO_2 for storage/disposal purposes.[77] However, the current government now finds itself in the position of trying to catch up with the current policy of the European Union to support CCS.

In June 2007, the UK government published a consultation paper on the decommissioning of offshore installations on the United Kingdom continental shelf (UKCS). The paper was mainly about liability for decommissioning costs, which would need new legislation. However, the government lost an opportunity to consider decommissioning jointly with the concept of carbon capture either for the purpose of EOR or for storage in depleted oil or gas fields. As was stated in the report by the North Sea Basin Task Force,[78] 'Without near-term intervention, the required infrastructure will be removed and the opportunity for re-use for CCS lost'.

Further, the largest potential source of CO_2 in the United Kingdom is from the power generation sector, more particularly coal-fired power plants. However, the government has not taken any steps to create incentives for CO_2 emitters to capture CO_2. In addition, there is no mention of CO_2 capture and storage in the 'United Kingdom National Emission Reduction Plan for Implementation of the Revised Large Combustion Plants Directive (2001/80/EC)' published by Department for Environment, Food and Rural Affairs (DEFRA) in February 2006. With all three issues being considered within the same time frame, the opportunity for 'joined-up thinking' is evident.

The UK government should think seriously about requiring coal-fired power plants to capture CO_2 and SO_2 from flue gases under the revised Large Combustion Plants Directive and requiring the operating oil and gas companies to provide for facilities for EOR using either the CO_2 or SO_2, depending on the structure of the uneconomic field, or to facilitate the storage of these gases in depleted fields under the new legislation regarding decommissioning of offshore installations. This would then provide the 'near-term intervention' required by the North Sea Basin Task Force.

If the UK government introduced domestic legislation similar to that in Norway, including a similar liability regime, this would pave the way for an 80% reduction in UK CO_2 emissions. Alternatively, liability could be addressed in a similar manner to liability for decommissioning on the UKCS, which could include the posting of surety bonds, letters of credit, trust funds, or environmental liability insurance. However, operators would possibly need a finite liability cap. Operators and insurers would need the ability to predict the costs of their liability because if liability costs are significant, organisation and investment decisions may be influenced by a desire to minimise costs.

14.4.5 CCS and State Aid

In his report, Stern says that to achieve the necessary CO_2 emission reductions, a disproportionately large share of the burden will have to be borne by the power generation sector.[79] He also states that 'Policy to reduce emissions should be based on three essential elements: carbon pricing, technology policy, and removal of barriers to behavioural change'.[80]

With regard to the energy sector, the first two policy measures are aimed at influencing the producers, whereas the third is designed to influence consumers. Stern goes on to state that establishing a price for GHG emissions is essential '. . . so that people are faced with the full social cost of their actions'.[81] As far as very large emitters are concerned, the advantage of a trading scheme over a tax is that a trading scheme determines the quantity of emissions in advance and leaves the market to determine the price, whereas a tax sets the price in advance and the market then determines the quantity.

The national allocations of tradable allowances under Phase I of the EU ETS were free and so it could be argued that this may distort competition and interfere with the internal energy market, as it 'involves an element of State aid that has neither been formally notified to, nor cleared by, the Commission under the EC Treaty'.[82] The Guidance notes state that 'the normal State aid rules will apply', but so far the Commission has not taken any formal decision in respect of the National Allocation Plans and state aid. However, in Commission press releases[83] in 2006, it was stated that the Commission was investigating the proposed new Danish and Swedish CO_2 tax reductions based on the state aid rules. Under the 2008 climate change package, the Commission issued new guidelines for the use of state aid for environmental protection. The new guidelines provide increased incentives for Member State to adopt more environmentally friendly processes to invest in new technologies.

14.4.6 CCS and the EU ETS

The European Union is now committed to a number of targets by 2020, namely, an energy efficiency target of 20% ensuring that 20% of its primary energy consumption comes from renewable sources, and encouraging 12 large-scale CCS power plants to be built. Most forms of renewable energy are not expected to be competitive within the next decade without some form of continuing subsidies but CCS coal plants are commercially viable at a carbon price of €35/tonne.[84] To achieve its target of a 20% reduction of GHG emissions by 2020, the Commission may well have to reduce the level of the ETS cap again in Phase III. Should this happen, it is the assumption of Deutsche Bank[85] that a price of €35/tonne would be achievable over 2013–2020. Given this analysis, CCS plants would then become more economically attractive over any forms of renewable energy. In addition, Mark-C Lewis[86] expects that the Commission will interpret the supplementary criterion[87] governing the use of credits in the ETS more strictly beyond 2010 in order to reflect the increased reliance on the supply side targets for renewable energy and the construction of CCS plant, by allowing only 45% of each Member State's effort to be met by the use of Kyoto flexible mechanisms compared to 50% during 2008–2012. This will mean that the energy sector will have to assume a larger share of the burden in order that emissions are reduced significantly by 2050.

Although the Stern Report analysis and policy prescriptions are intended for a global audience, Stern also has more specific recommendations for the European Union and the ETS. He is particularly concerned with the ETS framework for Phase III in this respect. He considers that 'Decisions made now for Phase three provide an opportunity for the scheme to influence, and become the nucleus of future global carbon markets'.

14.4.7 Auctioning EU Allowances to Support CCS

Under Phase I of the EU ETS, the allowances were allocated by the Member State on a 'grandfathering' basis.[88] Under Phase II, Member States were able to auction up to 10% of the allowances. At a meeting with the UK Trade and Investment, Mark-C Lewis suggested that if more than 10% were auctioned, the money raised could be hypothecated to encourage an accelerated deployment of renewables, including CCS.[89] The repost from the Chair of the discussion group[90] was that this was not possible. However, a closer inspection is needed. Firstly, under the UK Greenhouse Gas ETS in its original manifestation, the allocations were auctioned.[91] Under the EU ETS, during the second phase, 10% of the allowances are permitted to be auctioned. Therefore there is nothing new in auctioning the allowances. It would take only a small move in policy to allow for a greater proportion of allowances to be auctioned in order to change behaviour. Although, as the Chair stated, there is no tradition of hypothecation of revenue in the United Kingdom, it is not without precedent. The UK Landfill Tax is hypothecated to support environmental projects within 3 miles of the landfill sites. The electorate voted to support a government in 2001 who increased national insurance contributions on the understanding that the extra monies were to be spent on the National Health Service. If only the allowances for the energy sector were auctioned from the beginning of Phase III, it is estimated that proceeds to the public purse across the European Union would be €27 billion per annum.[92] Should allowances for all sectors be auctioned, then this would raise an estimated €60 billion per annum.[93] With such large numbers involved, this suggestion deserves further investigation in order to provide incentives for commercial companies to sign up for further research and development into clean coal technology, including CCS and renewable energy sources.

Having considered incentives for coal-fired power plant operators the question remains, why should oil and gas producers make their depleted sites, onshore or offshore, available for the long-term storage/disposal of GHGs? Given the discussions on long-term liability and the costs of developing the technology, there is no commercial reason for such companies to make their depleted sites available. The Directive discusses financial incentives for operators of coal-fired power stations to embrace CCS but no mention is made of any incentive for or transfer of liability of the original owners of the site.

Once an oil or gas field is no longer economically viable, the well is decommissioned by capping the well and the operator installs the legally required monitoring and identification technology. The operator will then transfer his operations to another site, possibly in another jurisdiction.

The key provisions of the draft CCS Directive go a long way to address the problems identified earlier with the broad regulatory CCS storage permitting scheme administered by Member State but under the auspices of the Commission. The draft Directive provides for two types of permit; a 2-year exploration permit to allow the holder to identify suitable geological formations within the permitted area and thereafter a storage permit for the development and utilisation of geological formations as storage sites for CO_2. The Directive also provides in general terms for third-party access to the assets of storage operators and pipeline operators on a non-discriminatory basis. This part of the Directive may well cause disquiet within the industry given the need for major capital investment. The capture and transport of CO_2 are largely provided for within existing regulations, which aim to manage the associated risks within this part of the process, and the draft Directive does not address these activities. The capture phase would be regulated under the Integrated Pollution Prevention and Control Directive and the transport would come under the same regime as the transport of natural gas. The CCS Directive as a whole establishes general requirements and objectives necessary to ensure an overall level of environmental integrity for CCS across the European Union. Under subsidiarity, the actual implementation of the draft Directive will be left to Member State with the Commission to play a coordinating role.

The draft Directive does include provisions to impose liability for any damage as a result of leakage from any storage site. With regard to the local environment, the Environmental Liability Directive[94] will be applied. It is proposed that liability for actual climate damage as a result of leakage will be by inclusion of geological storage in the revised EU ETS Directive to provide that EU ETS allowances will be required to be surrendered to compensate for any leaked emissions.

The storage operator must take remedial measures for any leakage, remaining liable even after closure of a storage site until the handover criteria are met. Such handover criteria are yet to be established, and will create much comment from the CCS industry. From the point of handover, the relevant Member State takes over the responsibility and liability. Under the draft Directive, Member State are permitted to require operators to lodge financial security for their prospective liabilities, which will again draw much comment from the industry during the consultation phase.

The draft EU ETS Directive provides for incentives to the industry through the ETS from Phase III (2013–2020) but may be opted-in by Member State for Phase II (2008–2012). As stated previously, CO_2 which is captured in this manner will not be regarded as emitted and so power stations with CCS technology will not be required to surrender allowances to the value of the CO_2 that is stored.

The draft CCS Directive does much to remove the legal uncertainties surrounding the activity of CCS, and the proposed inclusion of credits to power stations with CCS, which may then be used in the EU ETS, goes some way to provide incentive to the operators to use the technology. Nevertheless, the oil and gas companies with fast-depleting sites remain unconvinced as to any benefit they may gain from making their depleted oil and gas sites available.

Guidelines have now been produced for handover criteria, which have been discussed in Chapters 8, 9, and 10 of this book.

ENDNOTES

1. See Timothy Swanson (1995), 'Economic instruments and environmental regulation: A critical introduction', *Review of European Community and International Environmental Law* 4(4), 287–295.
2. This opinion was expressed with some force by Prof. Richard Macrory in conversation with the author. See also Anthony Ogus (1994), *Regulation: Legal Form and Economic Theory.* Hart Publishing: UK.
3. See Note 1.
4. This approach was first suggested by Ronald Coase (1960) as a solution to externality problems in his seminal paper, 'The problem of social costs', *Journal of Law and Economics* 3, 1–44.
5. See Newbury (1994), 'The impact of EC environmental regulation on UK energy policy', *Oxford Review of Economic Policy* 9(4), 81.
6. 'Annual Health Costs of Air Pollution Reach $50 Billion', *Environment Reporter* 1648, 1990.
7. See Chapter 7, also Patricia Park (1996), 'The marketable permit programme in California', *Environmental Law and Management* 8, 26–29.
8. RECLAIM Rules 2011 (SO_x) and 2012 (NO_x) incorporate by reference the lengthy protocols for MRR for both pollutants.
9. (1983) 'SCAQMD, RECLAIM socioeconomic and environmental assessments', 3.
10. See table on transposition in ENDS Rep. 249, October 1995, p. 37, which shows that only Denmark enjoys 100% transposition, while the United Kingdom, which prides itself in claiming that its compliance with EC legislation is virtually blemish-free, is one place from the bottom with 82%. The report accompanying the statistics from the Commission shows that suspected breaches of EC laws are more numerous in the environmental field than in any other, except for suspected breaches of internal market legislation.
11. See Susan Rose-Ackerman. 'Public law versus private law in environmental regulation: European Union proposals in the light of united states experience', *Review of European Community and International Environmental Law* 4(4), 312–320.
12. European Environment Agency (EEA) (1996), *Environmental Taxes: Implementation and Environmental Effectiveness*. European Environment Agency: Copenhagen.
13. See ENDS Rep. 250, November 1995, p. 3. After dropping its initial proposal for an *ad valorem* tax, the government announced in August that it would proceed with a two-band, weight-based levy.
14. Verbal report from Kevin Whiteman (1995), Regional General Manager of the Southern Region Environment Agency to the Regional Environmental Protection Advisory Committee of which the author is a member.
15. See ENDS Report, and UKELA, 'Garner Lecture', November 1995.
16. This is the State Environmental Regulatory Body.
17. This meeting was attended by the author on 12 January 1996, in Houston Texas. Barry Mcbee is a lawyer who worked for the Bush administration in Washington, DC; he was appointed to the Chairmanship of the Committee by Governor George T. Bush in September 1995 and has pledged to reduce the burden of environmental regulation for business within the state of Texas.
18. Holloway et al. (2006), 'Underground storage of carbon dioxide', Shackley and Gough, Editors, *Carbon Capture and Its Storage*. Ashgate: Aldershot.

19. DEFRA (2006), *Key Facts about Climate Change Emissions of Greenhouse Gases*, http://www.defra.gov.uk/environment/statistics/globatmos/kf/gakfo5.htm. Accessed 2009.
20. Sir Nicholas Stern, *The Stern Review on the Economics of Climate Change*, http://www.hm-treasury.gov.uk/independent_review. Accessed January 2010.
21. Simon Tysoe. 'CCS and clean coal: Legal barriers to development', Herbert Smith, *Energy Exchange*. Spring 2008.
22. Articles 55–57.
23. Article 194(2).
24. Part XI, Article 1(1). 'Area' means the seabed and ocean floor and subsoil thereof, beyond the limits of national jurisdiction.
25. Part XI, Article 1(2), 'Authority' means the International Seabed Authority.
26. Author's emphasis.
27. Part XI, Article 1(3).
28. Australia, co-sponsored by France, Norway, and the UK.
29. A new Subsection 4 details the circumstances when such streams should be considered for dumping.
30. Decision 2007/2.
31. OSPAR Guidelines for Risk Assessment and Management of CO_2 Streams in Geological Formations (Annex 7).
32. Professor David Johnson, Chief Executive of OSPAR Commission (2007), 'The Inaugural Carbon Capture and Storage Summit', Kingsway Hall Hotel, November 2007.
33. United Nations Framework Convention on Climate Change, 1992.
34. Article 4(1)(b).
35. Article 4(1)(c).
36. Article 4(1)(d).
37. Article 4(2)(a).
38. Articles 3(3), 4(1)(b), and 4(2)(c).
39. Article 1.8 defines a 'sink' as 'any process, activity or mechanism which removes a greenhouse gas, an aerosol or a precursor of a greenhouse gas from the atmosphere'.
40. Article 1.7 defines a 'reservoir' as 'a component or components of the climate system where a greenhouse gas or precursor of a greenhouse gas is stored'.
41. Kyoto Protocol Article 2(iv).
42. UNFCCC Article 4.1(c).
43. Kyoto Protocol Articles 5–8.
44. IPCC (2006), IPCC Guidelines for National Greenhouse Gas Inventories 1966 as revised in 2006.
45. Directive 2004/101/EC.
46. Directive 2003/87/EC.
47. EC Directive, Article 1.
48. As was Norway.
49. Europa.
50. Directive 2003/87/EC.
51. Energy Research Centre of the Netherlands (2007), Task 2: Choices for regulating CO_2 capture and storage in the EU, Task 2—*Policy Options paper v FINAL 27_04_07*.
52. 'Enhanced Oil Recovery: Enhancing energy security and emissions reductions', *Energy Exchange*, no. 26.
53. *Blackland Park Exploration Limited v Environment Agency* (2003), EWCA Civ 1795.
54. Consider the over-allocation of emission credits under the European National Allocation Plans.
55. Regulation no. 140/2002.
56. Coal Regulation, Article 7.
57. European Commission, *XXXIInd Report on Competition Policy*, 2002, 19.

58. Peter D. Cameron (2007), *Competition in Energy Markets: Law and Regulation in the European Union*, 2nd edition. Oxford University Press: New York.
59. Commission of the EC (2001), Community Guidelines on State Aid for environmental protection, (2001) O.J. C37/3.
60. *PreussenElektra AG v Schleswag AG*, Case C-379/98 (2001) E.C.R. I-2099.
61. (2001) O.J. C37/3.
62. Case NN 12/2004, (2005) O.J. C244/8.
63. Case C44/204, also Commission press release IP/05/1517, 'State aid: Commission closes formal investigation on CO_2 taxation system in Slovenia following changes to legislation', 1 December 2005.
64. Stern Report, p. xix.
65. Stern Report, p. xx.
66. Three joint industry projects have been set up by Det Norske Veritas (DNV) to develop industry guidelines and standards for capture, transmission, and storage of carbon dioxide.
67. European Commission, *Sustainable power generation from fossil fuels: Aiming for near-zero emissions from coal after 2020* {COM (2006) 1722}-{SEC (2006) 1723}-{SEC (2007) 12}.
68. Directive 2001/42/EC.
69. Directive 85/337/EEC as amended by 97/11/EC.
70. See EC cases—*Palin Granit, Alvesta Polarit, Saetti, Kingdom of Spain*, in which it was ruled that substances other than the primary product of a production process will be a by-product and not a waste. Also, *Blackland Park Exploration Limited v Environment Agency* (2003) EWCA Civ 1795.
71. 2004/35/EC.
72. See Note 73.
73. 'Carbon Capture and Storage' fact sheet published on 18 April 2007 by Norwegian Government, www.regjeringen.no/en/dep/oed/subject/carbon-capture-and-storage/. Accessed December 2007.
74. Pollution Control Act, 1981, Chapters 7, §§48 and 49.
75. Conference papers and unpublished reports by Raymond Scott Park and Dr John Walsky.
76. John Newman, 'Carbon dioxide recovery from power station flue gases and its use to enhance the recovery of oil', Trichem Consultants.
77. Unpublished report for Occidental Oil (UK) by Patricia Deanne Park (1990) on *Decommissioning of Oil and Gas Installations*.
78. *Storing CO_2 under the North Sea Basin*, A report by the North Sea Basin Task Force, June 2007.
79. Stern Report, p. xiii.
80. Ibid, p. xviii.
81. Ibid. p. xviii
82. Johnston (2006), 'Free allocations of allowances under the EU emissions trading scheme: Legal issues', *Climate Policy* 6, 115–136, 132.
83. IP/06/1274 and IP/06/1525.
84. Mark Lewis (Director of Research), *Carbon Emissions: Banking on Higher Prices* (Deutsche Bank, 23 July 2007).
85. See Note 84.
86. Director of Research, Deutsche Bank.
87. The principle of supplementarity is enshrined in the Kyoto Protocol and mandates that all states that have ratified Kyoto must ensure that any use they make of the flexible mechanisms to achieve their targets is supplementary to domestic action.
88. Industrial sectors were allocated allowances on their historical emissions.
89. Meeting at Church House on 5 June 2007.
90. Anthony Hobley, Partner, Norton Rose.

91. Patricia Park (2001), 'The UK Greenhouse Gas Emissions Trading Scheme: A brave new world or the result of hurried thinking', *Environmental Law & Management* 13(6), 292–299.
92. Deutsche Bank, *Environmental Carbon Emissions*, 23 July 2007.
93. Ibid.
94. EU Environmental Liability Directive 2004/35/EC.

Bibliography

'Annual health costs of air pollution reach $50 billion', *Environment Reporter* 1648, 1990.
(2011), '12th five year plan on greenhouse emission control', *Guofa* no. 41.
Coal Mining Award Scheme, *MPG3*, May 1999.
'Enhanced oil recovery: enhancing energy security and emissions reductions', *Energy Exchange*, issue 26.
(2011), 'Full scale CCS' St meld nr 9 (2010–2011)', www.reggjeringen.no/nb/dep/oed/dok/regpubl/stmeld/2010-2011/meld-st-9-20102011.html?id=635116. Accessed January 2012.
(1989), *Response to a Radioactive Materials Release Having a Transboundary Impact*, IAEA Safety Guides, Safety Series no. 94, STUI/PUB/814.
35 *AJIL* (1941), 712ff; Gray, *Judicial Remedies in International Law*. American Society of International Law, New York.
Environmental Agency, 'Abandoned mines and the water environment', National Rivers Authority Water Quality Series no. 14, No SR 41, 1994.
ABARES (2011), *Australian Energy Statistics*.
Act 29 June 1990, no. 50, §6-1. Also appurtenant Regulation 7 May 2002, no. 448 concerning system responsibility in the power system.
Act no. 11 of 22 March 1985 relating to petroleum activities.
Act no. 12 of 21 June 1963 relating to exploration for and exploitation of submarine natural resources.
AEMO, 'Australian energy market governance structure', http://www.aemo.com.au/en/About-AEMO/About-AEMO/Who-is-AEMO.
Arnesen et al. (2001), 'Energy law in Norway', Roggenkamp et al., Editors, *Energy Law in Europe*, Oxford University Press: Oxford, pp. 729–830.
Attorney General's Reference no. 1, 1995 1 *WLR* 599 at 615.
Australian Government Department of Resources Energy and Tourism, Bureau of Resources and Energy Economics (February 2012), 'Energy in Australia 2012', www.bre.gov.au. Accessed April 2012.
Bates (1997), *Environmental Law in Australia*, 4th edition. Butterworths: London, UK.
Beier (1980), 'The significance of the patent system for technical, economic and social progress', *International Review of Industrial Property (IIC)* 11, 663.
Bhattacharya & Dutta (2011), 'Climate regulation 2012 India', HSA Advocates.
Bhattacharya & Dutta (2012), 'Climate regulation 2012 India', HSA Advocates, India.
Bhattacharya & Dutta (2012), 'Climate regulation 2012 India', HSA Advocates, http://login.westlaw.co.uk/maf/wluk/app/document?src=doc&linktype=ref&&context. Accessed April 2012.
Bial et al. (2000), 'Public choice issues in international collective action: Global warming regulation, http://www.ssrn.com. Accessed February 2012.
Birkenbus (1987), *International Organisation*, 41, 482.
Birnie & Boyle, *International Law and the Environment*, 2nd edition.
Bishop's Committee for Justice, Development, Ecology and Peace, Media release, September 2002.
Bjerregaard Hints at Treaty Reform on Environmental Taxes, ENDS Rep. 250, November 1995, p. 38.
Blackeney (1989), *Legal Aspects of the Transfer of Technology to Developing Countries*. ESC Publishing Ltd: Oxford.
Bosselman, Rossi & Weaver (2000), *Energy Economics and the Environment*. Foundation Press: New York, USA

Brussels Convention on Nuclear Ships, Articles II and VIII.
Bugge & Ueland (November 2011), 'Case studies on the implementation of Directive 2009/31/EC on the geological storage of carbon dioxide', www.ucl.ac.uk/cclp.
Bunger (2011), *Deficits in EU and US Mandatory Environmental Disclosure*. Springer: London, UK.
Cameron (2007), *Competition in Energy Markets: Law and Regulation in the European Union*, 2nd edition. Oxford University Press: New York.
Carroll (1996), 'Transboundary impacts of nuclear accidents: Are the interests of non-nuclear states adequately addressed by international nuclear safety instruments?', *Review of European Community and International Environmental Law* 5(3), 205–210.
Carson (1962), *Silent Spring*. Houghton Mifflin: USA.
Chequer & Costa (10 December 2010), 'Brazilian pre-salt: PSA regime approved in the Congress', Tauil & Chequer.
Circular no. 2/85: Planning control over oil and gas operations.
Cleg (October 1989), Can a carbon tax really stem the global warming problem? *Petroleum Times* Report.
Cm 9389 (1955), A programme of nuclear power. HMSO: London.
Coal Mines Regulation Act 1872.
Coase (1960), 'The problem of social costs', *Journal of Law and Economics*, 3, 1–44.
Constitution of India Article 245(1).
Control of Pollution (Landed Ships Waste) Regulations 1987 as amended.
Control of Pollution Act 1974 as amended by the Water Act 1989.
Crawford (1992), 'The constitution', Bonyhady, Editor, *Environmental Protection and Legal Change*. Sydney Federation Press: Sydney, Australia.
Cregut & Roger (1991), *Inventory of Information for the Identification of Guiding Principles in the Decommissioning of Nuclear Installations*, EUR 13642 EN, Petroleum Times Report.
Criminal Jurisdiction (Offshore Activities) Order, 1987, SI 1987 no. 2198.
Cummins & King (2011), 'Multilateral and bilateral investment agreements', Dimitroff, Editor, *Risk and Energy Infrastructure: Cross Border Dimensions*. Globe Law and Business.
Daintith & Terence, *United Kingdom Oil and Gas Law*, Oxford University Press:Oxford, UK.
Daintith & Willoughby (1984), *Manual of United Kingdom Oil and Gas Law*, pp. 1–1107.
DECC (2009), *The UK Renewable Energy Strategy*, DECC: UK, p. 30.
Decision of Norwegian Parliament of 2 June 1972.
DEFRA (2006), 'Key Facts about climate change emissions of greenhouse gases', http://www.defra.gov.uk/environment/statistics/globatmos/kf/gakfo5.htm.
Department of Trade and Industry – Oil and Gas Directorate (New Licensing), January 2000.
Derclaye (2009), Should patent law help cool the planet? An inquiry from the point of view of environmental, *European Intellectual Property Review*, 31, SSRN: http://ssm.con/abstract=1373148.
Deuch and Holdren. The future of nuclear power: An interdisciplinary MIT study, 2003.
Diemann & Betlem (1995), 'Nuclear testing and Europe', *National Law Journal*, 145, 1236.
Dimitroff (2011), *Risk and Energy Infrastructure*. Globe Business Publishing: London.
Douben (1998), *Pollution Risk Assessment and Management*, Wiley Publishing: chichester, UK.
Dow, 'Energy law in the United Kingdom', Roggenkamp, Editor, *Energy Law in Europe*. Oxford University Press: Great Britain, pp. 901–971.
Easo (2009), 'Licences, concessions, production agreements and service contracts', Pickton Turbervil, Editor, *Oil and Gas: A Practical Handbook*. Globe Law and Business: London, UK.
EC Directive 90/313. O.J. L 158/56 of 23 June 1990.

Eisenhower (1945), *Atoms for Peace address*, GAOR 8th Session, 470th meeting, paras 79–126; Agreed Declaration on Atomic Energy, Washington, 1945, 1 UNTS 123 (US, Canada, UK); UNGA Resolution 1(1).
Electricity (class exemptions from the requirement for a licence) order, 1997.
Electricity Act 1989, section 96(4).
Electricity Association (2001), *Electricity Industry Review* 5.
E&P Forum, 'Guidelines for the development and application of health, safety and environmental management systems', Report no. 6, 36/210, New York, July 1994.
ENDS Report 250, November 1995, p. 3. After dropping its initial proposal for an ad valorem tax, the government announced in August that it would proceed with a two-band, weight-based levy.
ENDS Report and UKELA, 1995, 'Garner Lecture', November 1995.
Energy Act §1-1.
Environmental Protection Agency (2009), 'Inventory of US greenhouse gas emissions and sinks: 1990–2007 Annex I at A-4 tbl, A-I, http://www.epa.gov/climatechange/emissions/dpwnloads09/AnnexI.pdf.
EPA Pub L. no. 109-58, 119 Stat. 594.
Ernst (1994), *Whose Utility? The Social Impact of Public Utility Privatisation and Regulation in Britain*. Open University Press: Buckingham, UK, p. 49.
Esmaeili (2001), *The Legal Regime of Offshore Oil Rigs in International Law*. Ashgate: London, UK.
European Environment Agency (1996), *Environmental Taxes: Implementation and Environmental Effectiveness*. European Environment Agency: Brussels.
EURATOM 90/641, Council Directive of 4 December 1990 on the operational protection of outside workers exposed to the risk of ionising radiation during their activities in controlled areas (O.J. L-349 of 13/12/1990, p. 21).
EURATOM 96/29, Council Directive of 13 May 1996 laying down basic safety standards for the health protection of the general public and workers against the dangers of ionising radiation (O.J. L-159 of 29/06/1996, p. 1).
European Parliamentary Assembly Rec. 1068 (1988).
European Commission, 'Towards sustainability (1993–2000)', Fifth Environmental Action Programme of the EC.
Fikentscher (1980), '*The Draft International Code of Conduct on the Transfer of Technology*', Munich.
Final Rule, Power Reactor Security Requirements, 74 *Fed. Reg*. 13, 926 (27 March 2009).
Ford (1982), Three Mile Island: Thirty minutes to meltdown, and report of the President's Commission on the accident at Three Mile Island (October 1979) Kemeny Commission Report.
Forman (2011), 'The uncertain future of NEPA and mountaintop removal', *Columbia Journal of Environmental Law* 36, 163–194.
Foster (1992), *Privatisation, Public Ownership and the Regulation of Natural Monopoly*, Blackwell: Oxford, UK.
Francioni (1975), 'Criminal jurisdiction over foreign merchant vessels in territorial waters: a new analysis', Benedeto Conforti, Editor, *Italian Yearbook of International Law*. Oxford University Press: Oxford, UK, vol. 1, 27.
A.M. Freeman (2003), 'Economics, incentives, and environmental policy', Vig & Kraft, Editors, *Environmental Policy*, 5th edition. CQ Press: Washington, DC.
FRG Atomic Energy Act, 1985.
Friedrich (ed.) (1962), *The Public Interest*, Chapter 17, Atherton Press: New York, USA.
Froggatt & Thomas (2011), 'The world nuclear industry status report 2010–2011: Nuclear power in a post Fukushima world, http://www.worldwatch.org/nuclear-power-after-fukushima. Accessed December 2011.

Gas Act 1986 section 19 (as amended).
Gas Act 1995.
German Interests in Polish Upper Silesia case. PCIJ Ser. A no. 17, 1922.
GESAMP (1990), *The State of the Marine Environment*, 2nd International Conference on the Protection of the North Sea, Quality Status of the North Sea, London, 1987.
Gold (1997), *Gard Handbook on Marine Pollution*. Gard Publishers: London, UK.
Government Response to Fourth and Fifth Reports of the Trade and Industry Committee, October 1998.
Grotius (1609), *Mare Librum*. Lodewijk Elzevir: Dutch Republic. But also see Seldon (1618), *Mare Clausum*.
Group of Experts on the Scientific Aspects of Marine Pollution (GESAMP) (1990), *The State of the Marine Environment*. UNEP: Nairobi, Kenya.
Handl, 69 *AJIL* (1975), p. 50.
Handl, 92 *RGDIP* (1988), 55.
Hansard, HC Debs, vol. 153, col. 464, (1989).
Hansen (1980), 'Economic aspects of technology transfer to developing countries', *International Review of Industrial Property (IIC)* 11, 429–430.
Hardin (1968), 'The tragedy of the commons', *Science* 1622, 1243–1248.
Harvey, 'Time to clean up? The climate is looking healthy for investment in green technology', *Financial Times*, 22 June 2005, p. 15.
Higgins (1994), *Problems & Process: International Law and How We Use It*, Oxford University Press: Oxford, UK.
Higgins (1998), *Problems & Process: International Law and How We Use It*. Oxford University Press: Oxford.
Holdgate (1979), *A Perspective of Environmental Pollution*.
Hoel (1991), 'Efficient international agreements for reducing emissions of CO_2', *The Energy Journal* 12(2), 93–108.
Hood (1983), *The Tools of Government*. Macmillan: London; Dantith (1989), 'A regulatory space agency', *Oxford Journal of Legal Studies* 9, 534–546.
Holloway et al. (2006), 'Underground storage of carbon dioxide', Shackley & Gough, Editors, *Carbon Capture and Its Storage*. Ashgate: Guildford, UK.
http://web.worldbank.org.
http://www.eea.europa.eu/themes/technology
http://www.encharter.org/index.php?id=330.
http://www.lungsa.org/assets/documents/healthy-air/coal-fired-plat-hazards.pdf. Accessed December 2010.
http://www.lungusa.org/assets/document/heathy-air/coal-fired-plat-hazards.pdf. Accessed April 2011.
http://www.statnett.no/Documents/Kraftsysternet/Nettutvikingsplaner/Statnett%20nettutviklingsplan%202010.pdf.
http://www.uccee.org.
http://yosemite.epa.gov/opel/RuleGate.nsf/byRIN/2050-AG60.
IEA (2010), *Energy Technology Perspectives*.
IEA (2011), *Carbon Capture and Storage Legal and Regulatory Review*. p. 17.
Igiehon & Park, *Evolution of International Law on the Decommissioning of Oil and Gas Installations*, IELTR 198, Sweet and Maxwell: UK, 2001.
ILO Convention (no. 115) concerning the protection of workers against ionizing radiations, June 1960.
IMO Guidelines, clause 1.1.
'Incremental electricity supply costs from additional renewable and gas fired generation in Australia' (July 2002), http://www.originenergy.com.au/about/files/MMAReport.pdf (accessed August 2003); Environmental Business Australia (July 2002), *The Business*

Case for Ratification of the Kyoto Protocol: Consultation Draft, Petroleum Times Report.
International Atomic Energy Association (1985), *Basic Standards for Radiation Protection*, Safety Series no. 72, 1082nd edition. IAEA: Vienna.
IPPC (1998), paragraph 2.
Jennings (1961), 'State contracts in international law', *British Yearbook of International Law* 37, 156.
Jessop (1990), 'Regulation theories in retrospect and prospect', *Economy and Society* 19, 153–216.
Johnson, Chief Executive of OSPAR Commission. 'The Inaugural Carbon Capture and Storage Summit', Kingsway Hall Hotel, November 2007.
Jones A. (1996), 'Regulation, crime and pollution from abandoned coal mines', *Journal of Environmental Law* 8(1), 43–85.
Jones E. (1988), *Oil: A Practical Guide to the Economics of World Petroleum*. Woodhead-Faulkener: Cambridge.
Jurgen (1988), 'The role of EURATOM', Cameron et al., Editors, *Nuclear Energy Law after Chernobyl*, Graham & Trotman, International Bar Association: London, UK, ibid.
Keaton (1968), 21 CLP, 94.
Kirgis (1972), *American Journal of International Law* 66, 290.
Kneese & Spofford (1986), The economics of integrated environmental management, *Proceedings of the Third Symposium on Integrated Environmental Control for Fossil Fuel Power Plants*. US Environmental Protection Agency: Pittsburgh, PA.
Kulichenko & Ereira (2011), *Carbon Capture and Storage in Developing Countries: A Perspective of Barriers to Deployment*. World Bank: Washington, DC.
Landcatch Ltd v The International Oil Pollution Compensation Fund (IOPCF) (The 'Braer'). (1998), 2 Lloyd's Rep. 552.
Legislation of FRG (1980), 19 *ILM* 1330; France (1982), 21 *ILM* 808; Japan (1983) 22 *ILM* 102; United Kingdom (1981) 20 *ILM* 1271; and USA (1980) 19 *ILM* 1003.
Lehmann (1985), 'The theory of property rights and the protection of intellectual and industrial property', 11, *IIC* 525, 526.
List 1, 7th Schedule of the Indian Constitution, entries 7 and 52.
Local Government Act 1972, Schedule 16.
Louka (2006), *International Environmental Law: Fairness, Effectiveness, and World Order*. Cambridge University Press: Cambridge.
Lov 26 Mar 2004 nr 17 om finansiell sikkerhetsstillelse.
Lyster & Bradbrook (2011), *Energy Law and the Environment*. Cambridge University Press: Cambridge.
Machlup & Penrose (1950), 'The patent controversy in the nineteenth century', *Journal* of *Economic History* 10, 1–29.
Macrory & Purdy (1997), 'The enforcement of E.C. environmental laws against member states', Holder, Editor, *The Impact of E.C. Environmental Law in the United Kingdom*. Wiley: Chichester, UK.
Malcolm, *Encyclopaedia of Planning Law*. Sweet & Maxwell: London, UK, Loose leaf folder.
Marc (2004), 'Protecting the environment at the margins: The role of economic analysis in regulatory design and decision-making', Edmond, Editor, *Expertise in Regulation and Law*. Ashgate: Guildford, UK.
MARPOL Convention, 1973 and 1978, UNCLOS 1982 Articles 211, 217, 218, and 220.
Matthew (1995), 'Will a market in air pollution clean the nation's dirtiest air? A study of the south coast air quality management district's regional clean air incentives market', *Ecology Law Quarterly* 22, 359–378.
McEldowney & McEldowney (1996), *Environment and the Law* Chapters 8 and 10, Edward Elgar Publishing: Cheltenham, UK.

Mechoir (2010), 'The Chinese renewable energy law', http://horizonmicrogrid.com/node/19.
Moore (2000), *A Practical Approach to Planning Law*, Blackstone; ETC: London.
MPG 6: Guidance for the mining of aggregates.
Muchilinski (2009), *Multinational Enterprises & the Law*. Oxford University Press: Oxford.
Myres, McDougal & Lasswell (1959), 'The identification and appraisal of diverse systems of public order', *American Journal of International Law* 53, 1–38.
National Academy of Science (2009), *America's Energy Future Technology and Transformation*. Petroleum Times Report.
National Mining Association, 'The US Department of Energy's clean coal technology: From research to reality, http://fossil.energy.gov/aboutus/fe_cleancoal_brochure_web2.pdf.
Newbury (1994), 'The impact of EC environmental regulation on UK energy policy', *Oxford Review of Economic Policy* 9(4), 81–102.
NGO Committee on Education, A/Conf. 48/14/Rev. 1, Rec. 75, 'Action plan for the human environment'.
Norwegian Ministry of Petroleum and Energy, 'Facts 2008—Energy and water resources in Norway', p. 78, www.regieringen.no/en/dep/oed/Documents-and-publications/reports/2008/fact-2008. visited September 2010.
Norwegian Shipowners Claims 1UNRIAA 307.
NRC, 10 C.F.R. §50.72, www.nrc.gov/what-we-do/regulatory/event-assess.html.
NRC Inquiry Group (1980), TMI Report to the Commissioners and the Public, Kemeny Commission Report.
Ogus (1994), *Regulation: Legal Form and Economic Theory*. Clarendon Press: Oxford, Chapter 2.
Park (1996), 'The Marketable Permit Programme in California', *E.L.M.* January 1996.
Park (1998), 'Some challenges for science in the environmental regulation of industry', Michael, Freeman & Helen Reece, Editors, *Science in Court*. Ashgate: Aldershot, UK p. 191.
Part XI of the 1982 Convention, UN Doc A/CONF. 62/122;(1982) 21 *ILM* 1261.
Pearce (1990), *Economics and the Global Environmental Challenge*, Henry Sidgwick Memorial Lecture. Cambridge University Press: UK.
Petroleum Act, section 10-6.
Petroleum and Submarine Pipelines Act 1975, section 23 and the Pipelines Act 1962, section 10.
Planning (Hazardous Substances) Act 1990 and DOE Circular 11/92.
Poustie (1995), 'Sparring at oil rigs: Greenpeace, *Brent Spar* and challenges to the legality of disposing of disused oil rigs at sea', *Judicial Review*, 6, 542–548.
Powers (2012), 'The cost of coal, climate change and the end of coal as a source of 'cheap' electricity', *University of Pennsylvania Journal of Business Law*, 12.2.
Pring et al. (1999), 'Trends in international environmental law affecting the mineral industry', *The Journal of Electronic Resources Librarianship* 17, 39–48.
Proclamation 2667—'Policy of the United States with respect to the natural resources of the subsoil and seabed of the continental shelf', 28 September 1945, Basic document no. 5.
Proclamation no. 2667, 3 C.F.R. 67, White House Press Release, 28 September 1945, and 13 Dept. State Bull. 484 (1945); Bernard Taverne (1994), *An Introduction to the Regulation of the Petroleum Industry*, Graham & Trotman Ltd.: New York.
Proposal for a Directive introducing a tax on carbon dioxide emissions and energy (1992) O.J. C 196/1.
Protocol to LDC, Annex 3.
Protocols of 1984, 1985, 1988, and 1991.
Rasband, Salzman & Squilliance (2009), *Natural Resources Law & Policy*, Chapter 7, Westlaw/Reuters: Eagen, USA.

RECLAIM Rules (2011), (SO_x) and 2012 (NO_x) incorporate by reference the lengthy protocols for monitoring, reporting and record keeping for both pollutants, Petroleum Times Report.
Redgwell (2001), 'International law and the energy sector', Roggenkamp, Editor, *Energy Law in Europe*. Oxford University Press: Oxford.
Reitze (2010), *Air Pollution Control Law: Compliance and Enforcement*, 2nd edition.
Renewable Energy (Electricity) Act 2000, (Cth) section 18.
Report of the ISBA Secretary General, 14th Session, ISBA/14/A/2, 2008.
Resolution 1803 (xvii). UN Doc. A/5217, 1962.
Resolution 36/193 (17 December 1981).
Resolution XV/90 of 1968, the OPEC.
Reyners & Lellouche (1988), 'Regulation and control by international organisations in the context of nuclear accident', Cameron et al., Editors, *Nuclear Energy Law After Chernobyl*.
Rose-Ackerman (1995), 'Public law versus private law in environmental regulation: European union proposals in the light of united states experience', *Review of European Community and International Environmental Law* 4(4), 312–320.
Rossi (2009), 'The Trojan horse of electric power transmission line siting', *Environmental Law* 39, 1015–1049.
Rothwell & Stephens (2010), *The International Law of the Sea*. Hart Publishing: Oxford.
Royal Commission on Environmental Pollution (1990), *Setting Environmental Standards*, Cm. 1200. HMSO: London.
Royal Decree of 8 December 1972 relating to exploration for and exploitation of petroleum.
Royal Decree of 9 April 1965 relating to exploration for and exploitation of petroleum.
Safety Series no. 78, 'Definitions and recommendations for the convention on the prevention of marine pollution', 1986, Vienna, adopted at the 10th Consultative Meeting of the LDC.
Sands (ed.) (1993), *Greening International Law*. Earthscan: Cambridge, UK.
Sands (1996) 'Observations on international nuclear law ten years after Chernobyl', *Review of European Community and International Environmental Law* 5, 199–204.
Sands (2008), *Principles of Environmental Law*. Cambridge University Press: Cambridge.
Sands (2009), *Principles of International Environmental Law*, 2nd edition. Cambridge University Press: Cambridge.
SCAQMD (1983), 'RECLAIM socioeconomic and environmental assessments', VII-3.
Schanze (1978), 'Mining agreements in developing countries', *Journal of World Trade Law* 12, 135–173.
Section 11(5) of the Petroleum Act 1998.
Sharma, Puranik, Arora. Khaitan & Co., India.
Shelton (1991), 'Human rights, environmental rights and the right to a decent environment', *Stanford Journal of International Law* 28, 103–115.
Simmons (2002), Centre for the Study of Regulated Industries, Industry brief. Regulation of the UK Electricity Industry.
Slade (1996), 'The 1995 NPT review and extension conference', *Review of European Community and International Environmental Law* 5:3.
Smith (2008), *Energy Exchange*. Spring: London, UK.
Solow (1957), 'Technological change and the aggregate production', *The Review of Economics and Statistics* 39, 312–320; Ewing (1976), 'UNCTAD and the transfer of technology', *Journal of World Trade* 10, 197–214.
Statement by Statnett on the Renewable Energy Directive given to the Ministry of Petroleum and Energy by letter on 24 March 2009.
Statnett (2010), Netutviklingsplan, p. 70, www.statnett.no/Documents/kraftsysternet/Nettutviklingsplaner/Statnetts%20nettutviklingsplan%2020.pdf.

Stern, The Stern review on the economics of climate change, http://www.hm-treasury.gov.uk/independent_review. Accessed July 2010.

Swanson (2006), 'Economic instruments and environmental regulation: a critical introduction', *Review of European Community and International Environmental Law* 4(4), 287–295.

Swanstrom and Jolivert (2009) 'Has the Energy Policy Act of 2005 succeeded in stimulating the development of new transmission facilities?' *Energy Law Journal* 30, 415–467.

Swanstrom & Jolivert (2009), DOE Transmission Corridor Designations & FERC Backstop Siting Authority, *Energy Law Journal*, 30, pp. 415–466.

(1994), *Sustainable Development: The UK Strategy*. HMSO: London.

Taggart (1980), *Ship Design Construction*. Society of Naval Architects and Marine Engineers: London, UK.

Task 2: Choices for regulating CO_2 capture and storage in the EU.

The 1994 Agreement, Annex, section 7.

The agreement between France and Belgium on Radiological Protection concerning the Installations of the Nuclear Power Station of the Ardennes, March 1967, 588 *UNTS* 227.

The Coal Industry Act 1994, section 43.

The Coal Industry Benefits Act 1992 524 US 498, 1998.

The Control Regulation of 11 March 1999, no. 302

The Countryside and Rural Economy (1992), PPG 7.

The Electricity Supply and Grid Services Regulation of 11 March 1999, no. 301.

The Energy Regulations of 7 December 1990, no. 959.

The Environment Agency and Sustainable Development. Statutory Guidance to the Environment Agency, Department of the Environment, 1996.

The European Commission, 'Nuclear safety and the environment', Supporting documentation for the preparation of an EC Communication on the subject of decommissioning nuclear installations in the EU. P. Vankerckhoven, DG XI.C2.

The Gas Act 1986, section 4 (as amended).

The Gas Act 1995 and the Utilities and Competition Act 1992, in addition to the Utilities Act 2000.

The Industry Concession Act of 14 December 1917, no. 16 and The Mines (Notice of Abandonment) Regulations 1998, DETR April 1998, SI 1998, no. 892.

The Planning and Building Act of 27 June 2008, no. 71 and the Competition Act of 5 March 2004, no. 12.

The Prospects for Coal, Cm. 2235, 1993.

The Renewable Energy Directive 2009/28/EC.

The Review of Energy Sources for Power Generation, Cm. 4071, 1998.

The Rio Declaration, 1992.

The Universal Declaration of Human Rights, 10 December 1948.

The Watercourse Regulation Act of 14 December 1917, no. 17.

This Common Inheritance, Britain's Environmental Strategy. Cm. 1200, HMSO, 1990.

Tietenburg (1995), 'Economic instruments for environmental regulation', *Oxford Review of Economic Policy* 6 (1), 17–33.

Tomain (2011), *Ending Dirty Energy Policy: Prelude to Climate Change*, Chapters 3 and 4. Cambridge University Press: UK.

Town and Country Planning (Assessment of Environmental Effects) Regulations, 1988.

Tromans (1999), Environmental Protest and the Law, O.G.L.T.R., p. 342.

Tudway (ed.) (1999), *Energy Law & Regulation in the European Union*. Sweet & Maxwell: London, p. 1402.

Tysoe (2010), 'CCS and clean coal: Legal barriers to development'. M. Roggenkamp, Editor. *European Energy Law Report*, Insentia Publishers.

UK Department of the Environment, Quality status of the North Sea (1987), Clark, Editor, *Marine Pollution Oxford*, DECC, UK, 1986.

UK Environmental Protection Act 1990, section 7(7).
UK Nuclear Installations Act, 1965; Switzerland Act on Third Party Liability, 1983.
UN Commission on Trade and Development (2009), 'The role of international investment agreements in attracting foreign direct investment to developing countries', www.unctad.org/Templates/Download.asp?docid=1254&dang. accessed November 2011.
US DOE (October 2008), 'Program facts: Clean coal power initiative (CCPI)', http://fossil.energy.gov/programs/powersytems/cleancoal/ccpi/Prog052.pdf. Accessed April 2012.
Van den Burg (1995), *Beginning LCA: A Guide into Environmental Life Cycle Assessment.*
Venkat (2011), *Environmental Law and Policy*. PHI Publishing: London, UK.
Vienna Convention, Article II; Paris Convention, Article 3; Brussels Convention on Nuclear Ships, Article II.
Vienna Convention, Article IV(3)(b); Paris Convention, Article 9.
Vienna Convention, Article IV; Paris Convention, Articles 3 and 9.
Vienna Convention, Article VIII; Paris Convention, Article 13; Brussels Convention on Nuclear Ships, Article X.
Vienna Convention, Articles V and VI; Paris Convention, Articles 7 and 8; Brussels Convention on Nuclear Ships, Articles III and V.
Weiss (1989), 'The five international treaties: A living history', Weiss & Jacobson, Editors, *Strengthening Compliance with International Environmental Accords*. MIT Press: USA, pp. 89–173.
Wells (1993), *Corporations and Criminal Responsibility.*
White (1996), *The Law of International Organizations.*
Williams (1955), 'The definitions of crime', *Current Legal Problems*, 8, 107–130; Ogus & Burrows (1983), *Policing Pollution: A Study of Regulation and Enforcement*, DECC, UK, pp. 14–18.
Winter (1996), *European Environmental Law: A Comparative Perspective*. Dartmouth: London, UK.
Wood (1999), 'International seabed authority: The first four years', *Max Planck Yearbook of United Nations Law*, Martinus Nijhoff: Leiden, Netherlands, vol. 3, 173.
Writ Petition no. 888/1996 decided on 6 May 2005 by Honourable Supreme Court.
www.encharter/org/index.php?id=7. accessed November 2011.
www.iea.org/speech/2009/Tansaka/4th_OPEC_Seminar_speech.pdf. Accessed July 2010.
www.imo.org/home.asp.
www.indiaclimateportal.org/The-NAPCC.
www.info.ogp.org.uk/decommissioning.
www.nasdaqomxcommodities.com. accessed November 2011.
Yamin (2000), 'Joint implementation', *Global Environmental Change* 10(1).
Yamin & Depledge (2004), *The International Climate Change Regime.*

Index

O

T

U

V

W

CPSIA information can be obtained
at www.ICGtesting.com
Printed in the USA
LVHW081336280820
664441LV00005B/24